McGraw-Hill Series on Computer Communications (Selected Titles)

ISBN	AUTHOR	TITLE
0-07-005147-X	Bates	*Voice and Data Communications Handbook*
0-07-005669-2	Benner	*Fibre Channel: Gigbit Communications and I/O for Computer Networks*
0-07-005560-2	Black	*TCP/IP and Related Protocols*, 2/e
0-07-005590-4	Black	*Frame Relay Networks: Specifications and Implementation*, 2/e
0-07-006730-3	Blakeley/Harris/Lewis	*Messaging/Queuing using the MQI*
0-07-011486-2	Chiong	*SNA Interconnections: Bridging and Routing SNA in Hierarchical, Peer, and High-Speed Networks*
0-07-020359-8	Feit	*SNMP: A Guide to Network Management*
0-07-024043-4	Goralski	*Introduction to ATM Networking*
0-07-034249-0	Kessler	*ISDN: Concepts, Facilities, and Services*, 3/e
0-07-035968-7	Kumar	*Broadband Communications*
0-07-041051-8	Matusow	*SNA, APPN, HPR and TCP/IP Integration*
0-07-060362-6	McDysan/Spohn	*ATM: Theory and Applications*
0-07-044362-9	Muller	*Network Planning Procurement and Management*
0-07-046380-8	Nemzow	*The Ethernet Management Guide*, 3/e
0-07-049663-3	Peterson	*TCP/IP Networking: A Guide to the IBM Environments*
0-07-051506-9	Ranade/Sackett	*Introduction to SNA Networking*, 2/e
0-07-054991-5	Russell	*Signaling System #7*
0-07-057199-6	Saunders	*The McGraw-Hill High-Speed LANs Handbook*
0-07-057639-4	Simonds	*Network Security: Data and Voice Communications*
0-07-060363-4	Spohn	*Data Network Design*, 2/e
0-07-069416-8	Summers	*ISDN Implementor's Guide*
0-07-063263-4	Taylor	*The McGraw-Hill Internetworking Handbook*
0-07-063301-0	Taylor	*McGraw-Hill Internetworking Command Reference*
0-07-063639-7	Terplan	*Effective Management of Local Area Networks*, 2/e
0-07-065766-1	Udupa	*Network Management System Essentials*

To order or receive additional information on these or any other McGraw-Hill titles, please call 1-800-822-8158 in the United States. In other countries, contact your local McGraw-Hill representative.

KEY = WM16XXA

TCP/IP

Architecture, Protocols, and Implementation with IPv6 and IP Security

Dr. Sidnie Feit

Second Edition

McGraw-Hill

New York   San Francisco   Washington, D.C.   Auckland   Bogotá
Caracas   Lisbon   London   Madrid   Mexico City   Milan
Montreal   New Delhi   San Juan   Singapore
Sydney   Tokyo   Toronto

Library of Congress Cataloging-in-Publication Data

Feit, Sidnie.
    TCP/IP : architecture, protocols, and implementations with IPv6
and IP security / Sidnie Feit. — 2nd ed.
        p.   cm. — (The McGraw-Hill series on computer communications)
    Includes bibliographical references and index.
    ISBN 0-07-021389-5
    1. TCP/IP (Computer network protocol)   I. Title.   II. Series.
TK5105.585.F45   1996
004.6'2—dc20   96-18758
                                                                    CIP

McGraw-Hill

A Division of The **McGraw·Hill** *Companies*

        4 5 6 7 8 9 0   DOC/DOC   9 0 1 0 9 8 7

ISBN 0-07-021389-5

*The sponsoring editor for this book was Steven Elliot, the editing supervisor was
Nancy Young, and the production supervisor was Pamela A. Pelton. This book was
set in Century Schoolbook by McGraw-Hill's Professional Book Group
composition unit.*

Printed and bound by R. R. Donnelley & Sons Company.

McGraw-Hill books are available at special quantity discounts to use as premi-
ums and sales promotions, or for use in corporate training programs. For more
information, please write to the Director of Special Sales, McGraw-Hill, 11 West
19th Street, New York, NY 10011. Or contact your local bookstore.

This book is dedicated to my husband and friend, Walter.

Contents

Preface

This book provides a general introduction to TCP/IP. It also contains a body of specific reference information for those who will continue to use or support TCP/IP. The book is intended to be a practical guide to TCP/IP and contains detailed information on how to get started on a real network—how to tie together existing local and wide area networks, how to choose system names, where to get network addresses, how to get advice on administration and security, and what to expect from current TCP/IP products.

The book is intended for those who need to learn about TCP/IP, either as planners, network managers, network administrators, software developers, or technical support or as end users who like to understand their operating environment.

This book explains TCP/IP terminology, concepts, and mechanisms. It describes the standards that define the TCP/IP protocol suite. It contains many interactive on-line dialogues that show how to use TCP/IP applications, how to peek at what is going on in the background, and how to check up on the health of network resources. For those who need a truly detailed level of understanding, there are traces that show the byte-by-byte structure of network messages and the interactive flow of these messages. The standard socket programming interface is explained, and several sample client/server programs are included.

For this second edition, the amount of material that is covered has been expanded significantly. An entire chapter now is devoted to describing the commonly used routing protocols. A new chapter describes domain name system internals. There are chapters on the World Wide Web, news, and gopher applications. IP "next generation" (version 6) addressing and protocols are presented. Security technologies are explained early on, and security issues are discussed throughout the book. There also is a chapter on integrating security into IP.

Sidnie M. Feit

Acknowledgments

I would like to thank the Yale University mathematics department for their invitation to spend a term as visiting faculty, which led to the first edition of this book. The opportunity to use the diverse computer systems on the Yale computer network and to study their documentation was invaluable. Since that time, H. Morrow Long, Manager of Development for the Yale Computer Science department, has continued to point out security problems and solutions, interesting new software, and significant happenings on the Internet. Morrow also has been a reliable source for hard facts about real implementations.

Graham Yarbrough read the book from cover to cover and provided insights based on his vast knowledge of the real world of business computing and network equipment. Michael MacFarland commented on several chapters in great detail and made many helpful suggestions.

Jay Ranade, editor of this series, always has offered positive encouragement.

Netmanage, Inc., provided the latest version of *Chameleon NFS* (including a *Personal Web Server*) as well as *NewtWatch* monitoring software. Network management scenarios were executed with the help of *HP Open View for Windows Workgroup Node Manager*. Network General provided a *Sniffer* monitor, extensive documentation, monitoring tips, and many megabytes of protocol traces. Ashmount Research Ltd. provided a Windows-based *NSLookup* program.

Many other vendors contributed product and technical information. FTP Software, Inc., contributed software and documentation for their Windows product. Vendors, including, among others, Cisco Systems and Bay Systems, answered questions diligently and responded quickly with product information.

Trademarks

AT&T is a trademark of AT&T. Banyan VINES is a trademark of Banyan, Inc. BBN is a trademark of Bolt, Beranek, and Newman. Cabletron and Spectrum are trademarks of Cabletron Systems. Chameleon and NewtWatch are trademarks of NetManage, Inc. Cisco is a Trademark of Cisco Systems, Inc. Clarinet is a trademark of Clarinet News Service. Cylink is a trademark of Cylink Corporation. DEC, VAX, VMS, ULTRIX, DIGITAL, and DECnet are trademarks of Digital Equipment Corporation. EIT is a trademark of Enterprise Integration Technologies. Ethernet is a registered trademark of Xerox Corporation. Fetch is a copyright of Dartmouth College. Gauntlet is a trademark of Trusted Information Systems. GES is a trademark of Global Enterprise Systems. Hewlitt-Packard, HP, and HP OpenView are registered trademarks of Hewlitt-Packard Company. IBM, VM, MVS, and OS/2 are trademarks of International Business Machines Corporation. IBM PC and LAN Server are registered trademarks of International Business Machines Corporation. Intel Inside is a trademark of Intel Corporation. Lanrover is a trademark of Shiva, Inc. LAN Workplace is a trademark of Novell, Inc. MCI is a trademark of MCI, Inc. Macintosh is a trademark of Apple Computer, Inc. MacTCP is a copyright of Apple Computer, Inc. Microsoft, Microsoft Windows, and Windows 95 are trademarks of Microsoft Corporation. MS DOS and LAN Manager are trademarks of Microsoft Systems, Inc. Netscape and Netscape Navigator are trademarks of NetScape Communications Corporation. NetWare, Novell, and UNIX are trademarks of Novell, Inc. NSLookup for Windows is copyrighted by Ashmount Research Ltd. OnNet and PC/TCP are trademarks of FTP Software. PC DOS is a trademark of Internatinal Business Machines Corporation. PC Quote is a trademark of PC Quote. Pkzip is a trademark of PKWARE. Reuters is a trademark of Reuters News Service. RSA is a trademark of RSA Data Security, Inc. Shiva and Shiva LANRover are trademarks of Shiva, Inc. Sun is a trademark of Sun Microsystems. UNIX is a technology trademark of X/Open Company, Ltd. WAIS is a trademark of WAIS, Inc. XEROX is a trademark of Xerox Corporation. All trademarks and copyrights mentioned are the properties of their respective owners.

Introduction

I.1 BACKGROUND

In the early days of computing, hosts exchanged information with directly attached devices, such as card readers and printers. Interactive use of computers first required local and then remote attachment of end-user terminals. When an organization had acquired several computers, often there would be a need to transfer data between computers or to allow users attached to one computer to access another computer.

Computer vendors responded to these requirements by developing communications hardware and software. Unfortunately, this hardware and software:

- Was proprietary and worked only with the vendor's equipment
- Supported only a limited number of local and wide area network types
- Sometimes was extremely complex, requiring different software dialects for each device and each application
- Lacked the flexibility that would enable previously independent networks to be connected to each other easily and inexpensively

This situation changed with the help of Transmission Control Protocol/ Internet Protocol (TCP/IP) and the router technology that it spawned.

Today, an organization's computer network has become its circulatory system. Organizations have combined desktop workstations, servers, and hosts into Local Area Network (LAN) communities. They have connected LANs to other LANs and to Wide Area Networks (WANs).

It has become a necessity of day-to-day operation that pairs of systems must be able to communicate when they need to, without regard for where they may be located in the network.

1

I.2 TCP/IP APPLICATIONS

From the beginning, TCP/IP was packaged with several important generic application services:

- Terminal access to any host
- The ability to copy files from one host to another
- Exchange of electronic mail between any pair of users

Over the years, many other useful applications have been added to the TCP/IP protocol suite:

- Remote Printing
- Network File System
- Network News
- Gopher
- The World Wide Web (WWW)

In addition, the set of utilities that support network administration and maintenance has been expanding. A few are:

- Directory services for mapping user-friendly host names to network addresses
- Dynamic Host Configuration
- Network Management for hosts, routers, and other network devices

The TCP/IP family is alive, well, and growing. Its user community is expanding at an almost explosive rate, and new services are being developed and modularly integrated into TCP/IP product offerings.

The World Wide Web application suite has sparked a revolution in client/server computing and is changing the way that business is conducted.

As a by-product of the adoption of TCP/IP, thousands of organizations have connected to the Internet. This has caused some strains on a protocol family that originally was designed for internal military and government use. The problems are being addressed, and attention has been focused on evolving more efficient mechanisms for addressing and routing and finding robust solutions to security problems.

I.3 TERMINOLOGY

Like many technical disciplines, data communications has a language all its own. Everyone in the field seems to use a similar vocabulary. The only problem is that groups within the profession use the same words to mean different things, and different words to mean the same thing!

We have made an effort to select a fairly simple vocabulary and use it consistently within this text. In the sections that follow, we will describe some of the terminology and graphic conventions that will be used in this book.

I.3.1   Protocols, Units, Stacks, and Suites

A *protocol* is a set of rules governing the operation of some communications functions. For example, IP consists of a set of rules for routing data, and TCP includes rules for reliable, in-sequence delivery of data.

A *protocol data unit* (PDU) or *packet* is a formatted unit of data that is transmitted across a network. The information carried in a PDU is often called its *payload.*

A *protocol stack* is a layered set of protocols that work together to provide communication between applications. For example, TCP, IP, and Ethernet make up a protocol stack.

A *protocol suite* is a family of protocols that work together in a consistent fashion. The TCP/IP protocol suite encompasses a large number of functions, ranging from dynamic discovery of the physical address on a network interface card to a directory service that reveals how electronic mail should be routed.

I.3.2   Hosts

A *host* is a computer that runs applications and has one or more users. A host that supports TCP/IP can act as the endpoint of a communication. Note that personal computers (PCs), workstations, minicomputers, and mainframes all satisfy the definition of host, and all can run TCP/IP.

This book also uses the terms *station, computer,* and *computer system* synonymously with *host.*

I.3.3   Routers

A *router* routes data through a network. Back in the early days, TCP/IP standards writers adopted the word *gateway* for what the commercial marketplace now calls a router. Elsewhere in the communications world, the term *gateway* came to mean a system that performs some kind of protocol translation.

We will use the term *router* throughout this book. However, if you go back and read TCP/IP standards documents, keep in mind that they sometimes use the term *gateway.*

I.3.4   Internet

An *internet* (with a lowercase *i*) is a set of network facilities (local and/or wide area networks) connected by routers. The *Internet* (with a capital *I*) is a very special internet, connecting thousands of networks together.

I.3.5 Network Node, System, and Network Element

The terms *network node, system,* and *network element* will be used to refer to a communicating entity in a network without specifying whether it is a host, a router, or another device, such as a bridge. For example: *The goal of network management is to control and monitor all of the nodes in a network.*

I.3.6 LANs, WANs, and Links

A local area network is a data network intended to serve a relatively small area, usually a few square kilometers or less. A wide area network covers a large geographical area and usually is constructed using serial telephone lines and shared packet-switching facilities.

The more general term, *link,* includes any medium—local or wide area—over which nodes can communicate using a link layer protocol.

I.3.7 People

The term *hacker* is sometimes used in admiration, describing someone who has a high degree of computer or network skill. At other times, it is used disparagingly, to mean someone who tries to break into private computer resources. In this text, we will apply the term *cracker* to someone who tries to break into private computer resources.

I.3.8 Bytes and Octets

Most of us use the word *byte* to mean an 8-bit quantity. However, byte also has been used to denote the smallest addressable unit in a computer. At one time or another, computer designers have produced machines with all sorts of byte sizes.

Standards writers have finessed the problem by using the term *octet* to mean an 8-bit quantity. In this text, we will use octet and byte interchangeably and will use the term *logical byte* when the size of an addressable unit in a computer is other than 8 bits.

I.3.9 Big Endians and Little Endians

Some computers store data with the most significant byte first. This is called the *Big Endian* style of data representation. Other computers store data with the least significant byte first, in a *Little Endian* style.

Similarly, there are Big Endian data communications standards that represent transmitted data with the most significant bit and byte first.

Internet protocol standards writers are Big Endians. However, keep in mind that there are other groups (such as the Institute for Electrical and Electronics Engineers or IEEE) that represent transmitted data with the least significant bit or byte first.

I.4   IMPLEMENTATION IN A MULTIVENDOR ENVIRONMENT

Unlike the proprietary networking protocols used in the past, TCP/IP is implemented by dozens of computer vendors and independent software companies.

Implementations are based on written standards and on free software that has been made available by volunteers. Strong guidance is provided in additional *Host Requirements* and *Router Requirements* documents.

The degree of interworking that has been achieved is impressive, and end users generally find that their applications work very well.

However, if you look behind the scenes, you can find vendors who have taken shortcuts and have omitted features that would provide better performance or improve error recovery. Sometimes software developers simply have misunderstood some detail in the specification and, as a result, have not provided some beneficial feature.

Hence, you cannot assume that every mechanism described in this book is actually implemented in a given software package. When buying TCP/IP host software, it is a good idea to be armed with a copy of the *Host Requirements* documents, which are discussed in Chapter 1. Your vendor should be able to respond when asked whether particularly important features have been included. (These will be marked as MUST and SHOULD features in the Host Requirements.)

For readers of this book who are implementors, this text can help you to understand the *why* and *how* of many protocols. But no book should ever be used as a basis for implementation. TCP standards are free and are available on-line (see Appendix B). The documentation is continuously updated, reporting on new features and more efficient ways to implement functions and obsoleting features that are not useful.

I.5   DIALOGUES

There are many interactive demonstration dialogues in this text. The text dialogues were generated on several Sun Microsystems computers. Many demonstrations were run at *tigger.jvnc.net,* a host located in Princeton, New Jersey. *Tigger* is a large server operated by the *Global Enterprise Systems* (GES) Service Provider. GES formerly was called *JVNC,* which is why its hosts have names that end with *jvnc.net.* The network performance statistics generated at *tigger* are particularly interesting, since *tigger* communicates with Internet hosts all over the world. A few of the text-based demonstrations were run at Yale computers.

TCP/IP has been implemented with very similar text-based user interfaces and command sets across many types of computers. Hence, the dialogues are close to or identical to what you will experience across a wide range of systems. In the text dialogues, end-user input is represented in bold text, while computer prompts and responses appear without emphasis.

There also are screens showing Graphical User Interfaces (GUIs) for

TCP/IP applications running on a Windows PC and a Macintosh. Several Windows screens show Netmanage *Chameleon* applications and Netscape, Inc.'s, *Netscape Navigator*. There are some screens from Hewlett-Packard's *HP OpenView for Windows Workgroup Node Manager* and Ashmount Research Ltd.'s *NSLookup for Windows*. A Qualcomm *Eudora* Macintosh electronic mail screen is displayed.

I.6   RECOMMENDED READING

Like many works whose subject is data communications, this book is peppered with acronyms. Appendix A contains a list of acronyms and their translations. A separate Glossary contains full definitions.

Appendix B includes an extensive list of documents that define TCP/IP and related facilities. Appendix C describes the services provided by Network Information Centers (NICs) and indicates where they are located. Appendix D contains examples that show how IP addresses can be assigned very efficiently (by using variable-length subnet masks).

1

TCP/IP: What It Is and Where It Came From

1.1 INTRODUCTION

In the late 1960s, the Advanced Research Projects Agency of the U.S. Department of Defense, or ARPA (later changed to DARPA) began a partnership with U.S. universities and other research organizations to investigate new data communications technologies.

Together, the participants built the Advanced Research Projects Agency Network (ARPANET), the first packet-switching network. An experimental four-node version of the ARPANET went into operation in 1969. The experiment was a success, and the testbed facility evolved into a network spanning the United States from coast to coast. In 1975, the Defense Communications Agency (DCA) assumed responsibility for operating the network, which still was considered a research network.

1.1.1 The Birth of TCP/IP

The early ARPANET protocols were slow and subject to frequent network crashes. By 1974, the design for a new set of core protocols was proposed in a paper by Vinton G. Cerf and Robert E. Kahn.[1] The Cerf/Kahn design provided the basis for the subsequent development of the *Internet Protocol* (IP) and the *Transmission Control Protocol* (TCP). Starting in 1980, it took 3 years to convert the ARPANET hosts, then numbering over 100, to the new protocol suite.

The versatility of the new protocols was illustrated in a 1978 demonstration in which a terminal in a mobile van driving along California's Highway 101 transmitted data via packet radio to a node at SRI International, across the continent via ARPANET, and then over a satellite network to a host in London (see Figure 1.1).

---

[1] "A Protocol for Packet Network Interconnections," *IEEE Transactions of Communications,* May 1974.

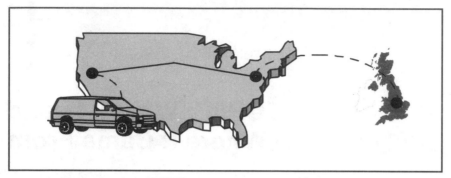

Figure 1.1   Demonstration of TCP/IP across a mixture of technologies.

During the early 1980s, the ARPANET was converted to the new protocols. By 1983, the ARPANET included over 300 computers and had become an invaluable resource to its users. In 1984, the original ARPANET was split into two pieces. One, still called ARPANET, was dedicated to research and development. The other, called MILNET, was an unclassified military network.

1.2   ACCEPTANCE OF THE PROTOCOLS

In 1982, the U.S. Department of Defense (DOD) issued a policy statement adopting a single set of communications protocols based on the ARPANET protocols and created the Defense Data Network (DDN) as the parent entity uniting its distributed operational networks.

In 1983, the Department of Defense adopted the TCP/IP protocol suite as its standard. Acceptance of TCP/IP spread to other government departments, creating a large market for the technology.

1.3   TCP/IP CHARACTERISTICS

TCP/IP has some unique characteristics that account for its durability. The TCP/IP architecture glues clusters of networks together, creating a larger network called an *internet*. To a user, an internet simply appears to be a single network, composed of all of the hosts connected to any of the constituent networks.

The TCP/IP protocols were designed to be independent of host hardware or operating system, as well as media and data-link technologies. The protocols were required to be robust, surviving high network error rates and supporting transparent adaptive routing in case network nodes or lines were lost.

1.3.1   Availability of TCP/IP

When the TCP/IP protocols became a requirement in computer procurements by the U.S. Department of Defense and other government agencies, vendors needed to implement TCP/IP in order to compete in government bids. Figure 1.2 illustrates how different systems, local area networks (LANs), and wide area networks (WANs) can be combined in a TCP/IP environment.

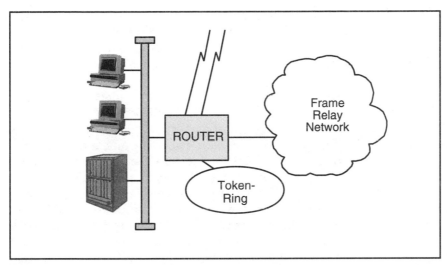

Figure 1.2 TCP/IP multivendor and multinetwork environment.

The U.S. Department of Defense encouraged the availability of TCP/IP by funding Bolt, Beranek, and Newman (BBN) to implement TCP/IP for Unix, and the University of California at Berkeley to incorporate the BBN code into the Berkeley Software Distribution (BSD) 4.2 Unix operating system. This operating system and its successors have been migrated to many hardware bases. Later, TCP/IP was added to AT&T's System V Unix.

In the 1990s, TCP/IP moved into the commercial world. It is the most universally available networking software. There has been rapid progress in bundling TCP/IP into LAN servers and desktop operating systems.

In addition, there is support for TCP/IP over an ever-increasing selection of transmission technologies. We will explore a number of these technologies in Chapter 4.

1.4 THE INTERNET

The ease of gluing TCP/IP networks together combined with an open door policy that allowed academic and commercial research networks to connect to the ARPANET spawned the supernetwork called the *Internet*.[2] Throughout the 1980s, the ARPANET was maintained as a backbone of this flourishing network.

Because of the characteristics of the TCP/IP protocols, Internet growth was steady and nondisruptive. The Internet became the world's largest network, comprising substantial government, military research, academic, and commercial networks, each containing hundreds of subnetworks. In 1985, a new backbone net, the National Science Foundation Net (NSFNET), was incorporated to

---

[2]Recall that, while an internet, with a small i, is any network created by joining two or more TCP/IP nets, the Internet is a specific network.

accommodate high-speed links to research sites and supercomputer facilities.

With the help of government support, an infrastructure of *regional Service Providers* mushroomed in areas all over the United States. Universities and research labs connected to the nearest regional provider, which in turn connected to the backbone.

The Internet spread across the globe, with Service Providers appearing in dozens of countries around the world.

By 1994, millions of computers were interconnected, and the Internet was ready for the commercial marketplace. The National Science Foundation (NSF) gave up its supporting role, and Service Providers in the United States connected to one another at large switching centers scattered across the country.

The Internet continues to be an incubator for new technology. Its mail, news, and bulletin board services provide a public forum in which ideas are debated and refined. Researchers, systems programmers, and network administrators exchange software bug corrections, solutions to interworking problems, and hints for improving performance. Software vendors publish free copies of their beta software at public sites and invite users to download, try, and comment on their new products.

1.5   THE INTERNIC

For many years, the Department of Defense retained an important coordination role for the Internet. Their DDN Network Information Center (DDN NIC) provided services to users, host administrators, site coordinators, and network managers.

In the spring of 1993, civilian Internet user support functions were turned over to the National Science Foundation, which currently funds two agencies:

- InterNIC *Registration Services,* located at Network Solutions, Inc., Herndon, Virginia
- InterNIC *Directory and Database Services,* operated by AT&T

Additional registration centers have been set up in other countries. Registration centers coordinate naming and addressing for TCP/IP computers.

InterNIC Directory and Database Services acts as a clearinghouse for Internet standards and other informational documents. All documents are available free of charge.

See Appendix C for more information about the InterNIC and other Network Information Centers.

1.6   IAB, IETF, AND IESG

Development of new TCP/IP protocols and maintenance of old protocols is coordinated by an organization called the Internet Architecture Board (IAB).[3]

---

[3]Formerly known as the Internet Activities Board.

The IAB identifies technical areas that need to be addressed. For example, in recent years, the IAB has spearheaded efforts to develop new network management protocols, more functional routing protocols, and a next generation version of IP.

In 1992, the *Internet Society* was formed, and the IAB was absorbed into the society. The purpose of the Internet Society is to promote the growth and successful operation of the Internet.

The IAB oversees some important groups. The *Internet Engineering Task Force* (IETF) writes and implements new protocols. Activities of the IETF working groups are supervised and reviewed by *the Internet Engineering Steering Group* (IESG).

1.6.1 Task Forces and Protocol Development

IETF members are volunteers. To tackle a specific problem, a working group whose members have appropriate technical expertise is formed. Participants in a working group use a methodology that combines theory with immediate implementation.

In fact, the validity and completeness of a protocol specification are tested by creating at least two independent implementations. An iterative *design-implement-experiment-review* process is used to evolve and enhance protocols and to improve the performance of the implementations.

This practical, hands-on approach to protocol development uncovers many flaws and oversights before a protocol is adopted. Features that make impossible demands on system resources or cause very poor performance are prevented from being incorporated into a protocol's architecture.

1.6.2 Other Sources of Internet Protocols

Although many of the protocols in the TCP/IP suite were designed and implemented by IETF working groups, university research groups and commercial organizations also have made significant contributions. To be accepted, contributions must be useful and usable.

Source code for new protocols often is placed in public databases on the Internet. Vendors can use this code as the starting point for new products. This has many benefits. Product development is rapid and less costly. Starting from a common source framework promotes interoperability.

Users also may copy and install public code on their own systems. Of course, when using free code, users do not get the benefit of the support and maintenance services provided by a vendor.

1.7 REQUESTS FOR COMMENTS

A new protocol's specification is circulated in a document called a *Request For Comments* (RFC). RFC documents are numbered in sequence. Hundreds of these documents have been written.

Users can obtain RFC documents from the InterNIC Directory and Database Services. RFC documents also have been stored at many other public computer directories at sites all around the world and may be copied from any one of these sites by means of electronic file transfer.

Sometimes protocol specifications are updated to correct errors, improve performance, or add new features. Updated protocols are published with new RFC numbers.

The InterNIC maintains an index of RFCs, and entries for obsoleted RFCs contain the number(s) of the superseding documents. For example, the index entry below announces that RFC 1098 obsoletes the original Simple Network Management Protocol standard and has itself been updated by a later document, RFC 1157:

```
1098 Case, J.D.; Fedor, M.; Schoffstall, M.L.; Davin, C.
Simple Network Management Protocol (SNMP). 1989 April; 34 p.
(Format: TXT = 71563 bytes) (Obsoletes RFC 1067; Updated by RFC 1157)
```

Not all RFCs describe protocols. Some just organize and present insights that have evolved within the Internet community. For example, there is an RFC that gives advice on selecting names for computers. Other RFCs provide guidance on how to administer a TCP/IP network and how to implement security procedures. There are RFCs that suggest implementation strategies for better performance, describe experimental algorithms, and discuss ethics on the Internet. After review, some of these are classified as *Best Current Practices* (BCP) documents.

1.7.1 State and Status of Standards

The IAB periodically publishes information on the progress of protocols as they move through several *states* which reflect their maturity level:

- Experimental
- Proposed
- Draft
- Standard

Some protocol documents are purely *informational,* and some have no current use and have been labeled as *historical.*

Protocols also are classified according to their requirement levels. Some protocols are required standards, while others are used where there is some special need. Some have outlived their usefulness and have been retired. The formal requirement *status* of a protocol is defined by one of the levels:

- Required
- Recommended
- Elective

- Limited Use
- Not recommended

The current state and status of Internet protocols is described in an RFC called *IAB Official Protocol Standards*. This document is updated periodically and released with a new RFC number.

1.7.2 Assigned Numbers

Network parameters, special network addresses, service names, and standard identifiers for terminals and computer systems are listed in an RFC called *Assigned Numbers*.

The Internet Assigned Numbers are administered by the *Internet Assigned Numbers Authority* (IANA), currently located at the University of Southern California's Information Services Institute. The *Assigned Numbers* RFC is updated every few months.

The *Assigned Numbers* RFC includes an address, telephone number, and electronic mail address that protocol and application developers can use in order to register parameter information with this authority.

1.7.3 RFCs that Promote Multivendor Interworking

The expectation that users would have needs that could not be met by a single computer architecture was a strong motivation for adoption of the TCP/IP communications standards by U.S. government organizations. These groups wanted to be able to purchase equipment in a competitive market, with several vendors able to satisfy requirements. They believed that the effort to establish and maintain standards would be repaid in lower costs and better service.

However, there are problems that can arise in a multivendor environment:

- Standards sometimes include optional features. By implementing different options, vendors can make interoperability very difficult.
- Vendors sometimes misunderstand the standards, and their products operate incorrectly.
- There are mistakes in standards specifications.
- Some implementations, although fairly accurate, are inflexible and don't allow a system administrator to tune configuration parameters to improve performance.
- A single system that uses poorly designed algorithms for pacing data transmission and retransmission can degrade performance for all of the systems on a network.

Two RFCs published in October of 1989 addressed many of these problems, correcting errors, clarifying definitions, specifying option support, listing configuration parameters, and identifying high-performance algorithms. Most importantly, these RFCs stated specific conformance requirements for host implementations. This was a major deficiency in the past. Correct operation, interworking, and performance are greatly improved by adherence to these RFCs, which are:

RFC 1122, Requirements for Internet Hosts—Communication Layers. This document deals with link layer, IP, and TCP issues.

RFC 1123, Requirements for Internet Hosts—Application and Support. This document covers remote login, file transfer, electronic mail, and various support services.

An RFC published in 1995 deals with equally important issues relating to the operation of routers:

RFC 1812, Requirements for IP Version 4 Routers

1.7.4 Related Documents

A series of RFCs that do not contain protocol specifications also were published as a separate set of *For Your Information* (FYI) documents. For example, RFC 1325 is an FYI: *Answers to commonly asked "new Internet user" questions.*

Another series, the *Internet Engineering Notes* (IEN) contains a set of discussion papers written in the early years of Internet protocol development.

1.8 OTHER INFORMATION RESOURCES

There are many World Wide Web servers and public file systems located at universities, research institutes, and commercial organizations attached to the Internet. These systems offer a variety of networking information, such as copies of RFCs, papers discussing new algorithms, performance test results, source code for protocols network management tools, free software, and product information. Any Internet user who can perform file transfer or operate a World Wide Web browser can copy documents or code from these sites.

Some networks are connected to the Internet by means of electronic mail gateways. Users at hosts on these networks cannot perform file transfers. Fortunately, many public file systems also support an electronic mail document distribution service.

1.9 OPEN SYSTEMS INTERCONNECTION

Open Systems Interconnection (OSI) was an international effort to create standards for computer communications and generic application services. OSI was an activity of the *International Organization for Standardization* (ISO), founded to promote trade and cooperative advances in science and technology. Standards promoting OSI are published as ISO documents.

The *OSI model* for computer communications is a standard part of any networking professional's education. It provides a framework for identifying where the functionality of various protocols fits into the overall scheme of things.

OSI protocols are used at quite a few European sites, and the IETF has published a number of RFCs that deal with internetworking between TCP/IP and OSI environments.

TCP/IP Suite Service Overview

2.1 INTRODUCTION

Why is the TCP/IP protocol family so widely used? Its ability to glue heterogeneous local and wide area networks together makes it a capable integrator. Equally important, it provides the foundation for peer-to-peer communications and then provides generic services on top of this foundation. Furthermore, TCP/IP was designed from the start to support client/server interactions.

2.2 APPLICATION-TO-APPLICATION COMMUNICATION

There are two styles of application-to-application interaction. *Connection-oriented* communication is appropriate when the applications need a sustained interchange of streams of data. In contrast, applications engaged in *connectionless* communication exchange stand-alone messages. Connectionless communication is well suited to interactions that are sporadic and involve small amounts of data.

2.2.1 TCP Connection-Oriented Communication

The TCP in TCP/IP stands for *Transmission Control Protocol,* which provides reliable, connection-oriented, peer-to-peer communications. Terminal login sessions and file transfers are carried over TCP.

2.2.2 UDP Connectionless Communication

Some data exchanges do not require a continuous interaction. For example, suppose that a database at a network server contains a table of names of company personnel and their telephone numbers. A telephone number might be looked up by sending a request message containing a person's name to the server. The server would respond with a message containing the matching telephone number. The *User Datagram Protocol* (UDP) supports this type of interaction.

2.2.3 Socket Programming Interface

Systems that implement TCP/IP usually provide a communications programming interface for software developers. Most of these are based on the *socket programming interface,* first defined for the Berkeley Unix operating systems.
The socket programming interface includes:

- Simple subroutines that create, transmit, and receive the stand-alone messages used in connectionless UDP communication
- Routines that set up TCP connections, send and receive data, and close the connection

2.2.4 Remote Procedure Call Programming Interface

Although not as prevalent as the socket programming interface, the *Remote Procedure Call* (RPC) client/server programming interface is quite widely available.[1]

A client using an RPC interface invokes a subroutine call that automatically causes a request to be sent to a server. The server executes the subroutine and returns the subroutine's output parameters to the client. This scenario is appropriately named a *Remote Procedure Call* because a locally invoked procedure is executed at a remote system.

For example, the telephone number lookup application described in Section 2.2.2 could be written using RPC routines.

2.3 BASIC SERVICES

Implementations of TCP/IP are expected to provide at least three application services: file transfer, remote login, and electronic mail. Many products include World Wide Web clients and servers. It also has become routine to offer a remote printing function.

2.3.1 File Transfer

File transfer was among the earliest services added to TCP/IP. The File Transfer Protocol (FTP) enables users to copy entire files from one system to another. FTP deals with simple types of files such as American National Standard Code for Information Interchange (ASCII) text or unstructured binary data. FTP also lets a user access a remote file system to perform housekeeping functions such as renaming files, deleting files, or creating new directories.

---

[1]The RPC interface introduced by Sun Microsystems has been ported to a very large number of platforms.

2.3.2 Terminal Access

In the early 1970s, most computer vendors built proprietary terminals that could be used only with their own computer systems. The U.S. Department of Defense (DOD) purchased systems from many different vendors but wanted every user to be able to connect to any host on their network from a single terminal. The *telnet* terminal access protocol was created to make this possible. Over the years *telnet* has been enhanced to work with a large assortment of terminal displays and operating system types.

2.3.3 Mail

Mail has attracted many end users to TCP/IP. Two aspects of mail are standardized:

- The format of the mail passed between users. There are formats for simple text and for multipart, multimedia messages.
- The mechanisms needed for direct or store-and-forward transfer of mail between hosts. The *Simple Mail Transfer Protocol* (SMTP) has been used to transmit mail since the earliest days of the Internet. Recent extensions have added new functionality.

Many proprietary mail systems have been linked to Internet mail, enlarging the community of potential mail partners.

Figure 2.1 illustrates interactions between hosts on a network. Note that TCP/IP truly is a peer-to-peer network architecture. Any host may act as a client, a server, or both.

2.3.4 World Wide Web Service

The World Wide Web is the most versatile of all of the TCP/IP client/server applications. Users can view attractive documents enhanced by images and sounds, navigate effortlessly from site to site with the click of a mouse, and search huge archives of information.

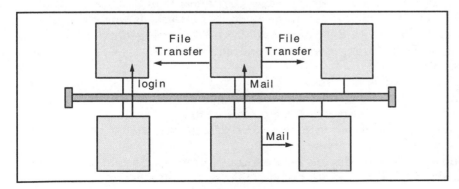

Figure 2.1 Application services on a TCP/IP network.

2.4   ADDITIONAL SERVICES

Other services have been added to the TCP/IP suite. The sections that follow describe those that are most popular and widely available.

2.4.1   File Access

File servers let users access remote files as if they are local. File servers first became popular in personal computer LAN environments as a means to share valuable disk resources and centralize maintenance and backup chores.

Many TCP/IP products include the Network File System (NFS). The products support one or both of the NFS roles:

File access client.   Lets a computer access remote files as if they are local. End users and local programs will be unaware of the actual location of these files.

File server.   Maintains directories that can be accessed by specified computers on a network.

2.4.2   News

The electronic news application started out as a way to support local bulletin board access and distribute bulletin board information to many sites.

Many organizations use free news software to publish internal information electronically. Others access Internet news groups that discuss topics ranging from sports to plasma physics. This software also is used to access commercial news wire services, such as Reuters, AP, and UPI.

2.4.3   Domain Name System Name Service

In order to use a network service, you must be able to identify remote computers. Users and programs can identify computers by names that are easy to remember and easy to type.

To set up communication with a host, its name must be translated to a numeric address. In earlier times, each TCP/IP host kept a complete list of all of the names and addresses of all hosts on its network. It was impossible to keep these lists up to date on a dynamically growing network like the Internet, with its hundreds—then hundreds of thousands—and then millions—of hosts.

The *Domain Name System* (DNS) was invented to solve this problem. The Domain Name System is a database of host names and addresses distributed across thousands of servers. DNS protocols enable a user to submit a database query to a local server and receive a response that may have been obtained from a remote server.

In addition to translating between host names and addresses, DNS servers also provide information that is needed to route electronic mail to its destination.

2.4.4   Commercial Software

Many third-party vendors have built applications that run on top of TCP/IP. For example, database vendors link desktop clients to their servers by means of TCP/IP.

2.4.5   Network Management

Over the years, many network management tools have been developed for use with the TCP/IP protocol suite. For example, there are commands that enable a network manager to see whether systems are active, view their current load, list logged-in users, and list services that are available.

These commands are very useful, but a lot more was needed to provide a consistent and comprehensive platform for centralized network management. The Internet community developed the *Simple Network Management Protocol* (SNMP) to manage everything from simple devices to host operating systems and application software.

2.4.6   Dialogues

The best way to become acquainted with TCP/IP services is to use them. We close this chapter with a few brief interactive dialogues that illustrate what some of these services are like. In all of the text dialogues, commands entered by the user are displayed in boldface. This convention will be followed throughout this book. The basic interactions are very simple.

Each service displays some messages as it does its work. End users ignore these most of the time. Later chapters will explain all of the mechanisms that underlie the services. We will discover that these messages are very informative when you want to know exactly what is going on.

2.4.7   Terminal Access Dialogue

Terminal access is the simplest service. Below we ask *telnet* to connect us to a host named *bulldog.cs.yale.edu*. After a TCP connection has been set up, *telnet* reminds us that CONTROL-] is a *hot-key* that will take us back to the local login session. The remote host then prints its *login:* prompt. From that point on, the session becomes a normal interactive login at that host.

```
> telnet bulldog.cs.yale.edu
Trying 128.36.0.3 ...
Connected to bulldog.cs.yale.edu
Escape character is '^]'.

login:
```

Although this was a very simple dialogue, quite a lot was going on behind the scenes. *Telnet* looked up the host name in a Domain Name System database and found out that the system's address is 128.36.0.3. *Telnet* used this address to connect to the remote system.

The TCP/IP naming and numbering schemes will be explained in Chapter 5. However, we already can observe that names consist of several words separated by periods, and addresses consist of four numbers separated by periods.

2.4.8   Looking Up a Name in the Domain Name System Database

Like many TCP/IP systems, our local host has a client application called *nslookup* (for "network server lookup"), which lets a user query a Domain Name System database interactively.

Below, the name and address of the local default server is printed in response to an *nslookup* command. A database query is entered simply by typing the name of the host to be looked up. The response repeats the identity of the server before providing the answer.

```
> nslookup
Default Server: DEPT-GW.CS.YALE.EDU
Address: 128.36.0.36

> bulldog.cs.yale.edu.
Server: DEPT-GW.CS.YALE.EDU
Address: 128.36.0.36

Name: bulldog.cs.yale.edu
Address: 128.36.0.3
```

2.4.9   File Transfer Dialogue

Next, we will use the File Transfer Protocol to copy a file named *chapter1* in directory *book* from a host named *plum.yale.edu* to the local host. Lines that start with numbers are messages from the file transfer server. The *cd* (change directory) command is used to change to directory *book* at the remote host. The *get* command is used to copy the file.

```
> ftp plum.yale.edu
Connected to plum.yale.edu.
220 plum FTP server (SunOS 4.1) ready.

Name : icarus
331: Password required for icarus

Password :
230 User icarus logged in.

ftp> cd book
250 CWD command successful.

ftp> get chapter1
200 PORT command successful.
150 ASCII data connection for chapter1 (130.132.23.16,3330) (32303 bytes).
226 ASCII Transfer complete.
32303 bytes received in 0.95 seconds (33 Kbytes/s)

ftp> quit
221 Goodbye.
```

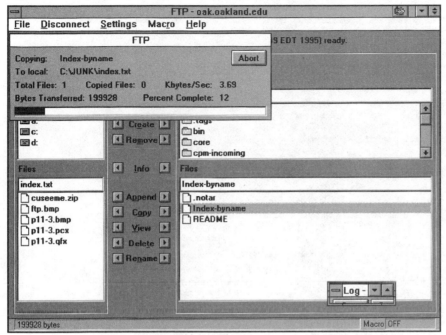

Figure 2.2 File transfer using a desktop application.

File transfer is even easier when you use a desktop product with a Graphical User Interface (GUI). Figure 2.2 shows a file being copied from a server to a PC using the Netmanage *Chameleon* file transfer application. The files listed on the right are located at a Unix server. The Unix file called Indexbyname has been selected. The name index.txt was typed in a window on the left and will be assigned to the PC copy of the file. Files are copied by clicking an arrow or by dragging and dropping.

2.4.10 World Wide Web

You can copy files from a World Wide Web server without even thinking about it. Figure 2.3 shows a *Netscape Navigator browser* screen. New documents are retrieved by clicking on an underlined phrase. They can be saved to the local disk using the File/Save menu.

2.4.11 News

Figure 2.4 shows a *Chameleon* news screen that lists some of the news groups that deal with scientific topics.

2.4.12 File Access Dialogue

Let's look at one last dialogue. In this example, we are using a Disk Operating System (DOS) PC which is attached to a TCP/IP network. We switch to device *d:* at our local host and list the contents of the root directory:

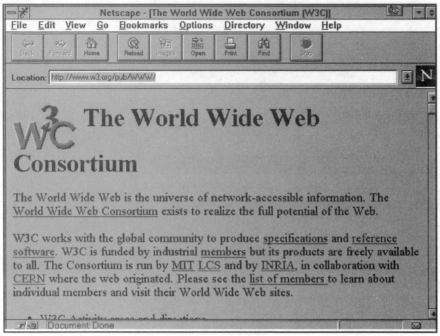

Figure 2.3   Using a *Netscape* browser.

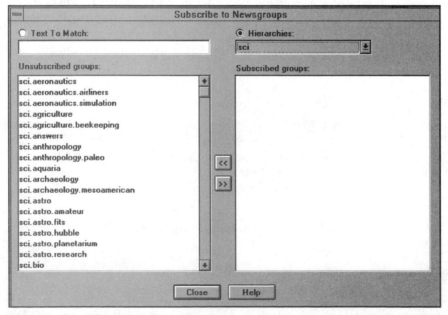

Figure 2.4   Scientific news groups.

```
d:
d:\> dir
 Volume in drive D is SERVER
 Directory of D:\

 .    <DIR>            10-25-95    8:03a
 ..   <DIR>            10-25-95    8:03a
 ALTBWPM        711     2-18-95   12:53p
 EGA512 FRS    3584     9-16-94    3:57p
 WPRINT1     344392    11-05-95   13:28p
 README WPD    5492     9-16-95    3:57p
 SPELL  EXE   40448     9-16-95    3:55p
 WP     EXE  252416    11-15-95    4:51p
 . . .
```

There is nothing about this that looks special—but that is the point. The files that appear to be on local device *d:* actually are read from a remote Network File System server.

3

TCP/IP Architecture

3.1 INTRODUCTION

TCP/IP was designed for an environment that was quite unusual in the 1970s but now is the norm. The TCP/IP protocols had to connect equipment from different vendors. They had to be capable of running over different types of media and data links. They had to unite sets of networks into a single *internet,* all of whose users could access a set of generic services.

Furthermore, the academic, military, and government sponsors of TCP/IP wanted to be able to plug new networks into their internets without interruption of service to the rest of the network.

These requirements shaped the protocol architecture. The need for independence of media technology and plug-and-play network growth led to the decision to move data across an internet by chopping it into pieces and routing each piece as an independent unit.

The functions that guarantee reliable data transmission were placed into source and destination hosts. Because of this, router vendors could focus their efforts on improving performance and keeping up with new communications technologies.

As it happens, the TCP/IP protocols turned out to scale very well, running on systems ranging from mainframes to PCs. In fact, a useful subset that supports network management routinely is ported to "dumb" network devices such as bridges, multiplexers, and switches.

3.2 LAYERING

In order to achieve a reliable exchange of data between computers, there are many separate procedures that must be carried out:

- Package the data
- Determine the path that the data will follow
- Transmit the data on a physical medium

- Regulate the rate of data transfer according to the available bandwidth and the capacity of the receiver to absorb data
- Assemble incoming data so that it is in sequence and there are no missing pieces
- Check incoming data for duplicated pieces
- Notify the sender of how much data has been received safely
- Deliver data to the right application
- Handle error or problem events

The result is that communications software is complicated! Following a layered model makes it easier to group related functions together and implement communications software in a modular manner.

The specific structure selected for the TCP/IP protocols was dictated by requirements that evolved in the academic and defense communities. IP does what is needed to glue different types of networks into an internet. TCP provides reliable data transfer.

The OSI data communications model was strongly influenced by TCP/IP's design. OSI layering and OSI terminology have become a standard part of the data communications culture.

Figure 3.1 shows the TCP/IP and OSI layers. Let's take a brief look what happens within each of the layers, starting from the bottom.[1]

---

[1]TCP/IP does not implement formal session and presentation layers.

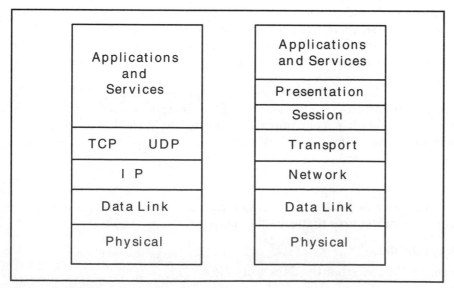

Figure 3.1   TCP/IP and OSI layers.

3.2.1 Physical Layer

The physical layer deals with physical media, connectors, and the signals that represent 0s and 1s. For example, Ethernet and Token-Ring network interface cards and cables implement physical layer functions.

3.2.2 Data Link Layer

At the data link layer, data is organized into units called *frames*. As shown in Figure 3.2, each frame has a header that includes address and control information and a trailer that is used for error detection.

A Local Area Network (LAN) frame header contains source and destination "physical addresses" that identify the source and destination network interface cards on the LAN. The header for a frame that will be transmitted across a frame relay Wide Area Network (WAN) contains a circuit identifier in its address field.

Recall that a *link* is a local area network, a point-to-point line, or some other wide area facility across which systems communicate by means of a data link layer protocol.

3.2.3 Network Layer

The Internet Protocol performs network layer functions. IP routes data between systems. Data may traverse a single link or may be relayed across several links in an internet. Data is carried in units called *datagrams*.

As shown in Figure 3.3, a datagram has an IP header that contains layer 3 addressing information. Routers examine the destination address in the IP header in order to direct datagrams to their destinations.

The IP layer is called *connectionless* because every datagram is routed independently and IP does not guarantee reliable or in-sequence delivery of datagrams. IP routes its traffic without caring which application-to-application interaction a particular datagram belongs to.

3.2.4 Transport Layer—TCP

The Transmission Control Protocol performs transport layer functions, TCP provides reliable data connection services to applications, TCP contains the mechanisms that guarantee that data is delivered error-free, without omissions, and in sequence.

Figure 3.2 Frame format.

Figure 3.3 IP datagram.

An application, such as file transfer, passes data to TCP. TCP adds a header, forming a unit that is called a *segment*.

TCP sends segments by passing them to IP, which routes them to the destination. TCP accepts incoming segments from IP, determines which application is the recipient, and passes data to that application in the order in which it was sent.

3.2.5 Transport Layer—UDP

An application sends a stand-alone message to another application by passing the message to the User Datagram Protocol (UDP). UDP adds a header, forming a unit called a *UDP Datagram* or *UDP message*.

UDP passes outgoing UDP messages to IP. UDP accepts incoming UDP messages from IP and determines which application is the recipient.

UDP is a connectionless communication service that often is used by simple database lookup applications.

3.2.6 Application Services

As we saw in Chapter 2, the TCP/IP protocol suite includes a set of standard application services such as terminal access, file transfer, Network File System (NFS) file server access, electronic mail, news, the World Wide Web (WWW), and Domain Name System address lookups.

3.2.7 Packaging Data for Transmission

Figure 3.4 shows how application data is packaged for transmission. The generic term for information combined with an appropriate layer header is *Protocol Data Unit* (PDU). For example, a TCP segment is a transport layer PDU, and an IP datagram is a network layer PDU.

3.3 PROTOCOL OVERVIEW

Figure 3.5 shows how the components of the TCP/IP protocol suite fit together.

Although text-based user interfaces for file transfer, terminal access, news, and Domain Name System name-to-address queries have not been formally

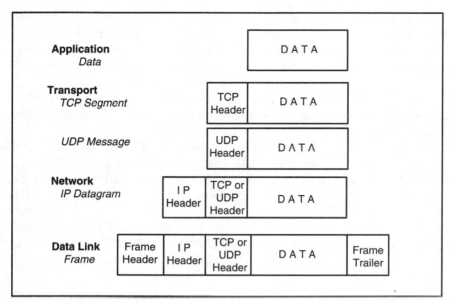

Figure 3.4 Packaging data for transmission.

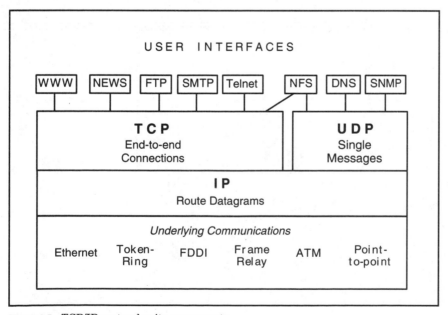

Figure 3.5 TCP/IP protocol suite components.

standardized, most vendors offer command sets that copy the Berkeley Software Distribution Unix end-user interfaces. Users who work in text command mode on two or more types of hosts find it very helpful that the user interface stays pretty much the same as they move from system to system.

There are many different graphical user interfaces (GUIs) available for Windows and Macintosh desktop systems. Although they differ in detail, they follow standard operating system conventions and generally can be used without special training.

World Wide Web, news, file transfer (FTP), mail (SMTP), and terminal access (*telnet*) clients communicate with their servers via reliable TCP connections. Most NFS file clients exchange UDP messages with their servers, although there are some NFS implementations that can run over both UDP and TCP.

Domain Name System directory lookups are based on UDP messages. Simple Network Management Protocol (SNMP) management stations retrieve information from network devices using UDP messages.

3.4 ROUTERS AND TOPOLOGY

The TCP/IP protocol suite can be used on stand-alone LANs and WANs or on complex internets created by gluing many networks together. Figure 3.6 illustrates stand-alone network links. Any hosts that are equipped with TCP/IP can communicate with one another across a LAN, point-to-point line, or wide area packet network.

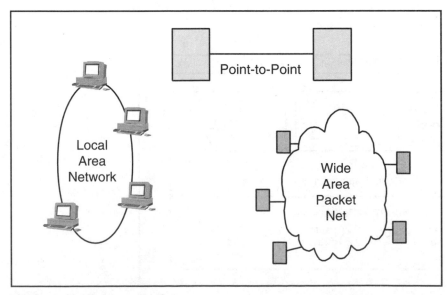

Figure 3.6 Stand-alone networks.

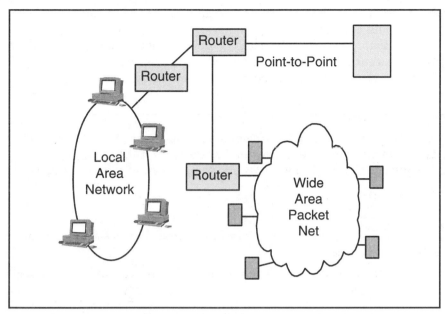

Figure 3.7   Gluing networks together with routers.

Networks are joined into an internet by means of *IP routers*. Figure 3.7 shows an internet that was created by connecting the stand-alone networks together via IP routers.

Modern router products are equipped with multiple hardware interface slots that can be configured with the combination of attachments that the customer needs: Ethernet, Token-Ring, Fiber Distributed Data Interface (FDDI), point-to-point synchronous, frame relay, or whatever.

Internets can be built up with arbitrarily messy topologies. However, when an internet has a coherent structure, it is easier for routers to do their job efficiently and to react quickly to a failure in some part of the network, altering paths so that datagrams avoid a trouble spot. An easy-to-understand logical design also helps network managers to diagnose, locate, and repair network faults.

The robust and competitive IP router market has helped to promote TCP/IP architecture. Router vendors are quick to implement new LAN and WAN technologies, widening their customers' connectivity options. The router price/performance ratio has decreased steadily over the past few years.

3.5   IP ROUTING

IP software runs in hosts and in IP routers. If the destination for a datagram is not on the same link as the source host, IP in the host directs the datagram to a local router. If that router is not directly connected to the destination link, the datagram must be sent on to another router. This continues until the destination link is reached.

IP routes to *remote* locations by looking up the destination network in a routing table. A routing table entry identifies the next-hop router to which traffic should be relayed in order to reach a destination.

3.5.1 Routing Protocols

In a small, static internet, routing tables can be entered and maintained manually. In larger internets, routers keep their tables up to date by exchanging information with one another. Routers can dynamically discover facts such as:

- A new network has been added to the internet.
- The path to a destination has been disrupted, and that destination cannot be reached at this time.
- A new router has been added to the internet. This router provides a shorter path to certain destinations.

There is no single required standard for router-to-router information exchange. The freedom to choose the most convenient protocol has stimulated competition and has led to great improvements in these protocols.

The network facilities under the control of one organization are called an *Autonomous System* (AS). An organization can choose any routing information exchange protocol that it likes within its own Autonomous System. A routing information exchange protocol used within an Autonomous System is called an *Interior Gateway Protocol,* or IGP.

The *Routing Information Protocol* (RIP) is a popular standard Interior Gateway Protocol. RIP is popular because it is simple and widely available. However, the newer *Open Shortest Path First* (OSPF) protocol has a rich set of useful features.

Although all routers support one or more standard protocols, some router vendors also provide a proprietary protocol for router-to-router information exchange. Many router products can run several routing protocols at the same time.

3.6 TCP ARCHITECTURE

TCP is implemented in hosts. The TCP entity at each end of a connection must ensure that the data delivered to its local application is:

- Accurate
- In sequence
- Complete
- Free of duplicates

The basic mechanism for doing this has been used since the dawn of data communications. The sending TCP:

- Numbers each segment
- Sets a timer
- Transmits the segment

The receiving TCP has to keep its partner informed of how much correct data has arrived by means of acknowledgments (ACKs). If an ACK for a segment does not arrive within a timeout period, TCP resends the segment. This strategy is called *retransmission with positive acknowledgment*. Occasionally, a retransmission will cause duplicate segments to be delivered to the receiving TCP.

The receiving TCP must arrange incoming segments in the right order, discarding duplicates. TCP delivers data to its application in order, without any missing pieces.

So far we have made it sound as if one side sends and the other receives. TCP is a *full duplex* protocol; that is, both ends of a connection can send and receive at the same time, so there are in fact two streams of data being transmitted. TCP plays a sender role and a receiver role simultaneously.

3.7   UDP ARCHITECTURE

UDP is implemented in hosts. UDP makes no promise of guaranteed delivery, and it is up to the peer applications to exchange information that confirms that data has arrived safely.

An application that wants to send data via UDP passes a block of data to UDP. UDP simply adds a header to the block and transmits it.

An application participating in UDP communications may send and receive User Datagram messages at any time. It is up to the clients and servers that are built on top of UDP to keep track of any relationship between the User Datagrams that are exchanged.

3.8   SECURITY CONCEPTS

TCP/IP has succeeded very well in opening up communications between computers on a LAN, across a site network, and even globally. But connectivity gives rise to new concerns about the security of information.

The basic security issues in a networked environment are the same as those in a central host environment:

- Authenticating users
- Integrity—assuring that data is not changed
- Confidentiality—preventing unwanted disclosure of information

3.8.1 Authentication

One important aspect of computer security is knowing who is who. In the past we have relied on user identifiers (IDs) and passwords to identify interactive users. We have relied on the "From:" field in an electronic mail message to identify the sender. But passwords can be captured by eavesdroppers, and electronic mail can be forged.

If we are going to conduct any serious transactions on our TCP/IP networks, we will need some way to identify originators reliably. Reliable identification of an originator is called *authentication*.

3.8.2 Message Digest Technology

A simple but effective authentication technology is based on *message digests*. As shown in Figure 3.8, a message digest is a calculation performed on a message using a secret key. Message Digest 5 (MD5) is currently in wide use.[2]

The *Challenge Handshake* illustrates one way that message digests are used. Just as with conventional authentication, a user is given a password which is registered at a host. However, the password will never be sent across a network. Instead, the user's desktop system will perform an MD5 calculation using the password as the secret key. As shown in Figure 3.9,

1. The user sends a userid to a host.
2. The host sends a random message to the user.
3. The host and the user's desktop system both perform an MD5 calculation on the random message and the user's secret password.

---

[2]MD5 was designed by Ronald Rivest. See RFC 1321.

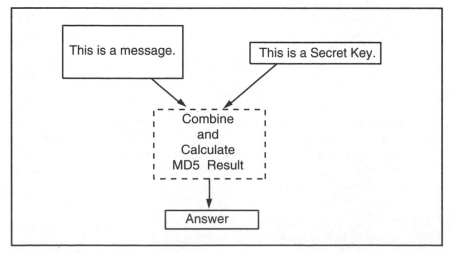

Figure 3.8 Using a message digest.

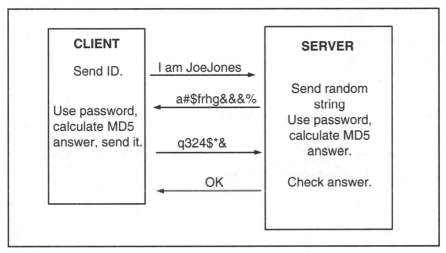

Figure 3.9   Using MD5 in a challenge handshake.

4. The user's system sends the answer to the host.

5. The host compares answers. If the user's system sent the right answer, the user is authenticated.

3.8.3   Message Integrity

MD5 and a shared secret key also can be used to detect whether data has been changed in transit. As shown in Figure 3.10,

1. An MD5 calculation is performed on the data and the secret key.

2. The data and the resulting message digest are sent to the partner.

3. The partner performs an MD5 calculation on the data and the secret key.

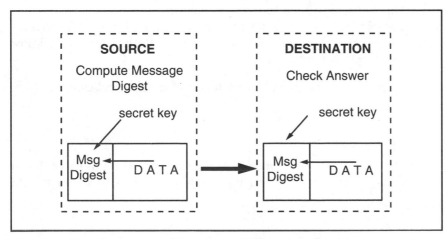

Figure 3.10   Protecting message data using an MD5 message digest.

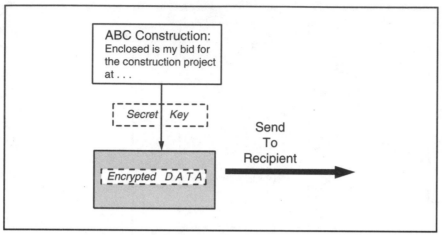

Figure 3.11 Symmetric encryption.

4. The partner compares the answer with the enclosed message digest. If they match, the data has not been changed.

Note that without knowing the secret key, a snooper cannot forge or change data. This mechanism is used for secure electronic mail and for client/server transactions that need to be protected.

3.8.4 Confidentiality via Symmetric Encryption

To prevent snoopers from reading and using your data, the data must be encrypted. The classical way to do this is for the sender and receiver to agree on a secret key. The data is encrypted using this key before it is sent. Often a message digest is added so that the receiver can check that the entire message is received exactly as it was sent. As shown in Figure 3.11, once the data has been encrypted, it looks like a string of gibberish.

This is the traditional *symmetric* method of encryption. Symmetric encryption uses the *same* key to encrypt and decrypt data. Both users must know the key and keep it secret. Disadvantages are:

- For safety, a separate key is needed for each pair of entities that communicate.
- Updating keys is difficult.

3.8.5 Asymmetric Public Key Encryption

Recently, methods that perform *asymmetric* encryption have been introduced.[3] Asymmetric encryption uses *different* keys to encrypt and decrypt data.

---

[3]The idea is due to Diffie, Hellman, and Merkle.

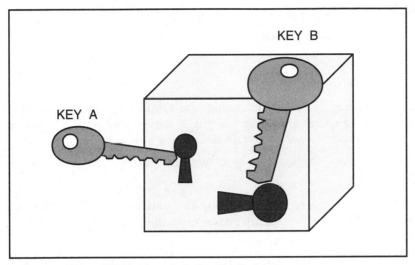

Figure 3.12 Using different keys to lock and unlock.

To understand this, suppose that you had a box with two different keys, A and B, as shown in Figure 3.12.

- If you lock the box using A, you must unlock it using B.
- If you lock the box using B, you must unlock it using A.

Asymmetric encryption also is called *Public Key* encryption because it enables you to manage your encryption keys in a very convenient way. Key A can be your *public key*. You can give it to your friends or print its value in a directory.

- All partners can use your public key to encrypt data sent to you.
- *No one else knows your private key, so no one else can decrypt the data that is sent to you.*[4]

Managing public/private keys is much easier than managing symmetric keys. But we still need a reliable registration authority so that we can be assured that a key published as "Jane Jones' Public Key" really belongs to Jane Jones, and not an impersonator.

Unfortunately, the asymmetrical encryption methods known today are very slow. A combined asymmetric/symmetric method is preferred.

---

[4]The public/private scheme that currently is in use is based on the fact that numbers that are the product of two large primes appear to be very difficult to factor. It took a worldwide team several months to factor a 129-digit number. However, computer speeds are increasing. Experts advise the use of 1024-bit keys for data that must remain secret for several years.

3.8.6 Combined Encryption

Combined encryption works as follows:

1. A random symmetric key is chosen.
2. Data is encrypted with this key.
3. Then the random key is encrypted using the recipient's public key and included in the message. (This is like putting the new random key inside a container that has been locked with the recipient's public key.)
4. The recipient decrypts the temporary random key and then uses this to decrypt the data.

As shown in Figure 3.13, the recipient's public key is used to put an envelope around the random key. The recipient is the only one that can unlock this envelope.

In later chapters, we will see that all of these methods are being introduced to TCP/IP communications and applications. The most ambitious effort is presented in Chapter 24, which describes how authentication and encryption are added to the IP layer for both classic version 4 IP and the new version 6 "IP, Next Generation."

3.9 LOOKING AHEAD

The topics that we have met in this brief overview will be visited in detail in the chapters that follow. The mysteries of the physical and data link layers are explored in Chapter 4. Chapter 5 explains naming and addressing. The IP protocol is mapped out in Chapter 6, while Chapter 8 focuses on routing.

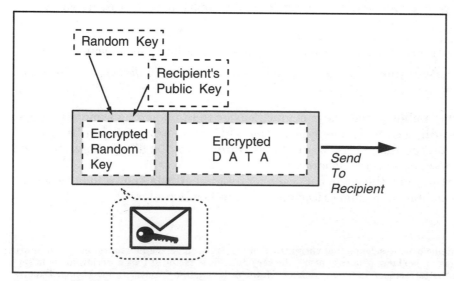

Figure 3.13 Enclosing a key used to decrypt a message.

Chapter 7 deals with some error handling and query facilities associated with IP. UDP is described in Chapter 9. Chapter 10, one of the longest in this book, is devoted to TCP.

Chapter 11 describes automatic configuration of TCP/IP systems via a boot server. Chapter 12 describes the Domain Name System. Chapters 13 through 19 describe the major application services.

Chapter 20 deals with network management. Chapter 21 describes the socket programming interface and examines a few sample programs.

Chapters 22 to 24 introduce IP version 6 and the new security architecture for IP networks.

4

Physical and Data Link Technologies

4.1 INTRODUCTION

During the past few years, an unprecedented number of innovative LAN and WAN technologies have been introduced and quickly absorbed into the marketplace. The use of twisted pair and fiber media have proceeded at a pace that no one could have predicted. Integrated Services Digital Network (ISDN), frame relay, T1, fractional T1, T3, SONET[1] fiber-optic lines, Switched Multimegabit Data Service (SMDS), cable connections, and Asynchronous Transfer Mode (ATM) promise wide area connections that are faster and cheaper.

As each new technology has emerged, the Internet Engineering Task Force (IETF) has responded quickly, writing specifications for running IP—along with other protocols—over the new medium. Then, with almost no delay, router vendors have produced hardware interfaces and software drivers that have enabled users to take advantage of the new technology.

The efforts of the IETF can be seen in the long series of Request For Comment (RFC) documents with titles like:

The Point-to-Point Protocol (PPP) for the Transmission of Multiprotocol Datagrams over Point-to-Point Links

Standard for the transmission of IP datagrams over IEEE 802 networks

Transmission of IP and ARP over FDDI Networks

Classical IP and ARP over ATM

---

[1]*Synchronous Optical Network*, a telephony standard for the transmission of information over fiber-optic channels.

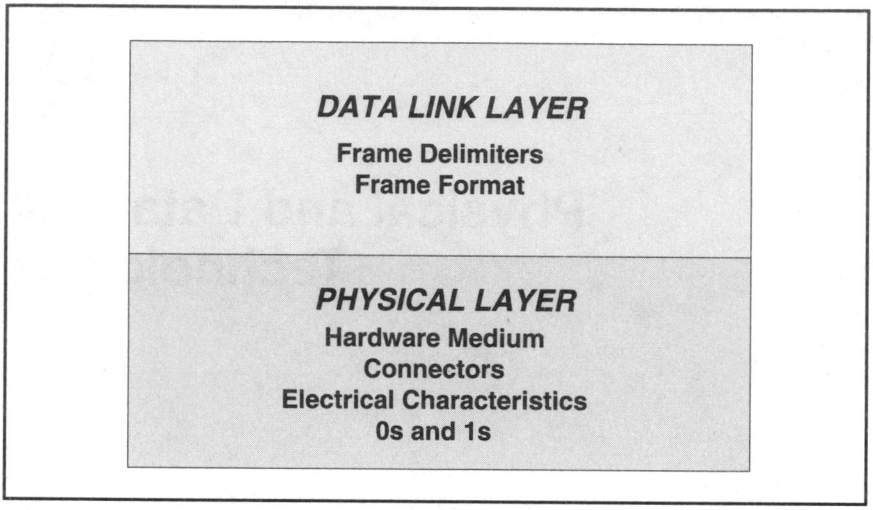

Figure 4.1   Lower layer functions.

4.2   PHYSICAL, MAC, AND DATA LINK FUNCTIONS

This chapter describes how IP runs on top of various lower-layer technologies. Let's take a moment to recall what happens at these lower layers, which are represented in Figure 4.1.

The physical layer specifies the cable, connectors, and electrical characteristics of a medium. The rules for impressing individual 0s and 1s upon the medium also belong to the physical layer.

To enable us to make sense out of the data that we transmit, we package it into units called frames.[2] A frame carries information across a single link. To reach its final destination, an IP datagram may need to be carried across several links.

The description of a frame's format belongs to the data link layer. The frame format differs depending on the underlying technology used for the link (e.g., a T1 line, frame relay circuit, or an Ethernet LAN). Each frame has a header that provides the information needed to deliver the frame across a link. The format of the header depends on the technology used.

4.3   NETWORK TECHNOLOGIES

Network technologies break down roughly into four categories:

1. Wide area point-to-point lines

2. LANs

---

[2]Note that some authors refer to these units as *packets*.

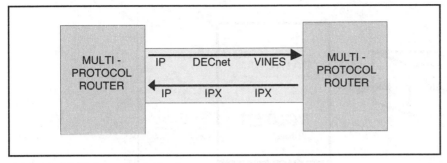

Figure 4.2 Multiple protocols sharing a medium.

3. Wide area packet delivery services
4. Cell switching services

For each technology, we need mechanisms to:

- Identify the destination when a single interface leads to multiple systems (for example, for a LAN interface)
- Detect errors when the data becomes corrupted in transit

Today, both local and wide area media are *multiprotocol* environments. As shown in Figure 4.2, a link often is shared by several protocols such as TCP/IP, Novell IPX/SPX, DECnet, Vines—and even bridged traffic. Multiprotocol hosts and routers need a way to sort out the various types of traffic, so we also need a mechanism to:

- Identify the protocol type for the Protocol Data Unit (PDU) carried within each frame.

Identifying a protocol type does not sound like it should be a difficult job. You just get a standards body to make a list of protocols, assign a number to each, and put that number into a field in the frame header.

And it is that easy—except that several standards bodies have done the job, and each uses different introducers for the fields and different numbers to identify the protocols. In this chapter, we will describe the various formats used for the major transmission technologies.

4.4 REPACKAGING

There is an Olympic event in which a competitor swims one link of a course, grabs a bicycle and pedals another, and runs a third. IP works the same way. Internet designers built IP so that datagrams could be relayed from one medium to another to reach their destination.

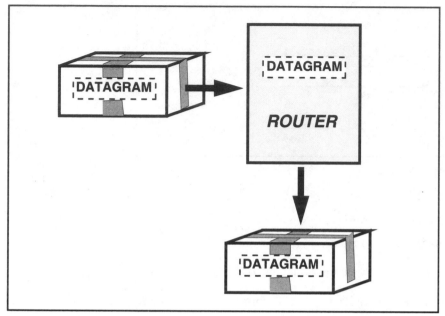

Figure 4.3   Repackaging datagrams.

Before a datagram is transmitted across a link, it is wrapped in frame packaging appropriate to the link. When a router receives a frame (see Figure 4.3):

- The router strips off the frame wrapping and extracts the datagram.
- The router looks at the datagram's destination IP address and chooses the next-hop medium.
- The router then repackages the datagram in a new frame wrapping for its trip across the next link and transmits the datagram on its way.

Now we are ready to get down to the specifics. In the sections that follow, we will discuss the way that data is packaged for the various types of network technology. We will start off with point-to-point links.

4.5   POINT-TO-POINT PROTOCOLS

IP datagrams can be sent across a point-to-point link between a pair of hosts, a pair of routers, or a host and a router. IP will transmit datagrams for many different TCP and UDP interactions across a single point-to-point link.

IP does not know or care about the identities of source or destination applications. Every time that IP is handed an outgoing datagram, it transmits the datagram as soon as it can. As illustrated in Figure 4.4, the traffic for many different client/server interactions shares a link—just as riders who have many different destinations share a subway ride.

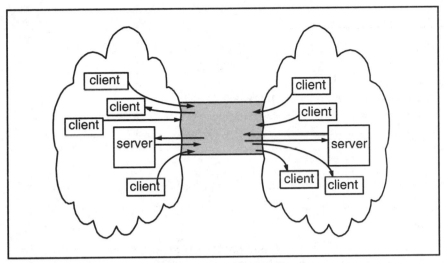

Figure 4.4   Multiple clients and servers sharing a link.

Today, IP traffic is carried across point-to-point links packaged in several different ways:

- Using one of the conventional versions of the High-level Data Link Control (HDLC) point-to-point protocol
- Via the Internet standard Point-to-Point Protocol (PPP)
- Using the Serial Line Interface Protocol (SLIP)

Little by little, implementations are migrating over to the Internet PPP standard, which offers many advanced features.

4.6   HDLC

The High-level Data Link Control (HDLC) protocol is an international standard for point-to-point links that was written in the 1960s. HDLC sends serial data as a clocked stream of bits which is partitioned into frames. Every frame is delimited by the special *flag* pattern:

0 1 1 1 1 1 1 0

In order to recognize this pattern, it is necessary to prevent the flag pattern from appearing within the user's data. To accomplish this, after transmitting the opening flag, the sending hardware will insert a 0 after any five consecutive 1s in the data. This procedure is called *zero-bit insertion* or *bit-stuffing*.

At the receive end of the link, after recognizing the beginning of a frame, the receiving hardware will remove any zero which appears after five consecutive 1s within the frame.

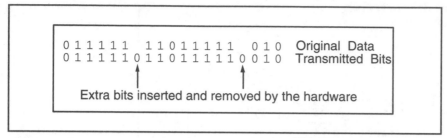

Figure 4.5   HDLC bit-stuffing.

Figure 4.5 shows some sample data before and after bit stuffing is executed.

4.6.1   HDLC Frame Format

The HDLC protocol established a basic pattern that has influenced all subsequent frame formats. As shown in Figure 4.6, an HDLC information frame is made up of a header, some data, and, at the end, a trailer that contains a *Frame Check Sequence* (FCS). Flag octets are used as delimiters at the beginning and end of the frame.

The frame check sequence is the result of a mathematical computation performed on the frame at its source.[3] The same computation is performed at the destination end of the link. If the answer does not agree with the value in the FCS field, some bits of the frame have been altered in transmission, and so the frame is discarded.

The use of a frame check sequence to detect transmission errors was a very successful idea. We will see an FCS field in all but one of the WAN and LAN frames.

The HDLC frame header contains a *destination address* field. This field is needed for *multipoint* versions of HDLC (such as IBM's Synchronous Data Link Control, or SDLC) that enable many systems to share a single line. Each system is assigned an address, and traffic is directed to a system by putting its address in the header.

---

[3]The calculation itself is called a Cyclic Redundancy Check (CRC). Some authors call the value placed in the trailer the CRC, instead of calling it the *frame check sequence*.

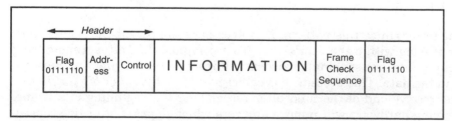

Figure 4.6   Format of an HDLC frame, with flag delimiters.

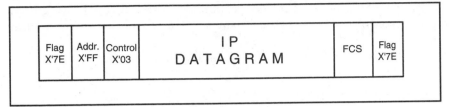

Figure 4.7   Format of an HDLC frame carrying an IP datagram.

IP does not make use of multipoint line technology, and IP datagrams are carried in an HDLC frame whose address is set to the binary value 11111111 (X'FF in hexadecimal) which is known as the *All-Stations,* or broadcast, address.

The HDLC frame header also contains a *control* field. Some link protocols[4] put frame numbers and acknowledgment numbers into the control field. These link protocols retransmit numbered frames that are not acknowledged within a timeout period.

Frames that carry IP—and many other protocols (including IPX and DECnet)—do not require numbering and acknowledgment. For IP (and these other protocols), the control field is set to X'03, which identifies an HDLC *Unnumbered Information* frame.

Thus, an IP datagram wrapped in an HDLC frame has the format shown in Figure 4.7.

To summarize, when an HDLC frame carries an IP datagram:

- The All-Stations address, X'FF, is used.

- The control field is set to X'03, meaning unnumbered information.

4.6.2   Problems with HDLC

The fact that HDLC is a "standard" does *not* mean that you can run a point-to-point line between any two HDLC interfaces and expect them to communicate successfully with one another.

There are many options defined within the HDLC standard—and many different "standard" versions of HDLC have been implemented. Just to confuse things even more, many vendors have implemented their own versions of HDLC point-to-point interfaces.

As a result, for a long time there was no single standard for point-to-point communication, making it difficult to interwork equipment from different vendors.

---

[4]For example, IBM's SDLC and the X.25 Link Access Protocol Balanced (LAPB) link layer are based on HDLC and use the control field for numbering, acknowledgment, and retransmission at the link layer.

HDLC was designed before the days of multiprotocol networking. Today, point-to-point lines often carry traffic for several protocols. This causes another problem.

An IETF committee was given the job of solving these problems.

4.7 THE INTERNET POINT-TO-POINT PROTOCOL

The IETF working group's solution was the *Point-to-Point Protocol,* more frequently called *PPP.* PPP can be used over any full-duplex circuit—either synchronous bit-oriented or asynchronous (start/stop) byte-oriented. It can be used on slow dialups, fast leased lines, ISDN, or even on SONET fiber-optic lines. And PPP was designed to carry PDUs for many protocols—IP, IPX, DECnet, ISO, and others. PPP even carries bridged data.

PPP includes several subprotocols. For example:

- The *Link Control Protocol* sets up, tests, configures, and closes down a link.
- *Network Control Protocols* are used to initialize, configure, and terminate use of a particular network protocol. A separate Network Control Protocol is defined for each of IP, IPX, DECnet, ISO, and so forth.

A typical PPP scenario is:

1. An originating PPP sends a *Link Control* frame to start things going. The partners exchange additional Link Control frames to establish the options to be used for the link.
2. *Network Control Protocol* frames are exchanged to choose and configure the network layer protocols to be used.
3. Data for the selected protocols is sent across the link in PPP frames. Each frame includes a header field that identifies what type of protocol data is enclosed.
4. Network Control and Link Control Protocol frames are used to close the link down.

A PPP frame header looks like an HDLC header that contains one extra field which identifies the next-layer protocol. Figure 4.8 shows the format of a PPP

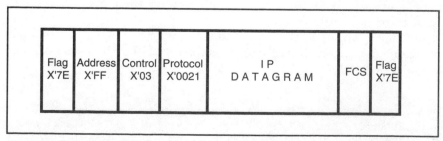

Figure 4.8 Format of a PPP frame carrying an IP datagram.

frame containing an IP datagram. The address field contains X'FF ("all stations") and the control field contains X'03 ("unnumbered information"). The additional *protocol field* contains X'00-21, a value that indicates that the frame is carrying an IP datagram. Protocol numbers to be used with PPP are published by the Internet Assigned Numbers Authority (IANA) in its *Assigned Numbers* RFC document.

4.7.1 PPP Compression

It may seem wasteful to include the same address and control octets in every frame. In fact, the partners at each end of the PPP link can negotiate to operate in a *compressed* mode that eliminates these fields.

The value in the protocol field indicates whether the information content is a Link Control message, a Network Control message, or information, such as an IP datagram. At PPP link setup time, the protocol field starts out as 16 bits in length, but the size of the protocol field used when information is transferred can be negotiated down to 8 bits. Hence, a datagram can be wrapped in the efficient packaging shown in Figure 4.9.

Van Jacobson compression is another PPP option that saves transmission bytes for TCP sessions. IP and TCP headers together account for 40 or more bytes of overhead. Van Jacobson compression reduces a typical 40-byte combination to 3, 4, or 5 bytes, which is a significant saving.[5]

4.8   ADDITIONAL PPP CAPABILITIES

The PPP working group tackled some other problems that can arise when using a point-to-point link.

4.8.1 Authentication

PPP often is used to connect a telecommuting or traveling user to an IP network via a dialup connection. Dialups sometimes are used by small branch office routers to connect a workgroup LAN to a headquarters site.

---

[5]The way that this is done is that both partners save the initial headers. These are updated during the TCP session by sending only the changes to header values. Since most header information is static, the saving is substantial.

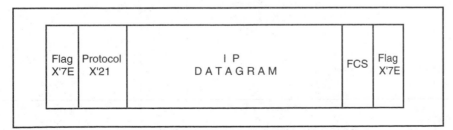

Figure 4.9 A PPP frame with compressed format.

Before permitting an external system to connect to your network via a dialup link, you might wish to authenticate that system! Currently, PPP supports two forms of authentication:

- The simple *Password Authentication Protocol* (PAP). A cleartext userid and password are packaged inside a frame sent across the link during the link setup procedure.
- The *Challenge Handshake Authentication Protocol* (CHAP).

The Challenge Handshake (which was described in Chapter 3) is quite clever. Recall, as shown in Figure 4.10, that:

1. A cleartext username is sent across the link.
2. The remote partner sends back a random challenge message.
3. The local system performs a message digest computation (using the challenge message and the user's password as inputs) and sends the answer back.
4. The remote partner looks up the password, does the same calculation, and compares the answers.

An eavesdropper will see different garbage bytes each time the link is set up. When a solid 16-byte password is used, it is virtually impossible to figure out the password by watching the link.

4.8.2 Automatic Link Quality Monitoring

PPP often is used between a pair of routers. Sometimes the quality of a link degrades for some reason. It would be helpful to have early warning of the

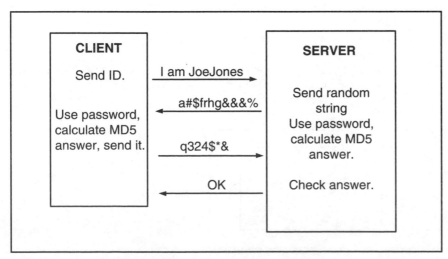

Figure 4.10 The PPP challenge handshake.

condition of the link so that some action could be taken automatically. For example, a router could terminate a dialup connection and redial. Or if the problem is occurring on a leased line, the router could send an alert to management personnel and possibly shunt traffic onto an alternative link temporarily.

PPP provides a very simple and effective way to check up on link quality. The link monitoring process simply counts the number of transmitted and received frames and octets. (Incoming discards and errors also are counted.) Periodically, a report is sent to the peer at the other end of the link.

This information gives a good picture of what is happening on the link. For example, if I have sent 100,000 octets during a time interval, but my partner reports that it has only received 50,000 of them successfully, something is very wrong with the link.

4.9 SERIAL LINE INTERFACE PROTOCOL

The Serial Line Interface Protocol (SLIP) was invented before PPP was available and provides a rudimentary method for transmitting IP datagrams across a serial link.

SLIP is surely the most primitive protocol ever invented. An IP datagram simply is transmitted, byte by byte, on a serial line. SLIP marks the end of a datagram with the delimiter byte, 11000000 (X'C0). What happens when X'C0 appears inside a datagram? The transmitting SLIP uses escape sequences that the receiving SLIP translates back to the data that actually was sent:

C0 in data — > DB DC

DB in data — > DB DD

SLIP typically is used to connect a PC, Macintosh, or Unix computer to an IP network via a dialup link. Note that SLIP provides no frame check sequence and leaves all error checking to higher layers. SLIP cannot carry any protocol other than IP.

Compressed SLIP (CSLIP) is an improved version of SLIP that compresses TCP/IP headers via the Van Jacobson algorithm. CSLIP provides much better link performance than SLIP.

SLIP can be used for host-to-host, host-to-router, or router-to-router communications. Figure 4.11 shows a communications server that supports both "dumb" ASCII terminal dialins and SLIP dialups. The device acts as an IP router for the SLIP traffic.

SLIP's most appealing feature is that it is widely available. Its most annoying feature is that the workstation user needs to write a script that will read prompts sent by the communications server and send a userid, password, and other information at appropriate points in the dialog. PPP is more functional, does not require scripts, and is gradually displacing SLIP.

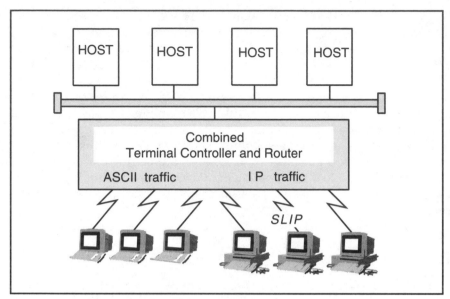

Figure 4.11 ASCII terminal and SLIP connections.

4.10   LOCAL AREA NETWORKS

Next, we will examine how IP and other protocols are packaged in frames that are sent across LANs. Classical LAN design includes several factors:

- Stations share a physical medium.
- There are *Media Access Control* (MAC) rules that determine when a station can transmit data.
- Data is carried in frames.

We'll look at the *Ethernet* technology first, since it provides a very simple example of a LAN implementation.

4.11   DIX ETHERNET

Ethernet LANs were the first to carry IP datagrams. Digital Equipment Corporation (DEC), Intel Corporation, and Xerox Corporation defined the original *DIX* Ethernet specification in 1980. This was revised as version 2 in 1982.

4.11.1   DIX Ethernet Media Choices

The traditional backbone medium for this technology was baseband coaxial cable. Originally, only a heavy half-inch 50-ohm cable was used. Later, a thinner, more flexible quarter-inch grade of coaxial, called *thinnet* or *cheaper-*

net, was introduced, and later still, many sites switched over to twisted pair wiring. A 10-megabit-per-second signaling rate was prevalent for quite a long time, but now speeds of 100 megabits per second are available. DIX Ethernet also runs on broadband and fiber-optic media.

To distinguish between the many different flavors of Ethernet implementations, the following notation is used:

```
[Data rate in megabits per second] [medium type] [maximum cable segment in
hundreds of meters]
```

Thus 10BASE5 means BASEband coax with a data rate of 10 megabits per second and a maximum cable segment length of 500 meters. The thin cable specification is 10BASE2, which means BASEband coax with a data rate of 10 megabits per second and a maximum cable segment size of 200 meters.

Similarly, 10BROAD36 is BROADband coaxial, 10 megabits per second, and a maximum cable segment length of 3600 meters. The twisted pair and fiber specifications are identified as 10BASET and 10BASEF, which do not quite match the pattern.

4.11.2 DIX Ethernet Media Access Control Protocol

DIX Ethernet uses a very simple Media Access Control procedure with a long title: *Carrier Sense Multiple Access with Collision Detection* (CSMA/CD).

An interface with data to send wraps the data in a frame and listens to the medium. As shown in Figure 4.12, if the medium is available, the interface transmits.

The frame header contains the physical address of the destination interface (which often is called the MAC address). The system with that physical address absorbs the frame and processes it. If two or more stations transmit at the same time, they detect the collision, back off for a random amount of time, and try again.

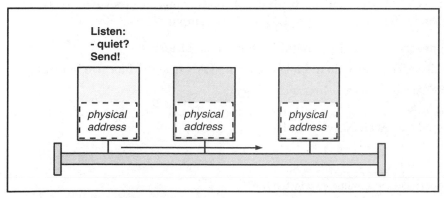

Figure 4.12 Ethernet media access control.

Figure 4.13 DIX Ethernet frame carrying an IP datagram.

4.11.3 DIX Ethernet Media Access Control Frames

The format of a DIX Ethernet frame that is carrying an IP datagram is shown in Figure 4.13.

The destination and source addresses are six octets long.[6] The *Ethernet type* code, X' 08-00, signals that the information content of this frame is an IP datagram.

There are Ethernet type codes that identify many other protocols.[7] The medium can be shared by several of these protocols because the Ethernet type code identifies the protocol for each frame, enabling the destination station to pass the frame's information to an appropriate procedure.

In order to operate correctly, the CSMA/CD protocol requires frames to be at least 64 octets in length. Therefore, it is necessary to add padding after a very short datagram.

4.12 802 NETWORKS

After DIX Ethernet and other LAN technologies had proved their usefulness in the marketplace, the IEEE established the *802 committee,* which was given the task of designing and publishing standards for LAN technologies.

Standards in the 802 series have guided many vendor implementations, and these standards have been recognized by ISO and republished with ISO document numbers.

The 802 standards deal with physical media, media access controls, and frame formats for many types of LANs. For example:

- 802.3 describes a slightly modified version of Ethernet.

- 802.4 describes a broadband token-passing LAN designed for use in factories.

- 802.5 describes Token-Ring technology.

- 802.6 describes a Distributed Queue Dual Bus subnetwork of a Metropolitan Area Network (MAN).

---

[6]These addresses are administered by the IEEE.

[7]See the IANA *Assigned Numbers* document.

4.13   802.2 LLC HEADER

A separate IEEE 802.2 standard defines a *Logical Link Control* (LLC) header to be used with all 802 LAN technologies. The LLC header has two jobs:

- For OSI frames, it identifies the source and destination protocols for the frame.
- It includes a control field.

How are these fields used? The IEEE description involves a lot of formal language, but the purpose of each element is very simple.

ISO destination and source protocol entities for the frame are defined by *Destination Service Access Point* (DSAP) and *Source Service Access Point* (SSAP) codes.

DSAP/SSAP values have been assigned to ISO protocols but not to IP or to a flock of other protocols that are in daily use. For IP and the other common protocols, the DSAP and SSAP are set to X'AA, meaning that another header follows that will tell you what kind of protocol data the frame is carrying. The additional header is called a *Subnetwork Access Protocol,* or *SNAP,* subheader.

What does the SNAP subheader contain? An introducer, followed by our old friend, the Ethernet type code. The introducer has a fancy title— *Organizationally Unique Identifier* (OUI). The OUI identifies who was responsible for assigning the protocol numbers.

The OUI introducer for Ethernet type codes is (as shown in Figure 4.14) X' 00-00-00. A separate OUI of X' 00-80-C2 is used to introduce protocol numbers for various bridge protocols.

4.13.1   802.3 and 802.2

The 802.3 standard includes specifications for Ethernet media, the CSMA/CD Media Access Protocol, and a MAC frame format. According to 802 committee standards, an 802.2 header must be included within an 802.3 MAC frame.

Figure 4.14 shows the result of putting an IP datagram into an 802.3/802.2 frame.

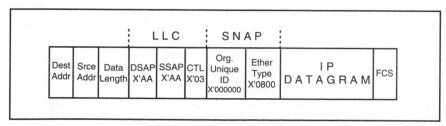

Figure 4.14   802.3 frame with 802.2 LLC and SNAP subheader.

■ Note that unlike DIX Ethernet, the third field of the 802.3 frame header contains the *length* of the information that follows[8] (exclusive of padding) instead of an Ethernet type code. We will see later that an IP header includes a datagram length field, so, for IP, this information is redundant.

■ The DSAP and SSAP are set to X'AA, signaling that a SNAP subheader will follow.

■ The control field is X'03, meaning unnumbered information—just as in HDLC.

■ The X'00-00-00 introducer in the SNAP field indicates that an Ethernet type follows. The Ethernet type is X'08-00.

Other protocols such as IPX and DECnet have similar frames—you just insert the appropriate value for their Ethernet type.

Note that 8 octets of overhead have been added without any change in functionality for IP. For this reason, there are many implementations that still use the older DIX Ethernet format. Ethernet Network Interface Cards and their software drivers usually support both protocols—customers can select their preferred configuration.

People frequently use the term *Ethernet* indiscriminately for either the older DIX or newer IEEE 802.3/802.2 implementations. Sometimes it is important to know which you are talking about. A system configured to talk DIX cannot communicate with a system configured to talk 802.3/802.2.

4.14 LAYERING FOR 802 NETWORKS

Let's spend a little more time exploring the IEEE view of the world. With the advent of 802 LANs, the IEEE divided layer 2, the link layer, into two sublayers, as shown in Figure 4.15.

The MAC sublayer provides the rules for accessing the medium—such as "listen and send" for 802.3 or "wait for the token" in 802.5. The MAC layer also defines the first part of the frame header, which includes the destination and source physical addresses.

The Logical Link Control sublayer defines the format of the LLC header. It also defines fairly complicated rules for communicating when frame numbering, acknowledgments, flow control, and retransmission are used at the data link level. A link that provides these capabilities is called a *Type 2* link. There are several protocols, including SDLC, LAPB, and LAPD, that perform Type 2 communication across a LAN.

Of course, for IP datagrams, the only requirement is that the Logical Link sublayer should indicate that an IP datagram is enclosed in the frame. IP normally runs over a *Type 1* link protocol.

---

[8]The length includes the 8 octets in the LLC and SNAP fields.

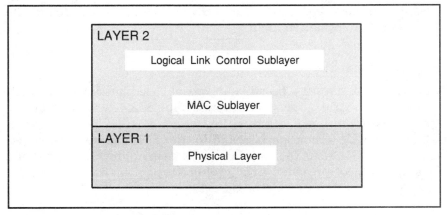

Figure 4.15   Layering for 802 LANs.

4.15   OTHER LAN TECHNOLOGIES

Token-Ring, token bus, and Fiber Distributed Data Interface (FDDI) LANs follow IEEE conventions and *must* include an 802.2 LLC header and SNAP subheader in order to carry IP and the other protocols identified via Ethernet type codes. There is no shortcut format for these LANs.

The same LLC and SNAP fields used in Figure 4.14, namely:

X'AA-AA-03-00-00-00 (*Ethernet type*)

are used to identify the enclosed protocol.

4.15.1   802.5 Token-Ring Configuration and Media

Token-Ring LANs were introduced by IBM, and the IEEE later published a standardized version of the protocol as 802.5. Stations on a Token-Ring are configured as a physical ring.

4.15.2   802.5 Media Access Control Protocol

The idea behind token-based Media Access Control for Token-Rings is simple. A special frame called the *token* is passed from station to station, around the ring. When a station receives the token, it has the right to transmit data for a limited period of time. When that time expires, the token-holder must pass the token to the next station.

Although the basic idea is straightforward, a ring protocol needs many more mechanisms than Ethernet does. In particular, the MAC layer protocol for 802.5 includes procedures for joining or leaving the Token-Ring, identifying neighbors, detecting a dead station or lost token, preventing data from cycling forever, and for signaling problems. There are different MAC layer headers defined for the various 802.5 functions. The protocol type of a frame that carries data is identified via LLC and SNAP headers, which follow the Token-Ring Routing Information Field.

4.15.3   802.4 Token Bus

The 802.4 standard described a broadband coaxial-based bus LAN that used token-passing to control access to the medium. 802.4 was part of the *Manufacturing Automation Protocol* suite, which was devised for use in industrial facilities. Signals on a broadband coaxial medium are not disrupted by the electronic emissions common in a factory environment. The use of a token-passing protocol provided predictable scheduling of LAN access. However, 802.4 never attained widespread use.

4.15.4   Fiber Distributed Data Interface

FDDI 100-megabit-per-second LANs frequently are used as backbone networks, interconnecting slower local area networks.

- FDDI primarily is intended for use with fiber-optic cable, although twisted pair cables also can be part of the network.

- As shown in Figure 4.16, the core of an FDDI network consists of a single or double ring called the *trunk*. Stations can connect directly to the trunk or to concentrators attached to the trunk. Optionally, tree-shaped networks can sprout from concentrators on the trunk.

- When the trunk is a dual ring, the LAN can be set up to recover from a break in the cable. Normally, traffic circulates on one ring. If there is a

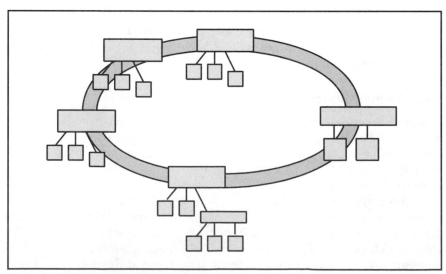

Figure 4.16   Topology of an FDDI network.

failure, the second ring is used to shunt traffic around the fault and keep the network running.

FDDI media access is based on token-passing. In fact, the MAC protocol is modeled closely on the 802.5 Token-Ring.

An FDDI frame has a MAC header and trailer and, when used to carry IP, uses the same 802.2 LLC and SNAP headers that were described earlier to identify the fact that a frame contains an IP datagram.

4.16 USE OF HUBS

Ethernets, Token-Rings, and FDDI LANs started out with very different cabling topologies, but over time, most organizations have chosen to connect their systems to centralized hubs. Hubs simplify LAN administration and repair. Thus the actual physical topology of the various types of LANs has been converging to a single physical topology—a star or a chain of stars.

4.17 SWITCHING

All of the LAN technologies that we have been discussing have one feature in common—a frame sent across the LAN can be "heard" by any station on the LAN. Although the rules say that a frame should only be accepted by an interface with a particular physical address, a system owner often has the power to set the network interface to a promiscuous mode that captures all of the data that appears on the LAN segment.

A desire for improved performance as well as concerns about security have led to the introduction of *traffic switching*. Some intelligent hubs switch each message between its source and destination and hide the message from all other stations.

4.18 BROADCASTING AND MULTICASTING

A multiaccess LAN technology supports broadcasting. An all-1s destination physical address is used to indicate that every interface attached to a LAN should absorb a frame. The hexadecimal representation of a broadcast address may be written as:

X'FF-FF-FF-FF-FF-FF

An interface also can be configured to absorb frames sent to one or more physical *multicast* addresses. Multicasting allows frames to be sent to a selected set of systems. A multicast LAN address always has a 1 in the low-order bit of its first byte—that is, in position:

X'01-00-00-00-00-00

The other bits will be set to values that have been picked for some multicast-based service.

The Internet Assigned Numbers Authority has reserved a list of multicast physical addresses for a number of its services. For example, a multicast address can be used to send a message to every bridge on a LAN. In Chapter 5, we'll see how layer 3 multicasts across an IP network get mapped onto layer 2 LAN multicasts.

The term *unicast address* is used to distinguish a unique physical address assigned to a single interface from broadcast and multicast addresses. If a frame header contains a unicast address, the frame is supposed to be delivered to one specific interface.

Now it is time to leave LANs and look at several special wide area technologies.

4.19 PACKET NETWORKS

The packet-switching technology that was introduced in the experimental ARPANET has been reshaped and used in many types of data communications facilities. X.25 packet networks gained wide use in the 1980s. However, many users are adopting the newer frame relay packet-switching technology, which provides a wider range of bandwidth options.

4.20 X.25 NETWORKS

Our telephone network lets any telephone instrument place a call to any other phone in the world. There is an international standards organization that is responsible for uniting national telephone networks into a global network. For a long time, this organization was called the *International Telegraph and Telephone Consultative Committee,* or *CCITT.* The name has since been changed to the *Telecommunication Standardization Sector of the International Telecommunications Union,* or, more simply, *ITU-T.*

During the 1970s, the CCITT started work on a set of recommendations intended to create a global *data* network. These recommendations reached maturity during the 1980s. The most important of these is X.25, which lays down the rules for connecting a computer to a data network. More specifically, X.25 defines the interface between a computer (called *data terminal equipment,* or *DTE*) and a network communications element (*data circuit-terminating equipment,* or *DCE*) that is part of a private or service provider data network.

X.25 sets up reliable data circuits between computers. These are called *virtual circuits* because, unlike the phone system, a fixed path for the exclusive use of the call is *not* reserved throughout a call. Real links are shared by many concurrent virtual circuits. However, the link sharing that takes place is invisible to the users of the circuits.

X.25 is popular worldwide, and there are many public X.25 data networks connecting computers around the globe.

X.25 data networks offer two types of circuits. *Switched virtual circuits* are data calls that are set up just like a phone call. Participating computers are assigned numbers.[9] A caller enters the number of the computer to be reached, and the call is put through. Alternatively, a customer may acquire *permanent virtual circuits* that behave like dedicated leased lines.

The CCITT recommendations do not place constraints on the *internal* structure of a regional X.25 data network. However, many X.25 data networks use an internal packet-switching technology.

4.20.1   X.25 Layering

X.25 is a three-layer protocol. Its link layer is called *Link Access Protocol Balanced* (LAPB), and its network layer is called the *X.25 Packet Level*. User premise DTE equipment sets up a link to an X.25 provider's DCE. This link is used to carry data for *multiple* layer 3 virtual circuits. A switched virtual circuit is initiated by sending a *Call Request* packet.

4.20.2   X.25 and IP

X.25 is one of the many wide area technologies used to transport IP datagrams. IP uses an X.25 virtual circuit in the same way that it uses a telephony point-to-point line. That is, IP traffic exchanged by hosts or routers is carried across an X.25 virtual circuit.

The X.25 link (layer 2) and packet (layer 3) protocols go to a lot of trouble to make sure that data is transmitted in order and free of errors. An X.25 circuit is intended to provide a reliable end-to-end data connection.

It might seem rather strange to run an unreliable IP datagram service on top of a hard-working protocol like X.25. It may seem even stranger when one realizes that both X.25 and IP provide layer 3 protocols. However, considerations of cost or convenience always override purity of layering. Layer 3 protocol units for VINES, DECnet, and Systems Network Architecture (SNA) also can be carried on X.25 circuits. Even layer 2 bridged data sometimes is carried on an X.25 circuit.

Figure 4.17 illustrates how IP traffic from multiple sources is routed across a single X.25 virtual circuit and forwarded to multiple destinations.

4.20.3   Multiprotocol over X.25

There are two methods of carrying multiprotocol traffic across an X.25[10] network:

1. Set up a *separate* virtual circuit for each protocol. During call setup, notify the partner of the protocol that will be carried.

---

[9]The 14-digit numbers used in X.25 calls are described in CCITT recommendation X.121.

[10]The same methods and formats are used for ISDN in the packet mode.

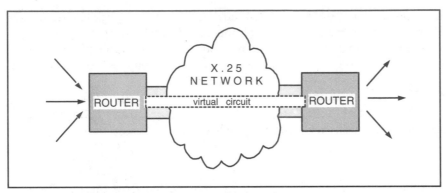

Figure 4.17   Using an X.25 network to route IP datagrams.

2. Set up a *single* virtual circuit that is shared by several protocols. During call setup, indicate that multiple protocols will appear. Notify the partner of the protocol carried in each packet by adding a header to each packet.[11]

The method that is selected depends on how much the service provider charges for additional circuits and on how long it takes to set up a new circuit.

Depending on the economics of the situation, a system may set up a switched X.25 connection on demand, when there is some traffic waiting to be forwarded to a remote site. The call will be closed after a period of inactivity. Call setup sometimes is a very slow process, which makes multiprotocol use of the circuit more attractive.

4.20.4   IP on a Separate X.25 Virtual Circuit

If IP traffic exclusively will be carried on a separate switched virtual circuit, this fact is indicated in the X.25 *Call Request* packet that initiates the circuit. There is an optional *Call User Data* field in the X.25 Call Request, and this is set to X'CC to indicate that IP traffic will be carried.

The value X'CC is a *Network Layer Protocol ID* (NLPID) that was assigned to IP traffic by the International Standards Organization.

4.20.5   Other Protocol on a Separate X.25 Virtual Circuit

A few other protocols have been assigned NLPID codes by ISO, but proprietary commercial protocols do not have ISO codes. However, as we have seen, many commercial protocols were assigned 2-byte type codes for the first multiprotocol environment—Ethernet. For example, AppleTalk traffic has Ethernet type code X'80-9B.

---

[11]The X.25 community calls its layer 3 protocol data unit a *packet*.

To run a single protocol that has an assigned Ethernet type code across a virtual circuit, the NLPID code X'80, followed by the SNAP subheader for the Ethernet type, is sent in the Call User Data field in the X.25 Call Request. For example, to set up a virtual circuit for AppleTalk traffic, send:

X'80-00-00-00-80-9B

4.20.6   Multiprotocol on a Virtual Circuit

If a virtual circuit will carry multiple protocols, the Call User Data field is set to X'00, and an extra header is placed in *each* packet in order to identify the protocol type for its contents. IP datagrams are identified quite efficiently by using the IP NLPID identifier, X'CC, as this extra header.

For protocols that must be identified by an Ethernet type code, the message header starts with an NLPID value of X'80 that indicates that a SNAP subheader follows. For example, *each* AppleTalk PDU on a multiprotocol circuit would be preceded by the header:

X' 80-00-00-00-80-9B

4.20.7   Packets versus Protocol Data Units

There is one small complication in the way that X.25 transfers information. Some X.25 networks transmit very small packets. However, they transmit entire higher-layer PDUs (such as IP datagrams) by sending them as contiguous *packet sequences* which are put back together into a single PDU at the other end of the circuit.[12] The protocol identifier is needed only in the header of the first X.25 packet of the sequence.

4.21   FRAME RELAY

X.25 networks provide reliable, in-sequence transmission of data. There is a great deal of overhead involved in assuring the level of quality offered by X.25. When IP traffic is streamed across an X.25 virtual circuit, much of the X.25 overhead is wasted effort.

The frame relay technology is better suited to TCP/IP use. Frame relay is a layer 2 protocol. When using frame relay, only a simple link layer header and error checking trailer are added to an IP datagram.

X.25 saves messages until they are acknowledged and retransmits if an Acknowledgment (ACK) is not received. Unlike X.25, frame relay does *not* save messages, it does *not* wait for ACKs, and it does *not* retransmit data. This results in the efficient use of the bandwidth that is available.

The initial frame relay standard defined service only for *permanent* virtual circuits. This meant that a user would contract with a service provider to

---

[12]A "more/nomore" flag is used to signal where a packet sequence ends.

obtain connectivity to prespecified sites at a set of agreed bandwidths. Many service providers offer bandwidths up to the T1 (1.544 megabits per second) rate.[13] Generally, a customer pays a fixed monthly fee based on a preagreed bandwidth.

Switched frame relay service enables systems that have been assigned global addresses to set circuits up dynamically—in much the same way that you would set up a switched telephone call. Supporting a switched service is more of a challenge because it is hard to predict how much traffic users will present to the service at any given time, and networks may occasionally be flooded with sudden bursts of traffic.

Frame relay offers good performance when compared to X.25 and has been well received. Some organizations have bought their own frame relay equipment and have built private networks.

As was done for the protocols discussed earlier, an IETF committee has specified the format that allows multiprotocol routed and bridged traffic to share a frame relay circuit. The encapsulation for an IP datagram is shown in Figure 4.18.

A frame relay address field usually is 2 octets long and includes a 10-bit *Data Link Connection Identifier* (DLCI) that identifies a specific circuit. A few bits in the address field are used to signal congestion and to indicate whether this frame should be given preferential treatment when frames are discarded due to congestion. (If a service provider needs more addresses, the address field can be extended to 3 or 4 octets.)

The control field is X'03, to denote unnumbered information. The protocol identifier X'CC indicates that this frame contains an IP datagram.

The frame is transmitted across the Service Provider's network. Frames whose frame check sequence values reveal that data has been corrupted are discarded.

For protocols (such as AppleTalk) that must be identified by an Ethernet type code, the message header has the format shown in the example in Figure 4.19. To improve the alignment of the message, an X'00 pad octet is inserted after the control field. The Network Layer Protocol ID value of X'80 indicates

---

[13]E1 rates of 2.048-megabit-per-second rates are supported at many locations outside of North America and Japan.

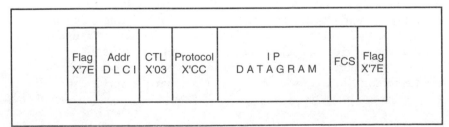

Figure 4.18   Frame relay encapsulation for an IP datagram.

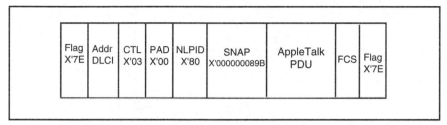

Flag X'7E	Addr DLCI	CTL X'03	PAD X'00	NLPID X'80	SNAP X'000000089B	AppleTalk PDU	FCS	Flag X'7E

Figure 4.19   Frame relay header with an Ethernet type code.

that a SNAP subheader follows. In this example, the SNAP subheader contains the Ethernet type code for AppleTalk.

Except for the pad byte, the header that is inserted is identical to the one used for multiprotocol X.25 circuits.

4.22 SMDS

The *Switched Multimegabit Data Service* (SMDS) is yet another public packet-switched data service. It was designed by the Regional Bell Operating Companies. The purpose of the service is to offer a large range of wide area bandwidth choices, including very-high bandwidth options (e.g., 155 megabits per second).

An interesting characteristic of SMDS is that data can be sent without opening a virtual circuit—it supports *connectionless* operation. In fact, a logical IP subnetwork can be constructed using wide area facilities, and (as shown in Figure 4.20) this logical wide area subnetwork will behave very much like a high-speed LAN. This makes SMDS an interesting option for a wide area backbone.

The *SMDS Interface Protocol* (SIP) is based on an IEEE standard, 802.6.

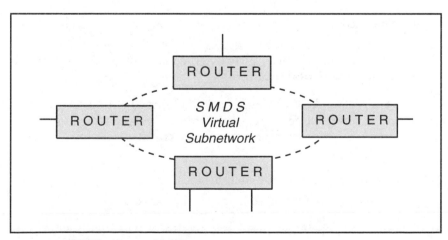

Figure 4.20   SMDS wide area backbone.

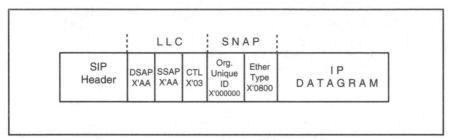

Figure 4.21 LLC and SNAP used to identify IP carried by SMDS.

4.22.1 IP over SMDS

Figure 4.21 shows the format of the header inserted after the SMDS SIP header, to signal the fact that the frame contains an IP datagram.

The format is identical to that used for IEEE 802 LANs. The first 3 octets are an IEEE 802.2 LLC header, and the SNAP subheader contains X'08-00, the Ethernet type code for IP.

4.23 ASYNCHRONOUS TRANSFER MODE

Asynchronous Transfer Mode (ATM) is a cell-switching technology suitable for use in both WANs and LANs. ATM combines the security benefits of switching with high performance and flexible choices of bandwidth. Characteristics of the technology include:

- ATM data is switched in 53-octet cells.

- Each cell contains a 5-byte header that includes cell routing information.

- Frames are split into cells at the source and combined back into frames at the destination by the *ATM Adaptation Layer* (AAL).

- There are several AALs, but the one that is relevant for IP datagram transmission is AAL5.

- The job of segmenting and reassembling frames for wide area transmission is done by a Data Exchange Interface (DXI), a piece of equipment analogous to a telephone line digital interface.

Like X.25 and frame relay, ATM communication is carried out by setting up a virtual circuit and sending frames across that circuit.

There are two methods of carrying multiprotocol traffic across an ATM network:

- Set up a separate virtual circuit for each protocol.

- Set up a single virtual circuit shared by several protocols.

The method chosen will depend on cost factors and the speed with which circuits can be set up and taken down.

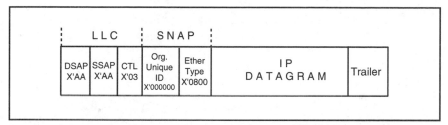

Figure 4.22 LLC and SNAP used to identify IP carried by ATM AAL.

If a separate virtual circuit is used for each protocol, just as for X.25, the protocol type for a switched circuit will be announced in a call request message.

When a single virtual circuit carries multiple routed protocols, as shown in Figure 4.22, the AAL5 frame will start with the familiar LLC and SNAP headers. The IP Ethernet type is enclosed in the SNAP subheader shown in the figure.

Note that the AAL5 frame does not have a header with destination and source address fields. This is because an end-to-end virtual circuit is established when the "call" is set up, and information needed to switch data from the source to the destination is included in the 5-octet cell headers.

The AAL5 trailer contains padding bytes (for alignment), a user data field, a *payload length* field, and a frame check sequence. The payload length covers the LLC and SNAP headers and the datagram.

4.24 MAXIMUM TRANSMISSION UNIT

Each of the technologies that we have discussed has a different maximum limit on the size of its frames. After subtracting away the size of the frame header, trailer, and the LLC and SNAP headers, if they are present, the residue defines the maximum size of the datagrams that can be carried across the medium. The maximum datagram size is called the *Maximum Transmission Unit,* or *MTU*.

For example, the maximum frame size for an 802.3 10BASE5 network is 1518 octets. Subtracting the MAC header and trailer (18 octets) and the Type 1 link control and SNAP headers (8 octets), we get a maximum datagram size of 1492 octets.

Table 4.1 summarizes maximum datagram sizes for a range of technologies.

Point-to-point lines are a special case. There really is no inherent limitation on datagram size for a point-to-point line. The optimum size depends on the error characteristics of the line. (If the error rate is high, better throughput can be attained by using shorter units.) The default maximum, 1500 bytes, is used most frequently.

SLIP originally was specified with a maximum datagram size of 1006 bytes. Some implementations can support a maximum of 1500 bytes, bringing SLIP in line with other point-to-point transmission formats.

TABLE 4.1   Maximum Transmission Units

Protocol	Maximum datagram octets (MTU)
Default for Point-to-Point	1500
Point-to-Point (low delay)	296
SLIP	1006 (original limit)
X.25	1600 (differs for some networks)
Frame relay	Usually at least 1600
SMDS	9235
Ethernet version 2	1500
IEEE 802.3/802.2	1492
IEEE 802.4/802.2	8166
16 Mb IBM Token-Ring	Maximum 17914
IEEE 802.5/802.2 4-Mb Token-Ring	Maximum 4464
FDDI	4352
Hyperchannel	65535
ATM	Default MTU 9180 Maximum possible 16K-1

The MTU values shown for Token-Rings are the largest allowed. Actual Token-Ring MTUs depend on a number of factors, including the token holding time for the ring.

4.25   TUNNELING

Adhering to a layered structure is a worthy goal, but sometimes the easiest way to get data from one place to another is to hitch a ride with another protocol. This process is called *tunneling,* probably because data temporarily disappears into the depths of another protocol until it pops out at some exit point.

Making tunneling work is not complicated—you simply wrap one or more headers for another protocol around your data unit, route it using the other protocol, and unwrap it at the destination.

In fact, we already have seen an example of tunneling. When IP datagrams are moved across an X.25 network, they are wrapped inside X.25 network layer headers. In this case, IP traffic is tunneled through X.25.

There are many other examples of tunneling in current use. Sometimes Novell NetWare IPX traffic is tunneled through an IP network. A NetWare message is wrapped in IP and UDP headers, routed through the IP network, and delivered to a remote NetWare server. A number of vendors offer products that tunnel SNA traffic through an IP network.

Tunneling always imposes a burden of extra overhead. Because it hides part of a network's path inside a foreign protocol, tunneling can degrade the ability to control and manage a network and sometimes creates bursts of traffic that are not subject to normal flow control.

4.26   SHARING A NETWORK INTERFACE

As we already have seen, it is not unusual to find that a LAN or WAN is being used for several protocols at once. In fact, a single node sometimes sends and receives a mixture of protocols on its network interface. How can this be done?

To keep the discussion simple, let's consider a specific interface—say for an Ethernet LAN. A PC or server may wish to use an Ethernet interface for TCP/IP, IPX, and DECnet. Can these protocols coexist?

We already have seen some evidence that they can. The link layer header will contain a field that identifies the network layer protocol for the message.

Figure 4.23 shows an Ethernet interface that is shared by TCP/IP, IPX, and DECnet protocol stacks. The intervening layer of *device driver* software hides the detailed hardware I/O interactions from the higher-level protocol stacks.

4.27   LINK LAYER ISSUES

The percentage of a datagram that is header information has an impact on throughput. Obviously, when bulk data is being transmitted, it is most efficient to carry as much data as possible in a datagram.

However, we have seen that there are different maximum datagram sizes for various network types. In Chapter 6, we shall see that IP provides a

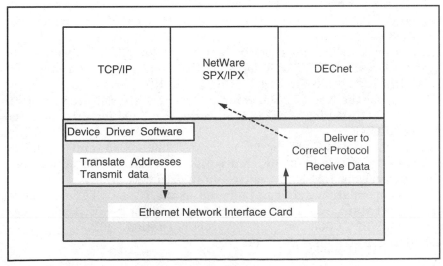

Figure 4.23   Protocols sharing a network interface.

mechanism for fragmenting large IP datagrams when passing data into a network with a small maximum datagram size. This feature assures that data can be delivered, even when an incompatible MTU size is encountered. However, as might be expected, the fragmentation and reassembly mechanisms slow down network response time.

When a pair of communicating hosts are attached to the same LAN, they will wish to optimize data transfers by using the largest possible datagrams. But when transmitting data to a remote host across unknown network types, some implementations drop down to a small maximum datagram size (sometimes 576 octets) in order to prevent fragmentation.

Later, in Chapter 7, we shall see that there is a procedure that automatically discovers the biggest datagram that can be used along a given path. This avoids fragmentation and enables bulk data to be transferred efficiently in optimally sized datagrams.

4.28 TRAILERS

A troublesome problem is the use of nonstandard protocol formats by some obsolete versions of TCP/IP. The Berkeley Software Distribution 4.2 implementation introduced a nonstandard format for Ethernet MAC frames that moved the frame type field and layer 3 and 4 header information into a *trailer*. The purpose of this rearrangement was to speed up the processing of incoming frames by reducing the number of times that data is copied. Some commercial products incorporated this feature.

The use of Berkeley trailers can lead to interworking problems. Fortunately, Berkeley trailers are becoming rare. However, if you need to use this feature, see RFC 1122 for advice on how to use them safely.

4.29 RECOMMENDED READING

RFC 1661 describes the Point-to-Point protocol. PPP authentication protocols are explained in RFC 1334, and automatic link quality monitoring is presented in RFC 1333.

There are several RFCs that describe how to transmit IP datagrams over lower-layer facilities: See RFC 1356 for X.25, RFC 1490 for frame relay, RFC 1209 for SMDS, RFC 1390 for FDDI, RFC 1577, 1932, 1626, and 1755 for ATM, RFC 1088 for the Network Basic Input Output System (NetBIOS), RFC 1055 for SLIP, RFC 1042 for IEEE 802 networks, RFC 894 for Ethernet, and RFC 1201 for ARCNET.

Information about HDLC can be found in ISO 3309, 4335, and 7809. The IEEE 802 series and ISO 8802 series describe physical, media access, and logical link protocols for LANs and metropolitan area networks.

CCITT Recommendation X.25 can be found in the 1984 CCITT red books. There are several standards documents relating to frame relay. A good place to start is with ANSI T1.606 and CCITT Recommendation I.122.

RFC 893 discusses trailer encapsulations.

Naming and Addressing

5.1 INTRODUCTION

Every node in a network needs to be given a name and address. How should this be done? It may not be much of a problem on a stand-alone LAN with a handful of hosts, but when dealing with hundreds or thousands of hosts, starting off with a good name and address plan saves a ton of headache remedies when hosts, routers, and networks are added, removed, or relocated.

Internet administrators have had to cope with name and address management for a worldwide internetwork whose size has doubled every year or so. They came up with a practical strategy—delegate.

The TCP/IP Internet scheme of name and address management:

- Makes it possible to delegate name and address assignment to someone in charge of all or part of a particular network
- Allows names to reflect the logical structure of an organization
- Assigns addresses that reflect the physical topology within an organization's network

We shall see that the Internet uses a hierarchical naming method that enables administrators to construct descriptive, easy-to-remember names.

5.2 EXAMPLES OF INTERNET NAMES

Some names are whimsical, such as a group of hosts at the Yale Medical school which have been called:

blintz.med.yale.edu

couscous.med.yale.edu

gazpacho.med.yale.edu

lasagne.med.yale.edu

paella.med.yale.edu

sukiyaki.med.yale.edu

strudel.med.yale.edu

Servers often are given names that make it easy for people to find them. For example:

www.whitehouse.gov

ftp.microsoft.com

gopher.jvnc.net

By the way, Internet node names are not case-sensitive. For example, *www.whitehouse.gov* could have been written *WWW.WHITEHOUSE.GOV* or *WWW.Whitehouse.Gov.*

Usually an end user will type host names in lowercase, while some tables list names in uppercase. In this text, you will see uppercase, lowercase, and mixed-case node names.

5.3   HIERARCHICAL STRUCTURE OF NAMES

It is easy to understand the hierarchical structure for these names. Each organization has a descriptive top-level name, such as *yale.edu, whitehouse.gov,* or *microsoft.com.* The organization then is free to design any convenient naming scheme. For example, Yale, like most universities, delegates naming responsibility to each department or division. Hence, there are names that end with:

cs.yale.edu

math.yale.edu

geology.yale.edu

Some departments create further subnames. For example, computers for Yale Computer Science majors are located in a room in the basement of the Computer Science (CS) building called *The Zoo.* The *Zoo* computers have names like:

lion.zoo.cs.yale.edu

tiger.zoo.cs.yale.edu

leopard.zoo.cs.yale.edu

All of the *Zoo* computers happen to be located on a single LAN. However, names can be assigned in any way that is *administratively* convenient. For example, another family of subnames at the Yale Computer Science department is:

hickory.theory.cs.yale.edu

pecan.theory.cs.yale.edu

olive.theory.cs.yale.edu

walnut.theory.cs.yale.edu

These machines are *not* located on a single LAN.

5.4 ADMINISTRATION OF NAMES

Using a hierarchical name structure makes it easy to ensure that all of your computer names are unique, while delegating the job of administering computer names to appropriate personnel. Note that:

lion	It is up to the administrator of the Computer Science *Zoo* facility to make sure that every *Zoo* computer is assigned a different name (*lion, tiger,* etc.).
lion.zoo	The Yale Computer Science department administrator must use different names for each departmental subgroup (*zoo, theory,* etc.)
lion.zoo.cs	As long as Yale's network administrator has assigned a different name to each department (*cs, math, geology*), every computer at Yale will have a unique name.
lion.zoo.cs.yale.edu	There are official registrars who make sure that every organization has a unique name (e.g., *yale.edu, microsoft.com*). Hence, each computer in the world can be given a unique name.

To ensure the worldwide uniqueness of names on the Internet, it was necessary to set up a registration service that assures that *every* business or organization uses a different name.

Initially, the Internet was sponsored by the United States Department of Defense (DOD). The DOD funded the *Department of Defense Network Information Center* (DDN NIC), which was responsible for administering the registration of all names and addresses.

In 1993, the National Science Foundation (NSF) assumed responsibility for nonmilitary names and addresses while the DDN NIC retained responsibility for military names and addresses.

The National Science Foundation funds the *InterNIC Registration Service,* which is the primary authority for worldwide naming and addressing. However, total centralization of these functions would create an unnecessary bottleneck. Therefore the InterNIC has delegated authority to two major Regional Registries:

- The Asia Pacific Network Information Center

- The RIPE[1] Network Coordination Centre (for Europe)

---

[1] Reseaux IP Europeens.

The InterNIC and each of the Regional Registries delegate naming and addressing authority to National and Local Registries within their regions.

See Appendix C for the postal addresses, telephone numbers, and electronic mail addresses of the InterNIC and the Regional Registries. Appendix C contains pointers to on-line archives of registration forms. It also includes information on accessing National and Local Registries.

5.5   FORMAL STRUCTURE OF NAMES

A name is made up of a series of labels separated by dots. It is not uncommon to see names made up of two, three, four, or five labels. All of the following are legitimate computer names:

bellcore.com

www.apple.com

ftp.ncsa.uiuc.edu

lion.zoo.cs.yale.edu

Names that are longer than this may be difficult for users to remember and type. However, Internet naming standards permit each label to have up to 63 characters, and a name can contain up to 255 characters.

5.6   WORLDWIDE NAMING TREE

Internet names are structured in a tree, as shown in Figure 5.1. Each node in the tree is assigned a label. Each node in the tree also has a name, called its *domain name*. A node's domain name is made up of the sequence of labels that lead *from* the node *to* the top of the tree. A node's domain name is written as the sequence of labels separated by dots.

Finally, a *domain* is defined to be a chunk of the naming tree that consists of a node and all of the nodes below it. In other words, a domain is made up of all names with a common ending. Examples of domains include:

- *edu* and all names under it (i.e., ending in *edu*)
- *yale.edu* and all names under it (i.e., ending in *yale.edu*)

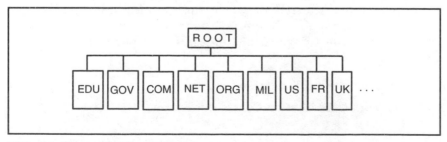

Figure 5.1   The worldwide naming tree.

- *cs.yale.edu* and all names under it (i.e., ending in *cs.yale.edu*)

The *top-level domains* (as shown in Figure 5.1) are:

edu	Four-year, degree granting institutions.
gov	United States federal government agencies.
com	Commercial organizations.
net	Internet network service organizations.
org	Not-for-profit organizations (*96olympics.org, npr.org*).
int	International organizations (*gopher.nato.int*). Rarely used and not shown.
mil	Military organizations (*army.mil, navy.mil*).
us	U.S. state and local government agencies, schools, 2-year colleges, libraries, and museums.
Countries	Two-character ISO country codes identify dozens of other top-level domains: *fr* for France, *uk* for the United Kingdom, *de* for Germany, and so forth. The structure of the tree under the country code is delegated to administrators for that country.

Domains *yale.edu, whitehouse.gov,* and *ibm.com,* shown in Figure 5.2, are called *second-level domains.*

There is one more formality. The label for the *root* of the naming tree is a period. Hence the complete name of the *lion* system in the Yale Computer Science network actually is:

lion.zoo.cs.yale.edu.

Most users (and authors) omit the final period when typing a name.

5.7 CONFIGURING A SYSTEM'S NAME

The way that a system's name is configured varies from system to system. Most often, the administrator either types the name into a menu or invokes a command.

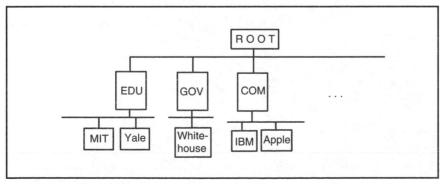

Figure 5.2 Second-level domains.

On *tigger,* which is a Unix system running *SunOS,* the *hostname* command is used to set or display the name of the host.

```
> hostname
tigger.jvnc.net
```

Some systems separate names into two parts—the initial label and the rest of the domain name. This is done so that automatic nicknames can be used for systems under the same domain node. For example, if a user working at a computer in domain *jvnc.net* types *mickey,* the name automatically is completed to *mickey.jvnc.net.*

Users of the *Chameleon* Windows desktop product enter their computer's name via two pop-up menus, as shown in Figure 5.3.

5.8 ADDRESSES

The IP protocol uses *IP addresses* to identify hosts and to route data to them. Every host must be assigned a unique IP address that can be used in actual communications. A host name is translated into its IP address by looking up the name in a database of name-address pairs.

When IP addresses were designed, no one dreamed that there would be millions of computers in the world and that many of them would want or need an IP address. The designers thought that they had to satisfy the needs of a modest community of universities, research groups, and military and government establishments.

They chose a design which seemed reasonable to them at the time. An IP address is a 32-bit (4-octet) binary number. Clearly, the address was chosen to fit conveniently into a 32-bit computer register. The resulting *address space* (which is the set of all possible address numbers) contains 2^{32} (4,294,967,296) numbers.

The *dot* notation was invented as a way of reading and writing IP addresses easily. Each octet (8 bits) of the address is converted into a decimal num-

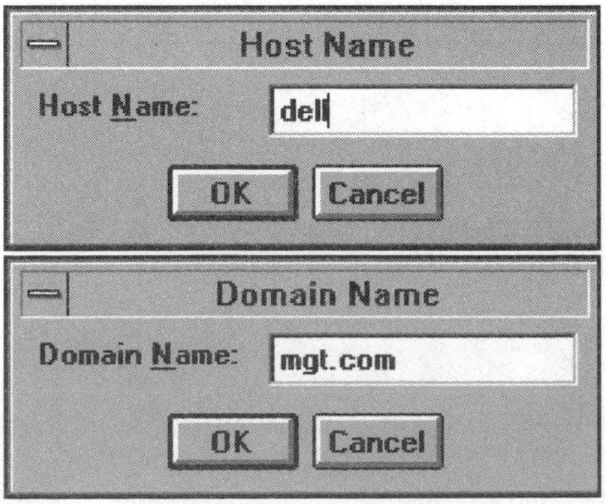

Figure 5.3 Configuring a system's name.

ber, and the numbers are separated by dots. For example, the address for *blintz.med.yale.edu* in 32-bit binary and in dot notation is:

10000010 10000100 00010011 00011111

130 . 132 . 19 . 31

Note that the biggest number that can appear in any position of a dotted address is 255, which corresponds to the binary number 11111111.

5.9   ADDRESS FORMATS

As shown in Figure 5.4, an IP address has a two-part format consisting of a *network address* and a *local address*. The network address identifies the network to which the node is attached. The local address identifies an individual node within the organization's network.

Every computer must have an IP address that is unique within the range of systems with which it will communicate.

5.10   ADDRESS CLASSES

An organization that plans to connect to the Internet must obtain a block of unique IP addresses. Addresses are obtained from a suitable registration authority.

For convenience, registration NICs delegate large blocks from their IP address space to Service Providers. This enables organizations to obtain their addresses from their Service Providers instead of a registration NIC.

For many years, there were only three address block sizes—large, medium, and small. There were three different network address formats for the three block sizes. The address formats were:

Class A for very large networks

Class B for medium-sized networks

Class C for small networks

These Class A, B, and C formats are displayed in Figure 5.5. Note that the address classes have the characteristics shown in Table 5.1.

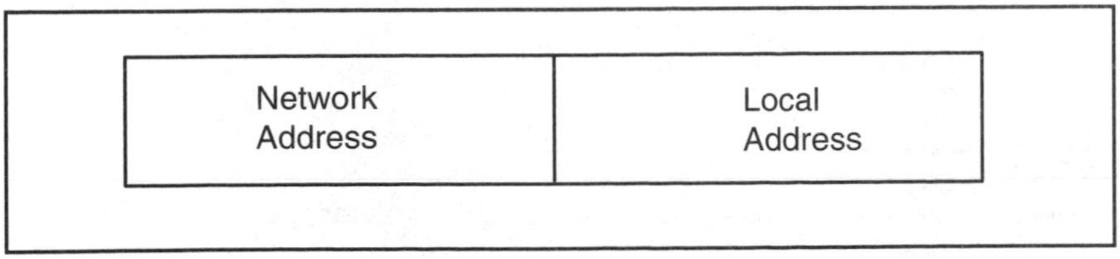

Figure 5.4   Format of an IP address.

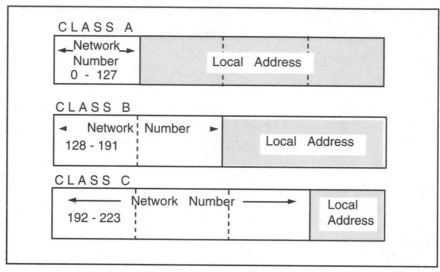

Figure 5.5   Traditional address classes.

In the early days of the Internet, organizations with very large networks—such as the United States Navy or Digital Equipment Corporation—were given Class A addresses. The network part of a Class A address is 1 octet long. The remaining 3 octets of a Class A address belong to the local part and are used to assign numbers to nodes.

There are very few Class A addresses, and most fairly big organizations have had to be content with a medium-sized Class B block of addresses. The network part of a Class B address is 2 octets long. The remaining 2 octets of a Class B address belong to the local part and are used to assign numbers to nodes.

Smaller organizations receive one or more Class C addresses. The network part of a Class C address is 3 octets long. This leaves only 1 octet in the local part that is used to assign numbers to nodes.

It is easy to spot the class of an IP address. You just look at the first number in the dotted address. The number ranges for each class are shown in Table 5.1 and Figure 5.5.

TABLE 5.1   Address Class Characteristics

Class	Length of network address (octets)	First number	Number of local addresses
A	1	0–127	16,777,216
B	2	128–191	65,536
C	3	192–223	256

In addition to Classes A, B, and C, there are two special address formats, Class D and Class E. Class D addresses are used for IP *multicasting*. Multicasting distributes a single message to a group of computers dispersed across a network. Multicast addresses, which support conferencing applications, will be discussed later in this book.

Class E addresses are reserved for experimental use.

- Class D addresses start with a number between 224 and 239.
- Class E addresses start with a number between 240 and 255.

5.11 ADDRESSES NOT CONNECTED TO THE INTERNET

Several blocks of addresses have been reserved for use for networks that will *not* be connected to the Internet and will never require connectivity to another organization. These addresses include:

10.0.0.0–10.255.255.255

172.16.0.0–172.31.255.255

192.168.0.0–192.168.255.255

Note that *many* organizations will be using these addresses. If your company merges with another company at some future date, or decides to communicate with clients or suppliers via TCP/IP, address conflicts could occur. However, you could register a Class C network and use it for any needed external communications. You can obtain proxy software that relays information between internal computers and the outside world via your registered Class C network.

The pros and cons of using these reserved addresses are discussed in RFC 1918, *Address Allocation for Private Internets*.

5.12 ADDRESSING EXAMPLES

In this section, we'll take a look at some examples of globally unique Class A, B, and C addresses. Later on, we'll examine the new *classless* method of assigning network addresses.

5.12.1 Assigning Class A Addresses to Networks

Some very large organizations have Class A addresses. In this case, the registration authority assigns a fixed value to the first octet of the address and the last 3 octets are managed by the organization. For example, the following addresses and host names belong to Hewlett-Packard, which was assigned Class A address 15.

15.255.152.2   *relay.hp.com*

15.254.48.2   *hpfcla.fc.hp.com*

Hewlett-Packard owns the numbers from 15.0.0.0 to 15.255.255.255. These numbers make up the organization's *address space.*

5.12.2   Assigning Class B Addresses to Networks

The registration authority assigns a fixed value to the first 2 octets of a Class B address. The last 2 octets are managed by the organization. For example, the following addresses and host names belong to the Global Enterprise Systems Service Provider, which was assigned the Class B address 128.121.

128.121.50.145	*tigger.jvnc.net*
128.121.50.143	*mickey.jvnc.net*
128.121.51.51	*camel-gateway.jvnc.net*

Global Enterprise systems owns the numbers from 128.121.0.0 to 128.121.255.255.

Class B addresses are very popular and many organizations have requested and received them. Unfortunately, although there are more than 16,000 possible Class B network identifiers, the supply is running out.

5.12.3   Assigning Class C Addresses to Networks

An organization with a small network that needs globally unique addresses is given one or more Class C addresses. This means that the registration authority assigns fixed values to be used in the first 3 octets of the organization's addresses. The organization has control of the last octet. For example, WAIS, Inc., was assigned the Class C address 192.216.46. Some of its addresses and host names are:

192.216.46.4	*ns.wais.com*
192.216.46.5	*webworld.wais.com*
192.216.46.98	*wais.wais.com*

WAIS, Inc., owns the numbers from 192.216.46.0 to 192.216.46.255.

5.13   TRANSLATING NAMES TO ADDRESSES

End users like to type easy-to-remember names, while IP needs to know the addresses of destination systems. Many computers are configured with a small file called *hosts,* which lists the names and addresses of local systems. Here is part of a *hosts* file stored at the Global Enterprise Systems host *tigger.jvnc.net:*

```
128.121.50.2      r2d2.jvnc.net    r2d2
128.121.50.7      nisc.jvnc.net    nisc
128.121.50.141    minnie.jvnc.net  minnie
128.121.50.143    mickey.jvnc.net  mickey
128.121.50.144    donald.jvnc.net  donald
128.121.50.145    tigger.jvnc.net  tigger
```

```
128.121.50.148   chip.jvnc.net    chip
128.121.50.149   bambi.jvnc.net   bambi
128.121.50.152   sleepy.jvnc.net  sleepy
```

The demonstrations in the next few sections of this chapter will be run at *tigger.jvnc.net*.

Recall that the distributed Domain Name System (DNS) database is used for global name-to-address translation. Below, the *nslookup* application sends name translation queries to a Domain Name Server called *r2d2.jvnc.net*. We will look up the address of a World Wide Server for the White House and a file transfer server at Novell, Inc.:

```
> nslookup
Default Server: r2d2.jvnc.net
Address: 128.121.50.2

> www.whitehouse.gov.
Server: r2d2.jvnc.net
Address: 128.121.50.2

Name: www.whitehouse.gov
Address: 128.102.252.1

> ftp.novell.com.
Server: r2d2.jvnc.net
Address: 128.121.50.2

Name: bantu.Provo.Novell.COM
Address: 137.65.1.3
Aliases: ftp.novell.com
```

The response to the second query indicates that the name *ftp.novell.com* actually is an *alias* for a computer called *bantu.Provo.Novell.COM*.

5.14 ALIAS NAMES

Often, it is convenient to assign some aliases (or nicknames) to a computer, in addition to its real name. For example, the host *nicol.jvnc.net* offers file transfer, gopher, and World Wide Web (WWW) services. For convenience, it has been assigned the nicknames:

ftp.jvnc.net

gopher.jvnc.net

www.jvnc.net:

```
> ftp.jvnc.net.
Server: r2d2.jvnc.net
Address: 128.121.50.2

Name: nicol.jvnc.net
Address: 128.121.50.10
Aliases: ftp.jvnc.net
```

```
> gopher.jvnc.net.
Server: r2d2.jvnc.net
Address: 128.121.50.2

Name: nicol.jvnc.net
Address: 128.121.50.10
Aliases: gopher.jvnc.net

> www.jvnc.net.
Server: r2d2.jvnc.net
Address: 128.121.50.2

Name: nicol.jvnc.net
Address: 128.121.50.10
Aliases: www.jvnc.net
>
```

If the load on *nicol* gets too heavy, one of its services—and the service nickname—could be transferred to a different host. This enables users to reach the service by means of the same name, even if the home site for the service has been changed. A host's true name is called its *canonical name*.

5.15   INEFFICIENCY CAUSED BY ADDRESS CLASSES

A Class A network spans 16,777,216 addresses, while Class B supports 65,536 and Class C only includes 256 numbers. The large gaps between these numbers led to very inefficient address block allocations and contributed to the depletion of the IP address space.

In Section 5.19, we will describe the more efficient *classless* method of allocating addresses to organizations.

5.16   TCP/IP NETWORKS AND SUBNETS

An organization that has a Class A or Class B network address is very likely to have a fairly complex network made up of many LANs and several WAN links. It makes sense to partition the address space in a way that matches the network's structure as a family of subnetworks. To do this, the local part of the address is broken into a *subnet part* and a *system part* in any convenient way, as illustrated in Figure 5.6.

The size of the subnet part of an address and the assignment of numbers to subnets is the responsibility of the organization that "owns" that part of the address space.

Subnet addressing often is done at a byte boundary. An organization with a Class B address such as 128.21 might use its third byte to identify subnets. For example:

128.121.1

128.121.2

128.121.3

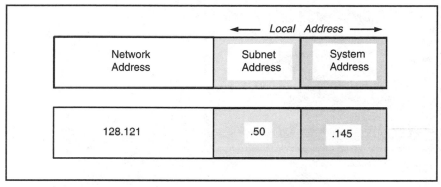

Figure 5.6 Breaking the local address into subnet and system parts.

The fourth byte would then be used to identify individual hosts on a subnet.

On the other hand, an organization with a Class C address has only a 1-byte address space. They might choose to do no subnetting at all or perhaps might use 4 bits for subnet addresses and 4 bits for host addresses, as shown in Figure 5.7. In the figure, the local address, 61, is expressed in binary as 0011 1101. The first 4 bits identify a subnet and the last 4 bits identify a system.

5.17 SUBNET MASKS

Traffic is routed to a host by looking at the network and subnet part of its IP address. A Class A, B, or C network part has a fixed size. But organizations can choose their own subnet field sizes, so how can hosts and routers recognize this field? The answer is that systems must be configured to know the size of the subnet part of the address. Figure 5.8 shows the *Chameleon* menu used to enter the size of the subnet field.

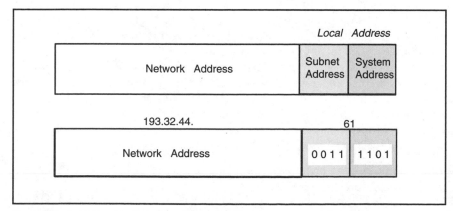

Figure 5.7 A 4-bit subnet part in a Class C address.

Figure 5.8 Configuring a subnet mask.

The subnet field size actually is stored in a configuration parameter called the *subnet mask*. The subnet mask is a sequence of 32 bits. The bits corresponding to the network and subnet fields of an address are set to 1, while the bits for the system field are set to 0.

For example, if we will use the third byte of addresses starting with 128.121 to identify subnets, the mask is:

11111111 11111111 11111111 00000000

Subnet masks often are expressed in dotted decimal notation. The mask above can be written as:

255.255.255.0

Sometimes the mask is written in a hexadecimal form, such as

X'FF-FF-FF-00

Hosts and routers connected to a subnet are configured with the mask for the subnet. It is common practice to use a single subnet mask throughout an organization's internet. There are exceptions to this practice, and some organizations use several different subnet sizes.

For example, if a network contains a large number of point-to-point lines, subnet numbers will be used up very wastefully since there are only two systems on each point-to-point subnet. An organization may decide to use 14-bit masks (255.255.255.252) for its point-to-point lines.

Table 5.2 shows the ways in which the local addresses in a Class B network can be partitioned into subnets. It also shows the number of subnets and hosts for each partition. The number of hosts and subnets each will be 2 less than you would expect, because of some restrictions that we will describe in the sections that follow.

For example, if 6 subnet bits are used, the subnet mask bit pattern is:

11111111 11111111 11111100 00000000

which translates to 255.255.252.0. We'll explain why 1/15 (1 subnet bit and 15 host bits) and 15/1 splits of the address bits are not allowed in the section that follows.

TABLE 5.2 **Subnets for a Class B Network**

Subnet bits	Number of subnets	Host bits	Number of hosts	Mask
0	0	16	65,534	255.255.0.0
1	—	15	—	Not allowed
2	2	14	16,382	255.255.192.0
3	6	13	8,190	255.255.224.0
4	14	12	4,094	255.255.240.0
5	30	11	2,046	255.255.248.0
6	62	10	1,022	255.255.252.0
7	126	9	510	255.255.254.0
8	254	8	254	255.255.255.0
9	510	7	126	255.255.255.128
10	1,022	6	62	255.255.255.192
11	2,046	5	30	255.255.255.224
12	4,096	4	14	255.255.255.240
13	8,190	3	6	255.255.255.248
14	16,382	2	2	255.255.255.252
15	—	1	—	Not allowed

In Appendix D, we provide examples that illustrate how you can use several different subnet masks within a single network. Using several mask sizes enables you to assign your addresses very efficiently.

5.18 SPECIAL RESERVED ADDRESSES

Not every number can be assigned to a subnet or to a host. For example, some addresses are used for broadcasting while others are reserved for use in routing tables. A good rule of thumb is: *stay away from using a block of 0s or a block of 1s in either your subnet field or your host field.* Also, there are no network numbers consisting of all 0s or all 1s.

5.18.1 Identifying Networks and Subnets

It is convenient to use a dotted address format to refer to a network. By convention, this is done by filling in the local part of the address with zeroes. For example, 5.0.0.0 identifies a Class A network, 131.18.0.0 identifies a Class B network, and 201.49.16.0 identifies a Class C network.

The same type of notation is used to identify subnets. For example, if network 131.18.0.0 uses an 8-bit subnet mask, 131.18.5.0 and 131.18.6.0 refer to subnets. This notation is used to represent destination networks and subnets in IP routing tables. The price of using this convention is that addresses of this form should not be assigned to hosts or routers. In addition, the use of 0 as a subnet number would make identifier 130.15.0.0 ambiguous. For this reason, the use of a zero subnet field is "forbidden" by the standards.[2]

5.18.2   Broadcasts to the Local Subnet

There are several IP address patterns that are used for broadcasts. A datagram can be broadcast to a set of systems within a particular scope.

IP address 255.255.255.255[3] (i.e., an address consisting of 32 ones) broadcasts a datagram to every system on the local link. This type of broadcast is used, for example, in the *BOOTP* and *DHCP* protocols, which a system invokes in order to obtain its IP address and other initialization data from a boot server. A client sends a boot request to 255.255.255.255 and uses the reserved address 0.0.0.0 as its source IP address.

A LAN broadcast is executed by wrapping the IP datagram in a frame whose header contains the all-1s physical broadcast address as its destination address.

5.18.3   A Broadcast Directed to a Subnet

A broadcast also can be directed to a specific subnet, which may be a directly connected subnet or may be a subnet that is remote from the source host. For example if 131.18.7.0 is a subnet of a Class B network, address 130.18.7.255 is used to broadcast a message to all nodes on this subnet.

If the destination subnet is remote, the result of sending an IP datagram to the broadcast address is that one copy of the datagram would be relayed to a router attached to subnet 131.18.7.0. Assuming that this subnet is a LAN, the router would then use a physical broadcast address in the destination field of the Media Access Control (MAC) frame to direct the message to all hosts on the subnet.

Note that this implies that no system could be assigned the reserved IP address 130.18.7.255.

5.18.4   Broadcasts to Networks

It is possible to send an IP datagram to every host on a selected remote network. This is done by setting the entire local part of the address to 1s. For

---

[2]See RFC 1122. In spite of the restriction, some sites use subnet 0 with a subnet mask.

[3]Berkeley 4.2 TCP/IP and some products based on it use 0s instead of 1s for broadcasting. This is a nonstandard practice, and, over time, these operating systems should be obsoleted or replaced.

example, suppose that an administrator wanted to send an announcement to all nodes on a Class C Ethernet network 201.49.16.0. The IP address used for the broadcast is:

201.49.16.255

This means that no host can be given the address 201.49.16.255.

Address 131.18.255.255 could be used to aim a message at every node in an entire Class B network. Note that if it were legal to assign subnet number 255 to one of the subnets, we would have a problem. It would not be clear whether a broadcast to 130.15.255.255 was intended for that subnet or for the whole network. The way to avoid this is never to assign an all-1s number (like 255) to a subnet.

5.18.5   Restrictions on IP Addresses

The set of usable IP addresses is diminished slightly because of the special broadcast and routing table formats described above. RFC 1122, *Requirements for Internet Hosts—Communication Layers,* mandates that:

- A network, subnet, or host field cannot consist entirely of 0s.
- A network, subnet, or host field cannot consist entirely of 1s.

Hence, to be usable, a field must contain at least 2 bits.

5.18.6   Loopback Address

At the opposite extreme to broadcasting are messages that never leave the local host. There are many hosts that contain both client and server processes. The local clients and servers communicate via IP within the host. To do this, they use a special address called the *loopback* address. By convention, any address starting with 127 is reserved for this purpose. In practice only the address 127.0.0.1 is used. Note that an entire Class A address space of 2^{16} numbers was reserved for loopback addresses.

It is easy to see the loopback address in action. For example, the *Chameleon* File Transfer client and File Transfer server can run concurrently under *Microsoft Windows.* We can start the server, whose screen is shown in Figure 5.9.

The client connects to the server at loopback address 127.0.0.1, as shown in Figure 5.10. Any "file transfers" that the client executes would simply copy files from one PC directory to another. The server log records the activities performed by the client at address 127.0.0.1, as shown in Figure 5.11.

5.18.7   Summary of Special Reserved Addresses

The various types of special addresses are summarized in Table 5.3.

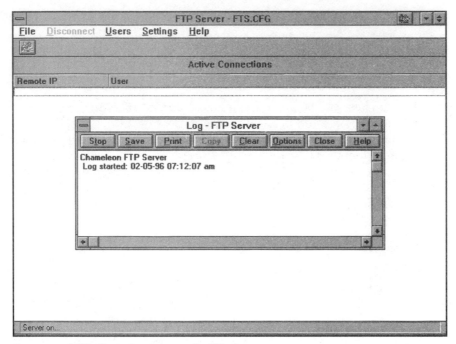

Figure 5.9 A File Transfer server under Windows.

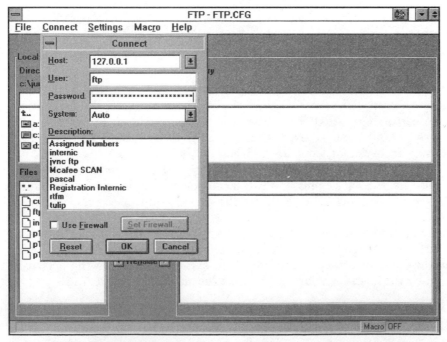

Figure 5.10 File Transfer client connecting to a local server.

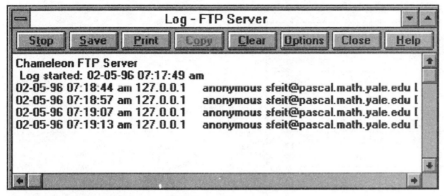

Figure 5.11   Client activities at a File Transfer server.

5.19   SUPERNETTING AND CIDR

The A, B, C method of handing out blocks of addresses was very inefficient. A Class C address only provided at most 254 usable addresses. (Recall that 0 and 255 cannot be used to address a node.) On the other hand, if an organization that needed a few hundred or a few thousand addresses was given a Class B address, many addresses would be wasted.

It would make better sense to assign organizations the number of bits that they really need. This is done very easily. For example, if an organization needs 4000 addresses, it is given 12 bits to use as the local part of its addresses. The remaining 20 bits are a fixed prefix, used as the new *supernetwork* or *prefix* part of the address. The conventional way to denote the size of this "classless" network part is */20.*

TABLE 5.3   Special Addresses

Addresses	Description
0.0.0.0	Used as source address in a boot configuration request. Also denotes the default route in a routing table.
127.0.0.0	Reserved.
127.0.0.1	Loopback. Client and server are in the same host.
127.0.0.2-127.255.255.255	Reserved.
128.0.0.0	Reserved.
191.255.0.0	Reserved.
192.0.0.0	Reserved.
255.255.255.0	Reserved.
240.0.0.0-255.255.255.254	Reserved.
255.255.255.255	Broadcast on locally attached LAN.

TABLE 5.4   CIDR Blocks from Class C Address Space

Size of network part	Number of local part bits	Equivalent number of class C nets	Number of addresses for the org.
/24	8	1	256
/23	9	2	512
/22	10	4	1,024
/21	11	8	2,048
/20	12	16	4,096
/19	13	32	8,192
/18	14	64	16,384
/17	15	128	32,768

The initial supernetted address allocations are being taken from the available Class C number space. Obtaining a 20-bit prefix is equivalent to obtaining 16 contiguous Class C addresses.

Table 5.4 shows the various address blocks that can be assigned from the Class C address space. To route to an organization, an Internet router needs to know:

- The number of bits in the network prefix
- The actual bit pattern assigned to the organization's network prefix

The router then can forward traffic to the organization using a single routing table entry. This is called *Classless Internet-Domain Routing* (CIDR).

Unused parts of the Class A numbering space could be divided up in exactly the same way. An organization could be assigned a string of bits as a network prefix and then could use the remaining bits to number its systems. All that is needed to make this work is to include the length of the network prefix in the routing information.

Internet routing has been made even more efficient by delegating large address blocks to Internet Service Providers. A provider then assigns subblocks to its customers. Traffic is routed to the provider using its block prefix. The provider then uses longer prefixes to route to its customers.

For example, a provider might be given the block starting with the 10-bit prefix 11000001 11, and one of its customers might be given the block starting with the 16-bit prefix 11000001 11011111.

5.20   THE NEED FOR IP, NEXT GENERATION

The introduction of classless, supernetted addresses and classless routing is a stopgap measure to extend the life of the current IP addressing scheme.

When IP addresses were originally designed, no one anticipated the computer technologies that would put computers on desks, into homes, and into commonplace devices—and network them together. The current addresses are inconvenient and inadequate for the job that needs to be done.

Unlike the hierarchical structure used for telephone numbers, the addresses were designed without country or area codes. This makes routing especially burdensome. Wide area routers store separate routing entries for tens of thousands of separate networks.

IP version 6 (*Next Generation*) is designed to solve these problems and also to prepare for new ways of using computers and computer networks. Version 6 will be described in Chapters 22 and 23.

5.21   IP ADDRESSES, INTERFACES, AND MULTIHOMING

Identifying networks and subnets within an IP address has many benefits:

- It simplifies the job of assigning addresses. A block of addresses can be delegated to the administrator of a particular network or subnet.

- It helps keep routing tables short. A routing table needs to contain only a brief list of networks and subnets, rather than a list including every host in an internet.

- It simplifies the job of routing. Table lookups of network or subnet numbers can be carried out quickly and efficiently.

These are important advantages. However, there is an important consequence of this addressing scheme. Consider Figure 5.12. The router in the figure has three different interfaces and is connected to two LANs and a leased line.

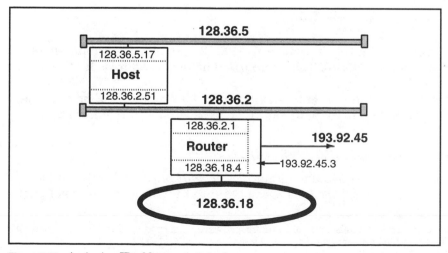

Figure 5.12   Assigning IP addresses to interfaces.

The router is connected to internal subnets 128.36.2 and 128.36.18. It also is connected to an external network, 193.92.45. What is "the" IP address of the router?

The answer is that *systems* do not have IP addresses, their *interfaces* do. Each interface has an IP address that starts with the network and subnet number of the attached LAN or wide area link. A router with three interfaces must be assigned three IP addresses.

A host also may be connected to more than one subnet or network. The host in Figure 5.12 has interfaces to two Ethernets. This host also has two IP addresses—128.36.2.51 and 128.36.5.17.

A system that is attached to more than one subnet is called *multihomed*. A multihomed host introduces some complications into IP routing. Data will be routed to a multihomed host differently depending on which of its IP addresses is chosen for communication. It may in fact be helpful to associate multiple names to the host, corresponding to its various interfaces. For example, (see Figure 5.12) users on LAN 128.36.2 might be told a different host name from the one told to users on LAN 128.36.5.

In spite of the drawbacks that result from multihomed hosts, the inclusion of network and subnet identifiers within an address has contributed greatly to the efficiency of routers and to the ease with which a TCP/IP internet can be enlarged.

5.22   CONFIGURING ADDRESSES AND SUBNET MASKS

As we saw earlier, the TCP/IP configuration user interface varies from host to host.[4] On *tigger,* the *ifconfig* command is used to set or display parameters associated with an interface. Below, the parameters associated with Ethernet interface 0 (*le0*) are displayed:

```
> ifconfig le0
le0:   flags = 63<UP,BROADCAST,NOTRAILERS,RUNNING>
       inet 128.121.50.145 netmask ffffff00 broadcast 128.121.50.255
```

The IP address for the interface is 128.121.50.145. Its subnet mask is expressed in hex, ffffff00. The appropriate broadcast address for the subnet is 128.121.50.255.

The same information is entered via menus for *Chameleon.* For example, a pop-up menu is used to configure an IP address, as shown in Figure 5.13.

5.23   RELATIONSHIP BETWEEN NAMES AND ADDRESSES

Users who look at a system name like *fermat.math.yale.edu* and its IP address in dot format, 128.36.23.3, can easily get the idea that parts of names

---

[4]At the time of this writing, up-to-date router software supported supernetted addresses, but hosts only understood conventional address classes.

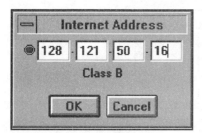

Figure 5.13   Configuring an IP address
via a menu.

actually correspond to numbers in the dotted address. This definitely is not
the case.

It is true that sometimes systems on a LAN are assigned names that
appear to match an address hierarchy. However:

- Completely unrelated names *may* appear on the same LAN
- Hosts with similar name structures *may* be located on different LANs or
  even on different networks.

For example, consider the following names and addresses:

macoun.cs.yale.edu   128.36.2.5
bulldog.cs.yale.edu   130.132.1.2

> *Addresses reflect network points of attachment and are bound to locations.*
> *But system names do not depend on their physical network attachments.*

An organization might populate a domain with names such as
chicago.sales.abc.com and *newyork.sales.abc.com*. These computers could be
located in the indicated cities.

Traffic is routed to a system based on its address, not its name, and a system's address always is looked up before data is sent to it. Thus, organizations are free to design a flexible naming plan that best meets their needs.

5.24   ADDRESS RESOLUTION PROTOCOL

Before a datagram can be sent between two stations on a LAN, it must be
wrapped up in a frame header and a frame trailer. The frame is delivered to
the Network Interface Card whose physical address matches the destination
physical address in the frame header.

> *Therefore, to deliver a datagram across a LAN, the physical address of the*
> *destination node must be discovered.*

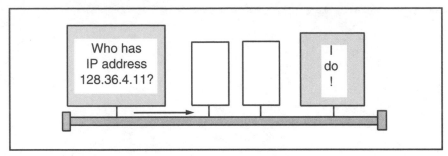

Figure 5.14   Finding the physical address of a system.

Fortunately, there is a procedure that automatically discovers physical addresses. The *Address Resolution Protocol* (ARP) provides a broadcast-based method for dynamically translating between IP addresses and physical addresses.

The systems on the local network use ARP to discover physical address information for themselves.[5] When a host wants to start communicating with a local partner, it looks up the partner's IP address in its ARP table, which commonly is kept in RAM memory. If there is no entry for that IP address, the host broadcasts an ARP request containing the destination IP address (see Figure 5.14).

The target host recognizes its IP address and reads the request. The first thing that the target host does is update its own address translation table with the IP and physical addresses of the sender. This is prudent because it is likely that the target soon will be conversing with the sender. The target host then sends back a reply containing its own hardware interface address.

When the source receives the reply, it updates its ARP table and is ready to transmit data across the LAN.

5.24.1   ARP Message Contents

ARP initially was used on Ethernet LANs, but its design is general, so it can be used with other types of networks such as Token-Rings, Fiber Distributed Data Interface (FDDI) LANs, and Switched Multimegabit Data Service Wide Area Networks (SMDS WANs). A variant of ARP has been designed for use with wide area virtual circuits (such as frame relay).

An ARP message is placed in the data field of a frame, following the lower-layer header(s).[6] The protocol type of the frame is identified to be ARP via Ethernet type code X'0806. Table 5.5 displays the ARP message fields.

---

[5] An administrator also can manually enter some permanent address translation entries into the ARP table, if desired.

[6] For example, for an Ethernet using DIX frames, an ARP message follows the MAC header, while for an 802.3 or 802.5 network, the ARP message follows the MAC header, Logic Link Control (LLC) header, and Sub-Network Access Protocol (SNAP) subheader.

TABLE 5.5   Format of an ARP Message

Number of octets	Field
2	Type of hardware address
2	Higher-layer addressing protocol
1	Length of hardware address
1	Length of higher-layer address
2	Type of message; 00 01 = request, 00 02 = response
*	Source hardware address
*	Source higher-layer (IP) address
*	Destination hardware address
*	Destination higher-layer (IP) address

The lengths of the last four fields depend on the technology and protocol in use. Hardware addresses for 802.X LANs contain 6 octets, and IP addresses are 4 octets long. The example in Table 5.6 shows the format of a message requesting translation of an IP address to an Ethernet address.

The source and destination roles will be reversed in the reply. For example, the *source* higher-layer address will be X'80-24-04-0B in the reply.

ARP use is not exclusively restricted to TCP/IP, since the second field identifies the protocol that is using ARP.

Since the original ARP request is broadcast, any system on the LAN could use the information in the ARP message to update its table entry for the

TABLE 5.6   Example of an ARP Request Message

Number of octets	Field	Description
2	00 01	Ethernet
2	08 00	IP
1	06	Ethernet 6-octet physical address length
1	04	IP address length
2	00 01	Request
6	02 07 01 00 53 23	Source hardware address
4	80 24 04 12	Source higher-layer address
6	00 00 00 00 00 00	Destination hardware address
4	80 24 04 0B	Destination higher-layer address

requester. However, usually a system enters an update only when it is the target of an ARP message.

5.24.2 ARP Table

Most systems provide a command that allows an administrator to:

- View the local ARP table
- Manually add or delete table entries
- Load a table with entries from a configuration file

The dialogue below uses the *arp* -a command to show how *tigger's* ARP table changes after a *telnet* connection is set up to host *mickey,* which is not currently in the table. Note that the output displays the name of each system, its IP address in dot format, and the 6 octets of its physical address as hexadecimal numbers separated by a : delimiter.

```
> arp -a
nomad-eth0.jvnc.net (128.121.50.50) at 0:0:c:2:85:11
r2d2.jvnc.net (128.121.50.2) at 8:0:20:a:2c:3f
jim-mac.jvnc.net (128.121.50.162) at 8:0:7:6f:a6:65
tom-mac.jvnc.net (128.121.50.163) at 8:0:7:ff:96:9e
chip.jvnc.net (128.121.50.148) at 0:0:3b:86:6:4c
nisc.jvnc.net (128.121.50.7) at 8:0:20:11:d2:b7
nicol.jvnc.net (128.121.50.10) at 0:0:3b:80:32:34
minnie.jvnc.net (128.121.50.141) at 8:0:20:7:b5:da
>

> telnet mickey.jvnc.net
Trying 128.121.50.143 ...
Connected to mickey.jvnc.net.
Escape character is '^]'.

SunOS UNIX (mickey.jvnc.net)

login:
. . .
logout

> arp -a
nomad-eth0.jvnc.net (128.121.50.50) at 0:0:c:2:85:11
r2d2.jvnc.net (128.121.50.2) at 8:0:20:a:2c:3f
jim-mac.jvnc.net (128.121.50.162) at 8:0:7:6f:a6:65
tom-mac.jvnc.net (128.121.50.163) at 8:0:7:ff:96:9e
chip.jvnc.net (128.121.50.148) at 0:0:3b:86:6:4c
nisc.jvnc.net (128.121.50.7) at 8:0:20:11:d2:b7
nicol.jvnc.net (128.121.50.10) at 0:0:3b:80:32:34
minnie.jvnc.net (128.121.50.141) at 8:0:20:7:b5:da
mickey.jvnc.net (128.121.50.143) at 8:0:20:7:53:8f
>
```

5.24.3 Reverse ARP

A variant of ARP called *reverse ARP* (RARP) was designed to help a node to find out its *own* IP address. It was intended for use by diskless workstations and other devices that need to get configuration information from a network server.

A station using the reverse ARP protocol broadcasts a query stating its physical address and requesting its IP address. A server on the network that is configured with a table of physical addresses and matching IP addresses can respond to the query.

Reverse ARP has been superseded by the BOOTP protocol and its improved version, the *Dynamic Host Configuration Protocol* (DHCP). These protocols are far more powerful and are used to provide a complete set of configuration parameters to a TCP/IP system. BOOTP and DHCP will be discussed in Chapter 11.

5.25 MULTIPLE ADDRESSES FOR ONE INTERFACE

Some router vendors allow you to assign *multiple* IP addresses to a single router interface. Why would anyone want to do this? Multiple subnet addresses might be needed for a LAN that has a very large number of systems. Or, separate subnet numbers might be used to apply different traffic filtering rules to systems belonging to two different workgroups. Each workgroup would belong to a separate *logical* subnet, although both share the same *physical* medium.

Figure 5.15 shows a LAN with two logical subnets, 128.36.4.0 and 128.36.5.0. The router's LAN interface has been assigned the two IP addresses, 128.36.4.1 and 128.36.5.1. Traffic *to* this LAN will be routed correctly. However, some extra work needs to be done to route datagrams originating at a LAN host correctly.

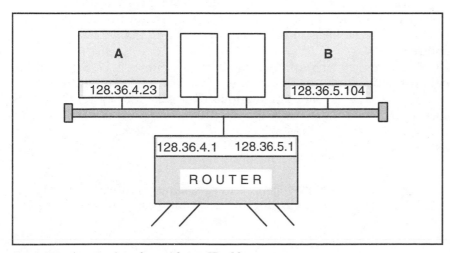

Figure 5.15 A router interface with two IP addresses.

Suppose that system A has an 8-bit subnet mask. When A wants to send a datagram to B, it will pass the datagram to the router. To avoid this, the hosts on the LAN could be configured with 7-bit subnet masks, since 4 corresponds to 0000 0100 and 5 is 0000 0101.

5.26 PROXY ARP

Suppose that adjacent subnet numbers were not available for use on the LAN. For example, suppose that 128.36.4.0 and 128.36.20.0 were sharing the medium. In this case, hosts on the LAN could be configured with the mask 255.255.0.0—which means "no subnetting." Hosts would then use ARP for *all* destinations on network 128.36. This will work perfectly for systems that share the medium, but what about traffic to subnets of 128.36 that are not on the LAN?

The LAN router will handle external traffic if it supports *Proxy ARP*. This means that whenever the router sees an ARP request for a destination that is external to the LAN,[7] the router sends an ARP response containing the *router's* physical address. The host would wrap a frame around the datagram and forward it to the router, which would unwrap the datagram and send it onward.

5.27 MULTICAST ADDRESSES

IP broadcasting causes a datagram to be delivered to every system on a network or subnet. A more restricted form of multiple delivery, called *multicasting,* causes a datagram to be propagated to a group of systems, as shown in Figure 5.16.

IP multicasting can be a very useful network tool. For example, a single message can be used to simultaneously update configuration data across a homogeneous group of hosts or to poll a group of routers for their status. Multicasting also is the basis of new applications that enable users to "tune in" to conferences.

Class D IP addresses, whose format is shown in Figure 5.17, are used for multicasts. A multicasting protocol standard has been defined, but the number of available hosts and routers that support the standard currently is limited. However, its use will grow over the next few years, so it is worthwhile to examine some of its features here.

5.27.1 Multicast Groups

A *multicast group* is a set of systems with an assigned multicast IP address. Members of the group still retain their own IP addresses but also have the

---

[7]If there are multiple routers on the LAN, the router that has the best path to the destination answers.

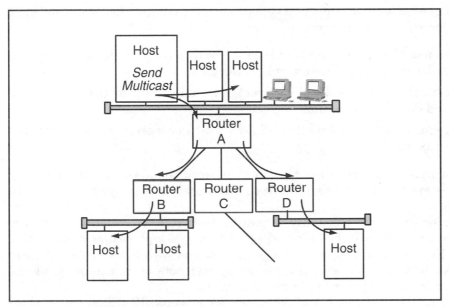

Figure 5.16 Propagating multicast datagrams.

ability to absorb data that has been sent to the multicast address. Any system may belong to zero or more multicast groups.

The Class D addresses that are used for multicasting start with numbers ranging from 224 to 239. Some IP multicast addresses are permanent and are listed in the Internet *Assigned Numbers* RFC. Some examples of permanently defined multicast IP addresses include:

224.0.0.1   All hosts on a local subnet

224.0.0.2   All routers on a local subnet

224.0.0.5   All routers supporting the Open Shortest Path First (OSPF) routing
             protocol

Multicast addresses also are assigned in an ad hoc manner to temporary groups that are formed and dissolved on an as-needed basis—for example, for an audio or video conference.

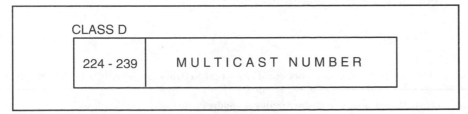

Figure 5.17 Class D format for IP multicast datagrams.

A number of functions must be supported within a host that is a member of one or more multicast groups:

- There must be a command that tells the host to join a multicast group and identifies the interface that should listen for this address.

- A host's IP layer must be able to recognize multicast addresses for incoming and outgoing datagrams.

- There must be a command that allows a host to cancel its membership in a multicast group.

Multicasting is not restricted to a local network. Routers with multicasting software are able to propagate multicast IP datagrams to systems across an internet.

In order to do this efficiently, a router needs to know whether there are hosts on its locally attached networks that belong to a particular multicasting group. Routers also need to exchange information with other routers so that they can find out whether there are group members on remote networks to whom they should forward multicast datagrams.

Hosts use the *Internet Group Management Protocol* (IGMP) to report their group memberships to neighboring routers that support multicast routing. The reports are sent to the IP multicast address that belongs to the group that the host is joining.[8]

To assure that their membership information is complete, the IGMP protocol enables routers to poll hosts periodically, asking for reports of their current memberships. The polls are sent to the all-hosts multicast IP address, 224.0.0.1.

5.27.2   Translating Multicast Addresses to Ethernet or FDDI Addresses

Physical interfaces to Ethernet and FDDI LANs optionally can be assigned one or more multicast addresses. These are logical assignments, and any convenient values can be selected. This makes it easy to translate an IP multicast address to a physical multicast address.[9]

The following translation rule can be used for Ethernet and FDDI LANs:

- The first 3 octets of the physical multicast address should be set to 01-00-5E.

- The next bit should be set to 0, and the final 23 bits should be set to the low-order 23 bits of the IP multicast address.

---

[8]Routers will not forward this report out of the local network, so it will be heard only by routers and by other local members of the group.

[9]Note that ARP is not used to translate multicast addresses.

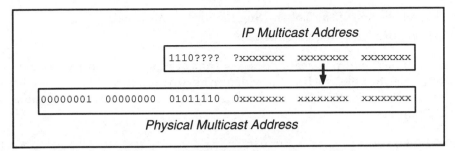

IP Multicast Address

| 1110???? | ?xxxxxxx | xxxxxxxx | xxxxxxxx |

| 00000001 | 00000000 | 01011110 | 0xxxxxxx | xxxxxxxx | xxxxxxxx |

Physical Multicast Address

Figure 5.18 Mapping part of the IP address into the physical address.

This mapping is illustrated in Figure 5.18:

- The final 23 bits in the IP multicast address have been marked "x." These bits are copied into the low-order bits of the physical multicast address.
- Positions marked "?" in the IP multicast address could be filled in with any bits. They are not copied into the physical multicast address.

This means, for example, that the three multicast IP addresses:

    11100000 00010001 00010001 00010001
    11100000 10010001 00010001 00010001
    11100001 10010001 00010001 00010001

all would map to the *same* physical multicast address:

    00000001 00000000 01011110 00010001 00010001 00010001

Interfaces on systems that belong to any of the three groups would capture multicasts for all of the groups. However, each host's IP layer would discard extraneous multicasts.

A good way to avoid this extra processing is to choose multicast addresses that set the ? positions to zero. This still leaves 2^{23}, or over eight million, multicast addresses.

5.27.3 Translating Multicast Addresses to Token-Ring Addresses

Unfortunately, the same scheme used above for Ethernet and FDDI cannot always be used for Token-Ring (at the time of writing) because many Token-Ring hardware interfaces cannot be configured with arbitrary multicast addresses. Therefore, any one of three translation methods can be used, depending on the hardware available at a site:

1. Embed 23 bits of the IP multicast address, just as above

2. Choose and use one of the Token-Ring *functional* addresses

3. Use a Token-Ring *all rings broadcast*

There are 31 *functional* physical addresses that are used to identify systems that have a specific role, such as a bridge, ring wiring concentrator, or ring error monitor. When method 2 is selected, multicasts are sent to the functional physical address:

03-00-00-20-00-00

When a station receives a frame that contains a multicast datagram, the IP address would have to be examined to see if this station really is a member of the multicast group.

Since one functional address is used for all multicast addresses, this method is not perfectly efficient. However, it is better than the last alternative, method 3, which broadcasts to all stations.

5.28 RECOMMENDED READING

The address classes are defined in the IP standard, RFC 791. Subnetting is defined in RFC 950, and supernetting is discussed in RFC 1519. Broadcasts are described in RFC 919 and RFC 922.

The Address Resolution Protocol is defined for Ethernets in RFC 826. Reverse ARP is described in RFC 903.

RFC 1112 describes IP multicasting. RFC 1390 describes the translation between IP multicast addresses and FDDI addresses. RFC 1469 discusses the translation between IP multicast addresses and Token-Ring addresses.

RFC 1178 contains both sound and entertaining advice on how to choose a name for your computer. RFCs 1034 and 1101 contain detailed explanations of domain naming. RFC 1035 describes the protocols used to build the Domain Name System and discusses the implementation of this system.

The Hosts Requirements standard, RFC 1122, provides further details about naming and addressing and corrects errors in the defining standards.

Chapter

6

Internet Protocol

6.1 INTRODUCTION

Recall that an internet is a set of networks connected by routers,[1] and the Internet Protocol is a network layer protocol that routes data across an internet. The researchers and designers who created IP were responding to U.S. Department of Defense (DOD) requirements for a protocol that could:

- Accommodate the use of hosts and routers built by different vendors
- Encompass a growing variety of network types
- Enable the network to grow without interrupting service
- Support higher-layer sessions and message-oriented services

The IP network layer architecture was designed to meet these needs.

It turned out that IP also gave network builders exactly what they needed in order to integrate the local area network (LAN) "islands" that had spread across their organizations. Furthermore, new islands could be plugged in without disrupting what already was in place.

These features eventually caused IP to become the network protocol of choice for government agencies, universities, and businesses.

6.2 IP DATAGRAMS

The IP protocol provides the mechanisms needed to transport units called *IP datagrams* across an internet. As shown in Figure 6.1, an IP datagram is made up of an IP header and a chunk of data to be delivered.

---

[1]Recall that many early RFC documents used the term *gateway* rather than *router*.

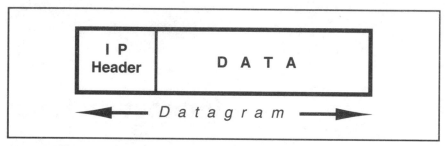

Figure 6.1 Datagram format.

IP is a "best effort" protocol. This means that IP does not guarantee that a datagram will be delivered safely to its destination. All that is guaranteed is that a best effort will be made (see Figure 6.2). A datagram may be destroyed along the way because:

- Bit errors occurred during transmission across a medium.
- A congested router discarded the datagram because of a shortage of buffer space.
- Temporarily, there was no usable path to the destination.

All of the features that assure reliability have been concentrated within the TCP layer. Recovery from destroyed data depends on TCP actions.

6.3 PRIMARY IP FUNCTION

The primary IP function is to accept data from TCP or the User Datagram Protocol (UDP), create a datagram, route it through the network, and deliver

Figure 6.2 IP best effort delivery.

it to a recipient application. Each IP datagram is routed independently. IP relies on two tools to help it to route datagrams:

- The *subnet mask*
- The IP *routing table*

6.4 HOW THE SUBNET MASK IS USED

Suppose that your computer has IP address 130.15.12.131 and is attached to a LAN. If you have data to be sent:

From: 130.15.12.131

To: 130.15.12.22

You might guess that both systems are on the same subnet. However, your computer needs to check whether this is true. This is done by checking the subnet mask. Suppose that your host has subnet mask:

255.255.255.0

This means that the mask consists of 24 ones and 8 zeros:

11111111 11111111 11111111 00000000

Recall that the 1s in the subnet mask identify the *network and subnet* portions of the address. Since the network and subnet part of both the source and destination address is 130.15.12, both are on the same subnet.

Your computer actually performs a logical AND between the mask and each of the IP addresses. The effect is that the 0s in the subnet mask zero out the host part of the address, leaving just the network and subnet parts.

For this example, routing is *direct*. This means that the datagram must be wrapped in a frame and transmitted directly to its destination on the LAN, as shown in Figure 6.3.

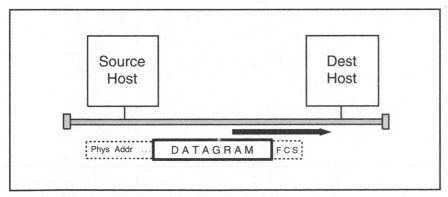

Figure 6.3 Framing and transmitting a datagram.

The destination address that is placed in the frame header must be the physical address of the destination system. The Address Resolution Protocol (ARP) table will be checked to see if there is an entry that provides the physical address for 130.15.12.22. If there is not yet an entry, the ARP protocol will be used to create one.

6.5   HOST IP ROUTING TABLE

Suppose that you have data to be sent:

From: 130.15.12.131

To: 192.45.89.5

A quick subnet mask check shows that this destination is *not* on the local subnet. In this case, IP must consult its local routing table.

A host's routing table usually is very simple. Figure 6.4 shows a LAN that is connected to remote sites by means of a single router. If a destination is not on the local network, a host does not have any choice. The only way to leave the local net is via that router.

Each host and desktop computer on this LAN contains a routing table that tells IP how to route datagrams to systems that are not connected to the LAN. To point the way to remote locations, this routing table needs the single entry:

default 130.15.12.1

In other words, *forward any nonlocal datagrams to the default router, which has IP address 130.15.12.1.* (Note that destination address 0.0.0.0 is used to mean *default* in routing tables.)

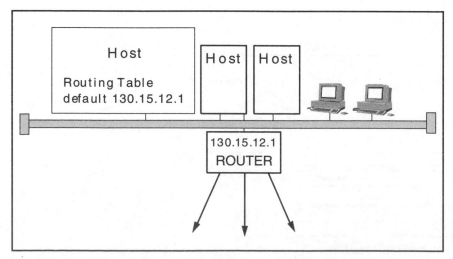

Figure 6.4 Forwarding traffic via a default router.

6.6 NEXT-HOP ROUTING

The reason that host routing tables can be kept this simple is that IP does not need to look at the complete route that will be followed to a destination. *It only needs to find out the next hop and forward the datagram there.*

To forward a datagram to the router interface at 130.15.12.1, the datagram must be wrapped in a frame whose header contains the physical address of the router's interface card.

When the router receives the frame, it will strip off the frame header and trailer and examine the IP datagram header in order to decide where it should go next.

6.7 ANOTHER HOST ROUTING TABLE EXAMPLE

Sometimes a host routing table is a bit more complicated. For example, there are two routers on the subnet 128.121.50.0 in Figure 6.5. The second router leads to a small LAN that is the home of several workstations.

Tigger has a route to such a LAN. We can view *tigger*'s routing table[2] with the *netstat -nr*[3] command.

```
> netstat -nr
Routing tables
Destination      Gateway          Flags   Refcnt        Use   Inter
                                                               face
127.0.0.1        127.0.0.1        UH      6           62806   lo0
default          128.121.50.50    UG      62        2999087   le0
128.121.54.0     128.121.50.2     UG      0               0   le0
128.121.50.0     128.121.50.145   U       33        1406799   le0
```

---

[2] Note the use of the term *gateway* rather than *router* in the display.

[3] Other computers may respond with tables that are formatted a little differently. They will contain similar but not necessarily identical information. For example, some systems may provide a column that displays a distance associated with each destination.

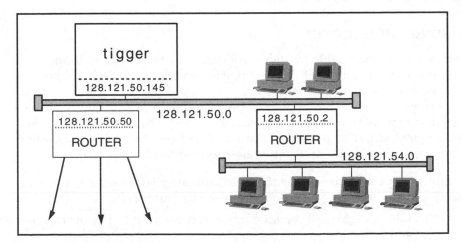

Figure 6.5 Routing decisions.

The *netstat* output discloses quite a lot of information about how and where *tigger*'s traffic is being routed:

- We can see that the first destination in the table is the *loopback* address, 127.0.0.1. It is a placeholder for traffic between clients and servers within *tigger*.

- The *default* entry is used to route to any destination that is not explicitly listed in the table. Traffic should be forwarded to the router interface at IP address 128.121.50.50.

- Datagrams to any system on subnet 128.121.54.0 should be forwarded to the router interface at IP address 128.121.50.2.

- The last entry is a placeholder that says "to route to any system on subnet 128.121.50.0, route via 128.121.50.145." But 128.121.50.145 is *tigger*'s own address, and 128.121.50.0 is *tigger*'s own LAN. Although this entry does not provide new routing information, it does let us view some interesting statistics about local traffic.

Netstat displays quite a lot of additional information:

- *Flags* tell whether a route is up (usable) and whether the next hop is a host (H) or gateway (G).

- *REFcnt* tracks the number of currently active uses of the route.

- The *Use* column counts the number of datagrams that have been sent on the route (since the last initialization).

- Interface *lo0* is a *logical* interface used for loopback traffic. All external traffic passes through the single Ethernet interface, *le0*.

Note that by including the local subnet, 128.121.50.0, in the report, we have discovered that more than twice as much traffic was sent to the outside world than to systems on the local area network.

6.8   RULE FOR ROUTING TABLE LOOKUPS

Each entry in a routing table provides information about routing to an individual destination. A routing table destination can be an individual host, a subnet, a network, a supernet, or *default*.

There is a general rule that applies to the way that IP uses a routing table, whether that table is in a host or a router. The entry chosen should be based on the *most precise match* to the destination IP address. In other words, when IP looks up the address of a destination host, conceptually, it is as if:

- The table is first searched to see if there is an entry that matches the complete IP address. If there is, this entry is used to route traffic.

- If not, the table is searched for an entry corresponding to the destination subnet network.

- If not, the table is searched for the destination network.
- If not, the table is searched to see if there is a routing prefix entry that matches.
- If this cannot be found, the default route is used.

Of course, a real implementation would search the table just once, throwing away a match when a more precise match was found.

6.9 ROUTER ROUTING TABLES

Host routing tables can be very simple, but the tables in routers often will contain a lot more information. A router has two or more interfaces, and each datagram must be transmitted through the appropriate interface. The router may need to record next-hop selections for many different subnets and networks, as shown in Figure 6.6.

6.10 BRANCH OFFICE ROUTING TABLE

Some routers have very simple routing tables. For example, the branch office router in Figure 6.7 directs incoming traffic from headquarters to the site's LAN and forwards all outgoing traffic across a wide area link to a router in company headquarters.

This router has two interfaces:

Interface	IP address
1	130.15.40.1
2	130.15.201.2

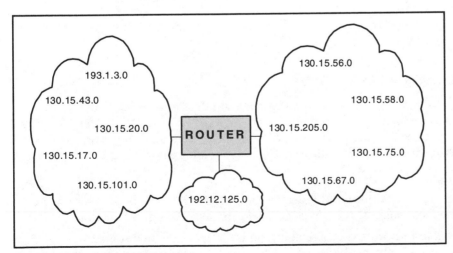

Figure 6.6 Routing to many locations.

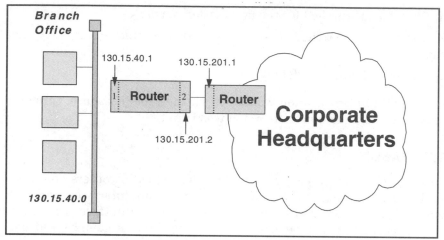

Figure 6.7 Branch office routing.

Routing table information would include:

Destination	Interface	Next hop	Type	Protocol
130.15.40.0	1	130.15.40.1	Direct	Manual
0.0.0.0	2	130.15.201.1	Indirect	Manual

The first entry just describes the direct connection to the locally connected subnet, 130.15.40.0. The subnet is reached *directly* through its own interface.

The second entry provides the default route to the rest of the network. The next-hop router is 130.15.201.1, and it is reached via interface 2. The headquarters destinations are reached *indirectly,* via the next-hop router. Both of these routes were entered manually.

6.11 GLOBAL ROUTING OPERATIONS

So far, we have been concentrating on a single routing decision. Figure 6.8 illustrates a global IP routing operation. When TCP or UDP at Host A wishes to send data to its peer at Host B, the sender passes its data to IP, along with the destination host's IP address. IP adds a header containing the destination IP address to the data.

- IP at Host A examines the destination address to see if the destination is on the local subnet. It isn't, so IP performs a routing table lookup.

- The table indicates that the next hop is Router X. The datagram is framed, and the LAN physical address for Router X is placed in the frame header.

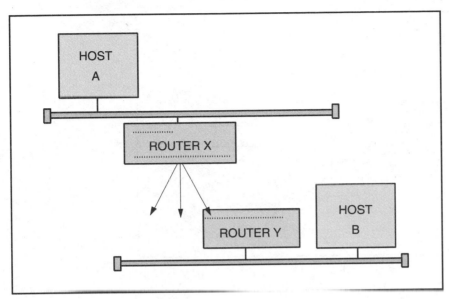

Figure 6.8   Global routing.

- When the datagram arrives at Router X, the framing is removed. IP at Router X compares the destination IP address with all of its own IP addresses (using the subnet masks) to see if the destination is on a locally connected subnet.

- It isn't, so IP performs a routing table lookup. The next hop is Router Y. The datagram is framed for serial transmission and sent to Router Y.

- When the datagram arrives at Router Y, the framing is removed. IP at Router Y compares the destination IP address with all of its own IP addresses (using the subnet masks) to see if the destination is on a locally connected subnetwork. It is, and Router Y frames the datagram for delivery to Host B.

This route from Host A to Host B consists of three hops: A to X, X to Y, and Y to B.

6.12   IP FEATURES

There are a number of features that contribute to IP's flexibility and ability to fit into many different environments. Among these features are *adaptive routing* and *datagram fragmentation and reassembly.*

6.12.1   Adaptive Routing

Ordinarily, datagram routing is *adaptive.* That is, the best choice for the next hop at any instant is made by checking the routing table at the current node. Routing table entries can change at any time, depending on network conditions.

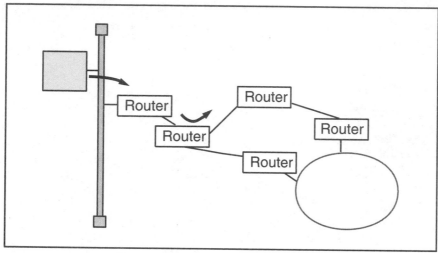

Figure 6.9 Adaptive routing.

For example, (see Figure 6.9), if a link goes down, datagrams will be switched to a different route, if one is available.

A change in network topology just causes datagrams to be rerouted automatically. Adaptive routing builds in flexibility and robustness.

On the other hand, an IP header *can* contain a strict source route to be followed to a destination. This might be done in order to route sensitive traffic along a secure path.

6.12.2 MTU, Fragmentation, and Reassembly

Before a datagram can be transmitted across a network hop, it must be encapsulated within the layer 2 header(s) required for the network technology, as shown in Figure 6.10. For example, to traverse an 802.3 or 802.5 network, a Logical Link Control (LLC) header, Sub-Network Access Protocol (SNAP) sub-header, Media Access Control (MAC) header and MAC trailer are added.

As we have seen in Chapter 4, each LAN and WAN technology imposes a different size limit on its frames. A datagram has to fit inside a frame, and so the maximum frame size restricts the size of the datagrams that IP can send across a medium.

The maximum datagram size for a medium is computed by subtracting the size of the frame header, frame trailer, and data link layer header from the total maximum frame size:

Max frame size – frame header size – frame trailer size – link layer header size

Recall that the biggest datagram size for a medium is called the *Maximum Transmission Unit,* or MTU. For example, DIX Ethernet has an MTU of 1500

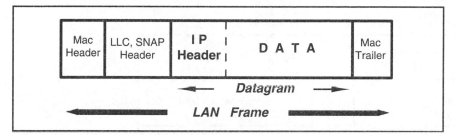

Figure 6.10 Transmission format for a LAN frame.

octets, 802.3 has an MTU of 1492 octets, Fiber Distributed Data Interface (FDDI) has an MTU of 4352 octets, and Switched Multimegabit Data Service (SMDS) has an MTU of 9180 octets.

In a large internet, an originating host may not know all of the size limits that a datagram will meet along its path. What happens if the source host has sent out a datagram that is too large for some intermediate network?

When the datagram arrives at the router that is attached to that intermediate network, IP solves the size problem by chopping the datagram into several smaller datagrams called *fragments*. It is up to IP *in the destination host* to gather up the incoming fragments and rebuild the original datagram.

Fragmentation most often is performed in a router. However, a UDP application might initiate a large message that causes the sending host to fragment a datagram.

6.13 IP PROTOCOL MECHANISMS

Now we are going to take a close look at what makes IP version 4 tick. We'll examine the formal protocol elements—the format of the IP header and the rules for handling the datagram as it traverses the network. See Chapter 22 for a description of IP version 6.[4]

6.13.1 Datagram Header

A datagram header is formatted as five or more 32-bit *words*. The maximum size for a header is 15 words (that is, 60 octets), but in practice, most datagram headers have the minimum length of 5 words (20 octets).

The header fields are displayed in Figure 6.11, which is arranged as a sequence of words. Note that the bits in a word are numbered from 0 to 31.

6.13.2 Destination, Source, and Protocol Fields

The most important fields in the header are *Destination IP Address, Source IP Address,* and *Protocol.*

---

[4]There is no IP version 5.

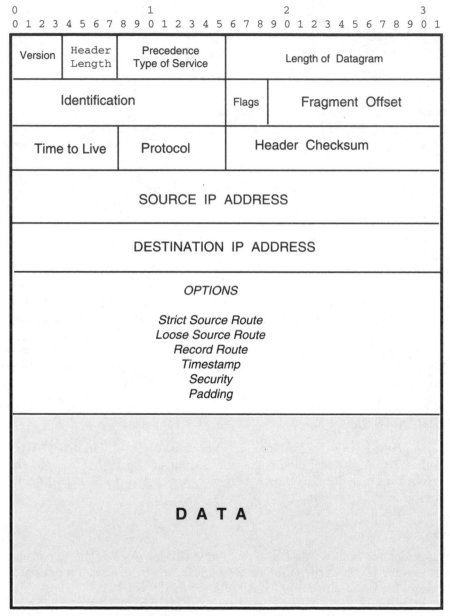

Figure 6.11 Format of an IP datagram.

The destination IP address enables IP to route the datagram. Once the datagram has reached its destination host, the *Protocol* field enables IP to deliver the datagram to the appropriate service—such as TCP or UDP. There are several other protocols besides TCP and UDP that send and receive datagrams. The Internet Assigned Numbers Authority (IANA) is responsible for coordinat-

TABLE 6.1   Common IP Protocol Field Numbers

#	Title	Protocol	Description
1	ICMP	Internet Control Message Protocol	Carries error messages and supports some network utilities
2	IGMP	Internet Group Management Protocol	Supports multicasting groups
6	TCP	Transmission Control Protocol	Supports sessions
8	EGP	Exterior Gateway Protocol	Older protocol used to establish routing to external networks
17	UDP	User Datagram Protocol	Provides delivery of stand-alone blocks of data
88	IGRP	Cisco's Interior Gateway Routing Protocol	Enables Cisco routers to exchange routing information

ing the assignment of TCP/IP parameter values, including values that can be used in the IP *Protocol* field. Some of the numbers that have been assigned for the IP *Protocol* field relate to proprietary, vendor-specific protocols.

Table 6.1 shows some of the most commonly used IP *Protocol* field numbers.

6.13.3   Version, Header Length, and Datagram Length

The currently deployed version of IP is 4. (The "Next Generation" version number is 6.)

The header length is measured in 32-bit words. If there are no options, the header length is five words (that is, 20 octets). If one or more options are included, the header may need to be padded with 0s so that it ends at a 32-bit word boundary.

The *Datagram Length* field measures the length in octets. The measurement includes both the header and the data portions of the datagram. This 16-bit field can express values up to a maximum of $2^{16}-1 = 65,535$ octets.

Network technologies are not the only reason to limit datagram size. The diverse types of computers that support IP have different limits on the sizes of the memory buffers that they use for network traffic. (The IP standard requires that all hosts must be able to accept datagrams of up to 576 octets.)

6.13.4   Precedence and Type of Service

The Department of Defense was the original sponsor of the TCP/IP protocol suite, and the ability to assign precedence levels to datagrams was important to the DOD. Precedence has been little used outside of the military and government, but that is changing. There are 3 precedence bits, providing eight different precedence levels.

The IP standard does not mandate the specific actions that are caused by the values of the precedence bits. It was originally intended that IP would map this setting onto similar options for the subnet to be crossed on the next hop. For example, access to a Token-Ring can be scheduled based on a precedence level. IP could map its precedence level to a corresponding Token-Ring precedence level.

The *Type of Service* (TOS) bits contain quality of service information that could affect how a datagram is handled. For example, when a router runs short of memory, it has to discard some datagrams. A router might consider a datagram whose reliability bit is set to 1 to be less eligible for discard than one whose reliability bit is equal to 0.

The position of the precedence and Type of Service bits in the field is:

Bits	Type	Description
0–2	Precedence	Levels 0–7. Level 0 is normal. Level 7 provides the highest priority.
3–6	TOS	Delay, reliability, throughput, cost, or security.
7	Reserved for future use	

The Type of Service values, as described in the current *Assigned Numbers* document, are shown in Table 6.2. Settings are mutually exclusive—only one TOS value can be requested in any one IP datagram. The *Assigned Numbers* standard recommends specific values to be used for various applications. For example, minimize delay for *telnet,* maximize throughput when copying a file, and maximize reliability when delivering network management messages.

Some routers ignore the Type of Service field entirely, while others can use the field in making routing decisions or in deciding which traffic should be protected against discard when memory is in short supply. It is believed that

TABLE 6.2   Type of Service Values

TOS Value	Description
0000	Default
0001	Minimize monetary cost
0010	Maximize reliability
0100	Maximize throughput
1000	Minimize delay
1111	Maximize security

TABLE 6.3   Recommended Type of Service Values

Protocol*	TOS value	Description
Telnet and other login protocols	1000	Minimize delay
FTP control session	1000	Minimize delay
FTP data session	0100	Maximize throughput
TFTP	1000	Minimize delay
SMTP command phase	1000	Minimize delay
SMTP DATA phase	0100	Maximize throughput
Domain Name Service UDP Query	1000	Minimize delay
Domain Name Service TCP Query	0000	No special handling
Domain Name Service Zone Transfer	0100	Maximize throughput
NNTP	0001	Minimize monetary cost
ICMP Errors	0000	No special handling
ICMP Requests	0000	Normally 0000 but sometimes sent with other value
ICMP Responses		Same as request being answered
Any IGP	0010	Maximize reliability
EGP	0000	No special handling
SNMP	0010	Maximize reliability
BOOTP	0000	No special handling

*FTP = File Transfer Protocol.
 TFTP = Trivial File Transfer Protocol.
 SMTP = Simple Mail Transfer Protocol.
 NNTP = Network News Transfer Protocol.
 IGP = Interior Gateway Protocol.
 EGP = Exterior Gateway Protocol.
 SNMP = Simple Network Management Protocol.

Type of Service will play a bigger role in the future. The *Assigned Numbers* document recommends using the values in Table 6.3.

6.13.5   Time-To-Live

When a topology change occurs in an IP internet—such as a link down or a new router initializing—some datagrams may wander around during the short period that new routes are being selected.

More serious misrouting problems can result from human error when routing information is entered manually. A mistake can cause datagrams to get "lost" or locked into circular paths for a long time.

The Time-To-Live (TTL) field limits the amount of time that a datagram will be allowed to remain in an internet. The TTL is set by the originating host and is decremented at each router that handles the datagram. A datagram that has not yet reached its destination host when the TTL reaches zero is discarded.

Although formally defined as a time in seconds, the TTL actually is implemented as a simple hop counter that is decremented—usually by 1—at each router. (Optionally, a larger decrement could be applied to a datagram that has just crossed a very slow link or has been queued for transmission for a long time.)

A recommended default value for the initial TTL is approximately twice the longest path in the internet. The length of the longest path is sometimes called the *diameter* of the internet.

6.13.6   Header Checksum

This 16-bit field contains a checksum that is computed on the fields[5] of the IP header. The checksum must be updated as the datagram is forwarded because the Time-to-Live field changes at each router. (Other header values also may change because of fragmentation or due to values written into option fields.)

6.14   FRAGMENTATION

The *Identification, Flags,* and *Fragment Offset* fields enable datagrams to be fragmented and reassembled. When IP needs to transmit a datagram whose size is larger than the MTU for the next link:

1. The first step is to check the *Flags* field. There is a "Don't Fragment" bit in the *Flags* field. If the "Don't Fragment" flag is set to 1, nothing can be done—the datagram must be discarded.
2. If the "Don't Fragment" flag is 0, the data portion is broken into pieces consistent with the next-hop MTU. Each break must be aligned on an 8-octet boundary.
3. Each piece is given an IP header similar to the header for the original datagram. In particular, each piece will have the same source, destination, protocol, and *Identification* field value. However, the following fields need to be set separately for each piece:
   a. Length of datagram. This is the length of the datagram holding the current piece.
   b. There is a "More" flag in the *Flags* field. This must be set to 1 in all but the final piece.

---

[5]The checksum is the 16-bit one's complement of the one's complement sum of all 16-bit words in the header. Prior to the calculation, the checksum field is set to 0.

c. The *Fragment Offset* field is set to indicate the position of this piece relative to the beginning of the original datagram. The start position is 0. The fragment offset is actually the true offset divided by 8.

d. Separate checksums must be calculated for each of the fragments.

Now we can look at the formal details.

6.14.1 Identification Field

The *Identification* field contains a 16-bit number. This number helps the destination host to recognize datagram fragments that belong together.

6.14.2 Flags Field

The *Flags* field contains three bits, as shown below:

Bit 0	Bit 1	Bit 2
0 = Reserved	0 = May Fragment 1 = Don't Fragment	0 = Last Fragment 1 = More Fragments

Bit 0 is reserved, and must be 0. The sender can set the next bit to 1 to prevent the datagram from being fragmented. If a datagram cannot be delivered without fragmentation and this bit is set to 1, the datagram would have to be discarded and an error message would be sent back to the source.

Bit 2 is set to 0 if this is the last—or the only—piece of a datagram. Bit 2 is set to 1 to indicate that this datagram is a fragment and more fragments follow.

6.14.3 Fragment Offset Field

An 8-octet chunk of data is called a *fragment block*. The number in the *Fragment Offset* field reports the size of the offset in fragment blocks. The *Fragment Offset* field is 13 bits long, so offsets can range from 0 to 8192 fragment blocks—corresponding to 0 to 65,528 octets in the entire datagram. For example, suppose that a router fragments a datagram (with ID 348) that carries 3000 bytes of data into three datagrams, each carrying 1000 bytes. Each fragment will include its own header and 1000 bytes (125 fragment blocks) of data. The contents of the *Identification, Flags,* and *Fragment Offset* fields would be:

Fragment	ID	Flags	Fragment offset
1	348	May Fragment, More	0 Blocks From Start
2	348	May Fragment, More	125 Blocks (1000 octets) From Start
3	348	May Fragment, Last	250 Blocks (2000 octets) From Start

On the other hand, if the datagram had been delivered without fragmentation, it would have had the following fields:

ID	Flags	Fragment Offset
348	May Fragment, Last	0 Blocks From Start

In this case, the recipient host can tell that the incoming datagram was not fragmented because an offset of 0 shows that it contains the start of the data, while the flag set to "last" shows that it contains the end of the data.

6.14.4 Reconstructing a Fragmented Datagram

A fragmented datagram is reconstructed at the recipient host. The pieces of a fragmented datagram can arrive out of order. When the earliest fragment arrives at the destination host, IP allocates some memory for reassembling the datagram. The *Fragment Offset* field indicates the byte boundary for the data in this fragment.

Fragments with matching *Identification, Source IP Address, Destination IP Address,* and *Protocol* fields belong together and are merged as they arrive. There is one inconvenient omission in the IP protocol; the recipient has no way of knowing how long the entire datagram will be until the end fragment arrives. The *Total Length* field in a fragment's header field reveals the length of only *that* datagram fragment.

This means that the recipient system has to do some guesswork about how much buffer space to reserve for an incoming datagram. Vendors handle this problem in different ways. Some allocate small incremental buffers to hold incoming fragments, while others use a single, fixed-sized buffer.

In any case, all implementations are *required* to handle at least incoming fragmented datagrams whose total length is up to 576 octets. Clearly, an implementation really should be able to handle fragmented datagrams whose total size is at least the MTU for the interface on which they arrive.

6.14.5 Reassembly Timeout

Now imagine the scenario:

- A datagram is transmitted.
- The sending process crashes.
- The datagram is fragmented.
- One of the fragments is destroyed along the way.

The receiving host could wait forever, but the missing piece will never arrive. Obviously, the receiving host needs to set a *reassembly timeout*. When the timeout expires, the destination host gives up, discards the received fragments, and sends an error message back to the originator. The reassembly timeout sometimes is configurable, and the recommended value is between 60 and 120 seconds.

6.14.6   To Fragment or Not to Fragment

After having gone to all of this trouble to support fragmentation, the procedure turns out to be a real performance bottleneck. As a result, most programmers carefully design their applications so that datagrams are small enough so that they will not be fragmented.

In Chapter 7, we'll find out about an MTU discovery protocol that prevents fragmentation while enabling use of the biggest MTU that can be carried to the destination.

6.15   LOOKING AT IP STATISTICS

We can get an idea of how IP behaves by checking some rough statistics. The *netstat -s* command provides counts of major IP events. The report that follows was run at *tigger.jvnc.net,* which is a server that is accessed by hosts all over the Internet. Note that the term *packet* is used in the report instead of the more correct term, *datagram.*

```
> netstat -s
ip:
13572051 total packets received
0 bad header checksums
0 with size smaller than minimum
8 with data size < data length
0 with header length < data size
0 with data length < header length
90 fragments received
0 fragments dropped (dup or out of space)
2 fragments dropped after timeout
0 packets forwarded
10 packets not forwardable
0 redirects sent
0 ip input queue drops
```

During the reporting period, there were no bad checksums, and *tigger* never had to drop datagrams because of a shortage of memory. Ninety fragments were received, for a grand total of 0.00066 percent. Two of the fragments were dropped after timeout.

There were 10 "nonforwardable" datagrams, possibly resulting from attempts to experiment with source routing via *tigger.*

6.16   OPTIONS

Up to 40 extra IP header octets are available to carry one or more options. The options that are included in a datagram are chosen by the sending applications. The use of options is fairly rare. The options consist of:

- Strict Source Route

- Loose Source Route
- Record Route
- Timestamp
- Department of Defense Basic Security
- Department of Defense Extended Security
- No Operation
- End of Option List (Padding)

Security options are used by the Department of Defense and some other government agencies. Several other options have been proposed.[6]

6.16.1   Source Routes

Two source routing options are provided. A *Strict Source Route* describes a complete path that must be followed to a destination. A *Loose Source Route* identifies milestones along the way. Any path can be used between identified milestones.

Strict source routes sometimes are used to improve data security. Unfortunately, as we shall see a little later, source routes also are part of the cracker's arsenal and are used in attempts to thwart network security.

Occasionally, source routes are used for network testing purposes. Loose Source Routes were intended to help out with routing to distant locations.

The mechanisms for strict and loose source routing are the same. The only difference is that *only* systems on the list may be visited when using a strict source route.

6.16.2   Reverse Route

When source routing is used, the traffic flowing back from the destination to the source must follow the same path (i.e., must visit the same set of routers in reverse order).

There is one complication; the source and destination views of a router's address are not the same. Figure 6.12 shows a path through two routers. The route from Host A to Host B traverses routers whose IP addresses are known to host A as 130.132.9.29 and 130.132.4.11. The route from Host B to Host A traverses routers whose IP addresses are known to Host B as 128.36.5.2 and 130.132.4.16. The addresses at each of a router's interfaces differ because the interfaces connect to different subnets, as shown in the figure.

The solution is simple. As each router is visited, the incoming address is replaced in the *Source Route* field by its outgoing address. The destination takes the resulting list, reverses its order, and uses it as the source route in the opposite direction.

---

[6]See the current *Assigned Numbers* and *Internet Official Protocol Standards* RFCs for the complete list of options and their status.

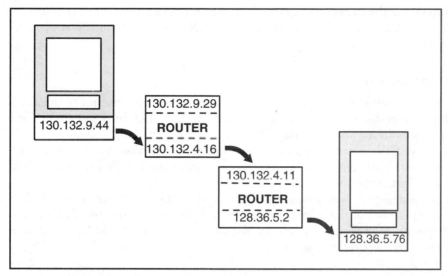

Figure 6.12   Routes from the point of view of host A and host B.

6.16.3   Describing the Route

You might expect the source route to be implemented by listing the routers *between* the source and destination, but this is not what is done. Table 6.4 shows the contents of the *Source IP Address, Destination IP Address,* and *Source Route* fields at each step along the path:

- At step 1, the *Destination IP Address* field contains the address of the first router. There is a pointer in the *Source Route* field that points to the next hop (shown in bold print).

- At step 2, the *Destination IP Address* field contains the address of the second router. The pointer in the *Source Route* field points to the next hop along the way. In this example, the next hop is the actual destination for the datagram.

TABLE 6.4   Source Routing

Step	Source IP address	Destination IP address	*Source Route* field
1	130.132.9.44	130.132.9.29	**130.132.4.11** 128.36.5.76
2	130.132.9.44	130.132.4.11	130.132.4.16 **128.36.5.76**
3	130.132.9.44	128.36.5.76	130.132.4.16 128.36.5.2

- At step 3, the datagram has arrived. Its *Source* and *Destination IP Address* fields have the true values, and the *Source Route* field lists the routers to be visited on the way back.

6.16.4 Source Routes and Security

Source Routes have become part of the network cracker's arsenal of burglary tools. They have been used to reach across the Internet into networks that administrators thought were safe.

Routers that filter traffic entering an organization have to be configured either to discard all source-routed traffic or to examine the *Source Route* field for the *real* destination of the datagram.

Another problem arises because multihomed hosts, which are connected to two or more subnetworks, can be targeted to carry source-routed datagrams, providing back door access for traffic. Multihomed hosts should be configured to discard source-routed datagrams.

6.16.5 Record Route

A *Record Route* field contains a list of IP addresses of routers visited by the datagram. Each router along the way will try to add its outgoing IP address to the list.

The length of the field is preset by the sender, and it is possible that all of the space will be used up before the datagram reaches its destination. In this case, the router simply forwards the datagram without adding its address.

6.16.6 Timestamp

There are three formats for a *Timestamp* field. It may contain:

- A list of 32-bit timestamps.
- A list of IP address and corresponding timestamp pairs.
- A list of preselected addresses provided by the source, each followed by space in which to record a timestamp. A node records a timestamp only if its address is next on the list.

Space may run out if the first or second format is used. There is an overflow subfield that is used to count the number of nodes that could not record their timestamps.

6.16.7 Department of Defense Basic and Extended Security

The *Basic Security* option is used to assure that the source of a datagram is authorized to transmit it, intermediate routers may appropriately relay it, and the destination should be allowed to receive it.

The *Basic Security* option parameters consist of a classification level that ranges from Unclassified to Top Secret and flags that identify the protection authorities whose rules apply to the datagram. Protection authorities include organizations such as the U.S. National Security Agency, Central Intelligence Agency, and the Department of Energy.

A datagram carrying the Basic Security option may also include an *Extended Security* option field. There are several different subformats for this option, depending on the needs of various defining authorities.

A host or router must discard information that it has not been authorized to handle. Secure systems are configured with the range of classification levels that they may transmit and receive and the authority or authorities that are valid. Note that there are many commercial products that do not support secure operation.

6.16.8 End of Option List and No Operation

The *No Operation* option is used as a filler between options. For example, it is used if it is desirable to align the next option on a 16- or 32-bit boundary.

The *End of Option List* option is used to pad the end of the options field to a 32-bit boundary.

6.16.9 Encoding Options

There are two single-byte options that are encoded as follows:

No Operation   00000001
End of Option List   00000000

The remaining options consist of several bytes. Each starts with a *type* octet and a *length* octet.

One issue that must be considered for these options is: Should the option be copied into the header of each fragment of a fragmented datagram? This must be done for *Security, Strict Source Routing,* and *Loose Source Routing. Record Route* and *Timestamp* fields appear only in the first fragment.

The type octet can be broken down as follows:

Bits	Function	Description
0	Copy flag	Set to 1 if copied on fragmentation
1–2	Option class	0 for datagram or network control 2 for debugging and measurement
3–7	Option number	Unique value for each option

Table 6.5 shows the value of the type octet, as well as its breakdown into Copy, Class, and Option Number, for each standard option.

TABLE 6.5   Copy, Class, and Option Number

Value	Copy	Class	Number	Name
0	0	0	0	End of Options List
1	0	0	1	No Operation
137	1	0	9	Strict Source Route
131	1	0	3	Loose Source Route
7	0	0	7	Record Route
68	0	2	4	Timestamp
130	1	0	2	Security
133	1	0	5	Extended Security

The formats for the common option fields are displayed in Figure 6.13.

6.16.10   Encoding a Strict Source Route

A Strict Source Route option contains a pointer and a list of addresses. The pointer contains the position of the next address to be processed. Initially, the pointer starts out with a value of 4. It is incremented by 4 at each hop.

6.16.11   Encoding a Loose Source Route

A Loose Source Route option contains a pointer and a list of addresses. Here again, the pointer starts out with a value of 4. The pointer is incremented when the next address on the list is reached.

6.16.12   Encoding Record Route

A Record Route option contains a pointer and space for addresses. Initially, the value of the pointer is set to 4. This is followed by unused space that is preallocated to hold addresses.

As each router is visited, its address is recorded at the location indicated by the pointer, and the pointer is incremented by 4. If all of the preallocated space gets used up, the datagram is routed to the destination, and no more addresses are recorded.

6.16.13   Encoding a Timestamp

A *Timestamp* option contains a pointer, an overflow subfield, and a flag subfield. The flag field indicates which of the three possible formats is to be used in this timestamp option.

If the flag subfield contains a 0, at each hop a timestamp will be recorded in the preallocated space and the pointer will be incremented by 4. If the preallocated space has been used up, the overflow field will be incremented by 1.

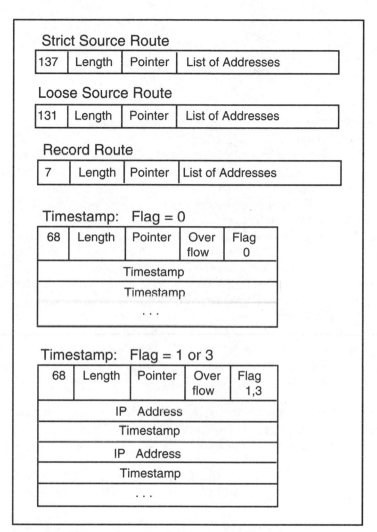

Figure 6.13 Formats for option fields.

What happens if the overflow count overflows? The datagram is discarded.

If the flag subfield contains a 1, at each hop an IP address and a timestamp will be recorded in the preallocated space and the pointer will be incremented by 8. If the preallocated space has been used up, the overflow field will be incremented by 1.

Suppose that the sender wants to record timestamps at a list of preselected nodes. In this case, the flag field is set to 3, and the sender fills in the selected internet addresses.

If the pointer is currently set at a router's address, the router fills in the timestamp value and increments the pointer by 8.

6.16.14   Encoding Basic and Extended Security Options

These options are under the control of military and government agencies. See RFC 1108 for further information.

6.17   SAMPLE IP HEADER

The display in Figure 6.14 shows a Network General *Sniffer* analysis of a DIX Ethernet MAC frame header and an IP header.

The MAC header starts out with the 6-byte physical addresses of the destination and source stations. Note that the *Sniffer* analyzer has replaced the first 3 bytes of each physical address with the name of the board manufacturer, which in this case is *Sun*. The type field contains the characteristic X'0800 code that says "deliver this information to IP."

In the display, an IP datagram follows immediately after the short DIX Ethernet MAC header. Recall that if this were an 802.3 frame, an 8-byte LLC header with a SNAP subheader would follow the MAC frame header.

The frame size is 61 bytes. This includes the 14-byte MAC frame header but does not include the 4-byte MAC trailer, so the complete frame originally was 65 bytes long. Ethernet or 802.3 frames on coaxial media must have a length of at least 64 bytes, so this frame barely exceeds the minimum size. The datagram in this frame has a total length of only 47 bytes.

Like most IP headers, this one carries no options and therefore has the standard 20-byte length. As is frequently the case, the *Type of Service* field has been set to 0.

We can tell that this datagram is not a fragment of a larger datagram because its *Fragment Offset* field is 0—showing that this is the start of a datagram—and the second flag is set to 0—indicating that this is the end of a datagram.

This datagram has 30 hops left in its *TTL* field. The *Protocol* field has value 6, which means that the datagram will be delivered to TCP at the destination host.

The *Sniffer* has translated the source and destination IP addresses into the convenient dot format.

The hexadecimal octets that made up the original MAC header and IP header are shown at the bottom of the display. The original *Sniffer* display of the hex fields has been altered to make it easier to match the hex codes to their interpretation.

6.18   DATAGRAM PROCESSING SCENARIOS

To get a better understanding of IP, it is useful to walk through the operations carried out when a datagram is processed at a router and at a recipient host. Figure 6.15 outlines the steps.

Problems or errors generally are handled by discarding the datagram and sending an error report back to the source. These reports will be described in Chapter 7, which discusses the *Internet Control Message Protocol* (ICMP).

```
DLC:  — DLC Header —
DLC:
DLC:  Frame 14 arrived at 10:26:10.5797; size is 61 bytes.
DLC:  Destination = Station Sun    076A03, Sun Atlantis
DLC:  Source      = Station Sun    07FD89, Sun Jupiter
DLC:  Ethertype   = 0800 (IP)
DLC:


IP:   — IP Header —
IP:
IP:   Version = 4, header length = 20 bytes
IP:   Type of Service = 00
IP:        000. .... = routine
IP:        ...0 .... = normal delay
IP:        .... 0... = normal throughput
IP:        .... .0.. = normal reliability
IP:   Total length = 47 bytes
IP:   Identification = 4458
IP:   Flags = 0X
IP:   .0.. .... = may fragment
IP:   ..0. .... = last fragment
IP:   Fragment offset = 0 bytes
IP:   Time to Live = 30 seconds/hops
IP:   Protocol = 6 (TCP)
IP:   Header checksum = 12F4 (correct)
IP:   Source address = [192.42.252.1]
IP:   Destination address = [192.42.252.20]
IP:   No options
IP:


HEX

MAC Header
08 00 20 07 6A 03      (Destination physical address)
08 00 20 07 FD 89      (Source physical address)
08 00                  (Protocol Type for IP)


IP Header
45 00 00 2F (Version, Hdr Length, Prec/TOS, Total Length)
11 6A 00 00 (Identification, Flags, Fragment Offset)
1E 06 12 F4 (Time to Live, Protocol, Header Checksum)
C0 2A FC 01 (Source IP Address)
C0 2A FC 14 (Destination IP Address)
```

Figure 6.14   Interpretation of MAC and IP headers.

ROUTER	HOST
Compute Header Checksum.	Compute Header Checksum.
Check parameters. Discard if invalid.	Check parameters. Discard if invalid.
Decrement the Time to Live. Discard if 0.	Check Destination. Here? Keep it.
Perform security screening checks. Discard if datagram fails any test.	Source routed? Check whether allowed to forward.
Select next hop. Discard if fragmentation is needed and "don't fragment" flag=1.	Not Fragmented? Deliver to higher-layer service.
Process options, if present. Update fields for options such as source routing.	Fragment? Insert. If datagram is complete, deliver to upper-layer service.
Update datagram header (or headers for fragments). Compute new header checksum(s).	Fragmentation timer expired? Discard fragments.
Transmit to next hop system.	

Figure 6.15 Datagram processing.

6.18.1 Router Processing

When a router receives a datagram, the first thing it does is to go through a series of checks to see if the datagram should be discarded. The header checksum is recomputed and compared to the checksum field.

The *Version, Header Length, Total Length,* and *Protocol* fields are screened to see that they make sense. The Time-to-Live value is decremented. A checksum error, parameter error, or zero Time-to-Live value cause the datagram to be discarded. Of course, the datagram also could get discarded if the router did not have enough free buffer space to continue processing the datagram.

The next step is to perform security screening. A series of preconfigured tests is applied to the datagram. For example, a router might restrict incoming traffic so that only a small number of destination servers were accessible.

Next, the router executes the routing procedure. A strict or loose source routing option will be consulted if it is present. An advanced router might take the Type of Service value into consideration. If the datagram cannot be routed without fragmentation and the "don't fragment" field is set to 1, the datagram will be discarded. If allowed and necessary, the datagram now is fragmented.

If options are present, they are processed. An updated header must be built for each datagram (or datagram fragment). Finally, the header checksum is recalculated, and the datagram is forwarded to its next-hop system.[7]

6.18.2   Destination Host Processing

At the destination host, the checksum is computed and compared to the *Header Checksum* field. The destination address is checked to make sure it is valid for this host. The *Version, Header Length, Total Length,* and *Protocol* fields are screened for correctness. A datagram will be discarded if any of these are in error or if the host does not have buffer space available to process the datagram.

If the datagram is a fragment, the host checks four fields: *Identification, Source Address, Destination Address,* and *Protocol.* Fragments with identical values in all of these fields belong to the same datagram. Next, the Fragment Offset value is used to position this fragment correctly within the whole.

Complete datagrams are delivered to the appropriate higher-layer service, such as TCP or UDP.

A host cannot wait indefinitely to complete the reassembly of a datagram. When the initial fragment arrives, a timer is set at a locally configured value—usually between 1 and 2 minutes. The fragments of the incomplete datagram are discarded when the timer expires.

6.19   FIREWALLING AND SECURITY

A site may need protection from the outside world. Everyone wants to enjoy the benefits of communication. But prudent network managers know that their computer resources must be protected from crackers. The *firewall router* has become the most popular defense weapon in a network manager's arsenal.

A firewall router is set up to filter traffic based on the security needs of a site. As shown in Figure 6.16, a firewall router can be configured to enable or disable traffic based on:

- Source IP address
- Destination IP address
- Protocol
- Application

For example, internal users might be permitted to send and receive electronic mail and to access external World Wide Web (WWW) servers. External users might be allowed to access a small selection of servers at the site.

---

[7]This is the common scenario for datagram processing at a router. However, there are times when a router will be the final destination for a datagram. For example, a request for network management information may be sent to a router.

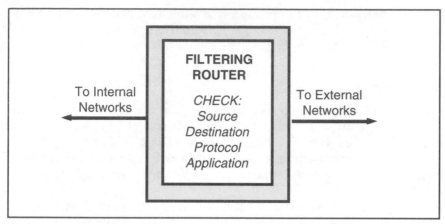

Figure 6.16 A firewall router.

Extra protection is provided by adding a smart filtering firewall host. For some implementations, internal users must connect to the firewall and authenticate themselves before they will be allowed to connect to the outside world. Users can be assigned privileges on an individual basis. All traffic from the outside world is filtered through the firewall host and can be screened carefully.

Some firewall hosts act as proxies. When an internal user requests information from the outside world, the proxy firewall actually communicates with the external system, gets the information, and then relays the information to the internal user.

To be well protected, a site can set up a "demilitarized zone" LAN that places a firewall host and all externally accessible application servers on a LAN that is protected by filtering routers. Figure 6.17 shows a demilitarized zone LAN that is used to protect a site from Internet intruders.

When a proxy firewall is used, private IP addresses can be assigned to the site's computers. Only the systems on the demilitarized zone LAN need unique public addresses.

6.20   IP PERFORMANCE ISSUES

The performance of an internet depends on the quantity of available resources in its hosts and routers and on how efficiently the resources are used. These resources are:

- Transmission bandwidth
- Buffer memory
- Central Processing Unit (CPU) processing

Perfect protocol mechanisms are unknown. Protocol design involves trade-offs between gains and losses in efficiency.

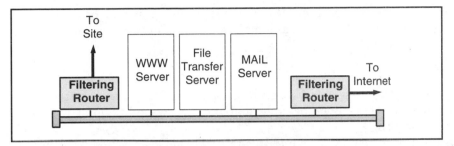

Figure 6.17   Protecting a site with a demilitarized zone.

6.20.1   Transmission Bandwidth

IP makes efficient use of bandwidth. Datagrams queued for transport to their next hop can be transmitted as soon as any bandwidth is available. There is no waste due to having to reserve bandwidth for specific traffic or waiting for acknowledgments.

Furthermore, there are new and very capable IP routing protocols that can split traffic over multiple paths and can choose routes dynamically so that they avoid a congested router or an overloaded link. Use of these protocols will help to maintain the best possible use of the available transmission resources.

There is little overhead due to control messages. ICMP error messages are the only source of control traffic.

There are some potentially negative features. When a load of traffic is directed from one or more high-speed LANs to a lower bandwidth point-to-point line, datagrams start to pile up in a queue at the router. Delivery time from source to destination increases, and some datagrams will be discarded. This will cause TCP to retransmit datagrams, increasing the load and decreasing the effective throughput.

Note that once a network becomes congested, datagram delivery becomes slower and less reliable. TCP retransmissions could have had the effect of keeping the network congested. Fortunately, some very effective algorithms cause TCP to respond to congestion immediately by throttling back the amount of data that is sent and slowing down the retransmission rate.

These algorithms have a significant impact on network performance and have become a required part of the TCP standard. They will be discussed in Chapter 10.

Router vendors are competing vigorously in offering ever more capable products, able to process tens of thousands of datagrams per second. To assure smooth performance, it is safest to configure the network so that the normal expected maximum memory load on a router is approximately 50 percent of capacity.

6.20.2   Buffer Utilization

Once an IP router has transmitted a datagram, its responsibility for that datagram is over. The buffer that was occupied by the datagram is available for immediate reuse. However, IP at a destination host will have to tie up some of its buffer space if it is reassembling a fragmented datagram.

6.20.3 CPU Processing

There is little CPU overhead in processing datagrams. Header analysis is straightforward. There is no need for elaborate software to manage timeouts and retransmissions.

Because it is connectionless and dynamic, IP requires routing to be executed at each hop. But this is accomplished by simple table lookups which can be accomplished quickly even for very large tables.

Security screening performed by routers can slow down processing, especially if there is a very long list of conditions to be checked for each datagram.

6.21 MORE ABOUT MULTICASTING

Recall (see Chapter 5) that there is a class of IP addresses that is used for *multicasting*—that is, the routing of IP datagrams from a source to a group of systems identified by a particular Class D multicast address. The technology and protocols to support multicasting applications, such as conferencing, are expected to change and grow over the next few years.

In this section, we will describe briefly some of the current mechanisms used to implement multicasting. First, some basic facts:

- A multicast sender does not necessarily have to be a receiving member of a group.

- Some multicast addresses are standard and permanent. They are registered with the IANA and published in the *Assigned Numbers* RFC.

- Temporary multicast addresses are selected through some ad hoc administrative process—their uniqueness cannot be guaranteed.

- The *all-hosts* multicast address, 224.0.0.1, is special. Datagrams sent to the all-hosts group are never forwarded out of the local link.

- The *Internet Group Management Protocol* (IGMP) provides the mechanisms that enable multicast routers to find out whether local hosts belong to multicast groups. IGMP is considered to be integral to IP. IGMP messages are carried in IP datagrams with protocol = 2.

- A *multicast router* is any system that executes special multicast routing software. This software may be running in an ordinary router or in a host that has been configured to perform multicast routing.

Now, let's walk through a scenario for a multicast host:

- A host that wants to join groups and receive multicasts starts out by listening to the all-hosts multicast address, 224.0.0.1.

- If a host wants to join a particular group, it must report this to all multicast routers on the local link. The host does this by sending an IGMP *report* message to the multicast address of the group that it wants to join.[8]
- The host then listens for datagrams sent to that multicast address.
- The host also listens for periodic queries from a local router and responds with a report when appropriate.
- To quit a group, a host simply stops listening for that group address and stops sending reports for that group.

The host actions are quite straightforward. Routing is more difficult and is still evolving. Let's take a look at a router scenario:

- A multicast router listens on each interface for host reports. For each of its interfaces, the router keeps a list of all multicast groups that have at least one active member on the subnetwork reached via that interface.
- The router must send other multicast routers a list of active group addresses for each of its attached subnetworks.
- Because hosts quit silently, the multicast router needs to check periodically to see if any local systems still want to belong to the group. The router periodically sends *queries* to the *all-hosts* address. Each host that is a member of some group waits a random amount of time. The first member to time out sends a response to the *group address*. The router—and all systems in the group—hear the response. Since the router now knows that someone is in the group, no one else needs to respond.
- When a router receives a multicast datagram containing data, it transmits the datagram onto each attached subnetwork that has a member in the group. The router may also need to forward the datagram to other multicast routers.

IGMP host messages have the format shown in Figure 6.18. Type = 1 is a Host Membership Query, and type = 2 is a Host Membership Report.

---

[8]The Time-to-Live on reports is set to 1, so they will not leave the local subnetwork.

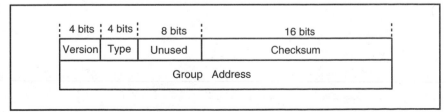

Figure 6.18 IGMP host messages.

6.22 RECOMMENDED READING

The IP protocol was defined in RFC 791. Updates, corrections, and conformance requirements are specified in RFC 1122. RFC 1812 details the requirements for IP version 4 routers and explains many details about the operation of these routers.

Department of Defense security options are discussed in RFC 1108. RFCs 1071, 1141, and 1624 discuss computation of the Internet checksum. RFC 815 presents an efficient algorithm for reassembling fragmented datagrams at a recipient host.

RFC 1112 specifies host extensions for IP multicasting.

7

Internet Control
Message Protocol

7.1 INTRODUCTION

IP has a simple, elegant design. Under normal conditions IP makes very efficient use of memory and transmission resources. But what happens when
things go wrong? After a router has crashed and disrupted the network, what
notice is given that datagrams are wandering around until their Time-to-
Lives (TTLs) expire? What warning is provided so that applications don't persist in sending information to an unreachable destination?

The *Internet Control Message Protocol* (ICMP) offers remedies for these ills.
ICMP also plays the role of network helper, assisting hosts with their IP routing
and enabling network managers to discover the status of network nodes.
ICMP's functions are an essential part of IP. All hosts and routers must be able
to generate ICMP messages and process the ICMP messages that they receive.
Properly used, ICMP messages can contribute to smoother network operation.

ICMP messages are carried in IP datagrams with ordinary IP headers (see
Figure 7.1) with the *Protocol* field set to 1.

7.2 ICMP ERROR MESSAGES

There are a number of situations that cause IP datagrams to be discarded.
For example, a destination may be unreachable because a link is down. The
Time-to-Live hop count may have expired. It might be impossible for a router
to forward a large datagram because fragmentation was not enabled.

When a datagram has to be discarded, ICMP messages are used to report
the problem to the address that sent the datagram. Figure 7.2 shows ICMP
messages being distributed to a datagram source.

ICMP notifies systems of problems quickly. ICMP is a very robust protocol
because error notification does not depend on the existence of a network management center.

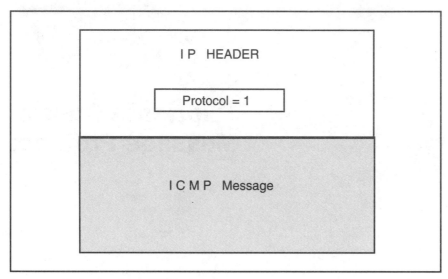

Figure 7.1   Packaging for an ICMP message.

There also are disadvantages. For example, if a destination is not reachable, messages will be propagated to sources all over the network, rather than to a network management station.

In fact, ICMP has no facilities for reporting errors to a designated network operations center. This is left up to the Simple Network Management Protocol (SNMP), which is presented in Chapter 20.

7.2.1 Types of Error Messages

Figure 7.3 summarizes the messages that routers and destination hosts send in order to report problems. Table 7.1 lists the formal names for the ICMP error messages.

7.2.2 Obligation to Send ICMP Messages

The ICMP protocol specifies that an ICMP message *should* or *may* be sent in each instance. It does not require that *every* error *must* result in an ICMP message.

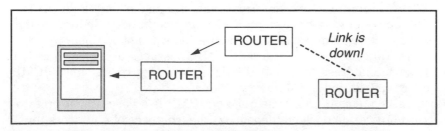

Figure 7.2   ICMP message directed to a traffic source.

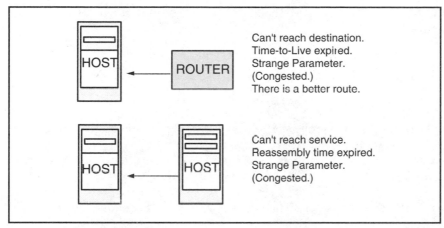

Figure 7.3 Types of ICMP error messages.

This makes very good sense. The first priority of a router on a network is to forward datagrams. And a congested recipient host should give more attention to delivering datagrams to its applications than to remote error notifications. It won't hurt if occasionally some discards are not reported.

7.2.3 Incoming ICMP Messages

What happens when a host receives an ICMP message? Let's look at an example. We'll try to connect to an address at one of the reserved—and therefore unreachable—networks:

```
> telnet 10.1.1.1
Trying 10.1.1.1 ...
telnet: connect: Host is unreachable
```

Note that we've been told exactly what happened.

TABLE 7.1   ICMP Error Messages

Message	Description
Destination Unreachable	A datagram cannot reach its destination host, utility, or application.
Time Exceeded	The Time-to-Live has expired at a router, or the Fragment Reassembly Time has expired at a destination host.
Parameter Problem	There is a bad parameter in the IP header.
Source Quench	A router or destination is congested. (It is recommended that systems should *not* send Quench messages.)
Redirect	A host has routed a datagram to the wrong local router.

To find out *which* router sent the ICMP message(s), we can use a handy tool called *traceroute:*

```
> traceroute 10.1.1.1
traceroute to 10.1.1.1 (10.1.1.1), 30 hops max, 40 byte packets
 1 nomad-gateway (128.121.50.50) 2 ms 2 ms 2 ms
 2 liberty-gateway (130.94.40.250) 91 ms 11 ms 78 ms
 3 border2-hssi2-0.NewYork.mci.net (204.70.45.9) !H !H !H
```

The New York router has sent *Destination Unreachable* messages, which are reported on the screen as the "!H" responses.

The *traceroute* function itself is based on ICMP *Time Expired* messages. The procedure is:

- A short UDP message is constructed. It is given an IP header whose Time-to-Live field is set equal to 1.

- The datagram is transmitted three times.

- The first router, (*nomad-gateway* in the example above), decrements the Time-to-Live value to 0, discards the datagram, and sends an ICMP *Time Expired* message back to the source.

- The *traceroute* function identifies the router that sent the messages and prints the three round-trip times.

- The Time-to-Live is set to 2, and the messages are sent again.

- The process is repeated, increasing the Time-to-Live at each step.

If the destination can be reached, eventually the full route will be displayed.

7.3   WHEN *NOT* TO SEND ICMP MESSAGES

We can expect ICMP to send error messages when a network is under stress. It is important to assure that the ICMP traffic does not flood the network, making the situation much worse. Some obvious limits have been imposed on the protocol. ICMP must not report problems caused by:

- Routing or delivering ICMP messages
- Broadcast or multicast datagrams
- Datagram fragments other than the first
- Messages whose source address does not identify a unique host (e.g., source IP addresses such as 127.0.0.1 or 0.0.0.0)

7.4   ICMP MESSAGE FORMAT

Recall that an ICMP message is carried in the data part of an IP datagram. Each ICMP message starts with the same three fields: a *Type* field, a *Code* field that sometimes provides a more specific description of the error, and a *Checksum* field. The format of the rest of the message is determined by the type.

ICMP error messages enclose the IP header and first 8 octets of the datagram that caused the error. This information can be the basis for problem-solving since it includes data such as the intended destination and the target layer 4 protocol. When we study TCP and UDP, we will discover that for these protocols, the extra 8 octets include information that identifies the communicating application entities.

The ICMP checksum is applied to the ICMP message (starting from its *Type* field).

7.4.1 Destination Unreachable Message

There are many stages at which delivery of a datagram can fail. Because of a broken link, a router may be physically unable to reach a destination subnet or to execute the next hop in a source route. A destination host may be unavailable because it is down for maintenance.

As we have seen in Chapter 6, modern routers have powerful security features. A router could be configured so that it screens the traffic entering a network. Thus, datagrams may be undeliverable because communication with the destination has been administratively prohibited.

The format of the *Destination Unreachable* message is shown in Figure 7.4. The *Type* (3 in this case) identifies this as a *Destination Unreachable* message. The *Code* indicates the reason. The full list of codes is quite extensive and is shown in Table 7.2.

7.4.2 Time Exceeded Message

A datagram may time out because the Time-to-Live reached zero while the datagram was in transit. Another kind of timeout occurs when a destination host's reassembly timer expires before all of the pieces have arrived. In either case, an ICMP *Time Exceeded* message is sent to the datagram source. The format of the *Time Exceeded* message is shown in Figure 7.5.

The code values (shown in Table 7.3) indicate the nature of the timeout.

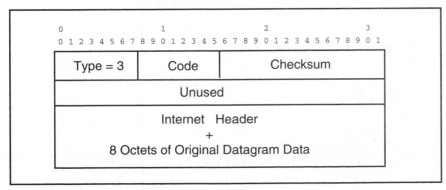

Figure 7.4 Format of the Destination Unreachable ICMP message.

TABLE 7.2   Destination Unreachable Error Codes

Code	Meaning
0	Network is unreachable.
1	Host is unreachable.
2	Requested protocol is not supported at the destination.
3	Port is unreachable. (Remote application may be unavailable.)
4	Fragmentation is needed but the "Don't Fragment" flag is set.
5	Source route has failed.
6	Destination network is unknown.
7	Destination host is unknown.
8	Source host is isolated.
9	Communication with destination net is administratively forbidden.
10	Communication with destination host is administratively forbidden.
11	Network is unreachable for specified Type of Service.
12	Host is unreachable for specified Type of Service.

7.4.3   Parameter Problem Message

The ICMP *Parameter Problem* message is used to report problems not covered by any of the other error messages. For example, there may be some inconsistent information in an options field that makes it impossible to process the datagram correctly, so the datagram must be discarded. Most often, parameter problems arise because of implementation errors at the system that wrote the parameter into the IP header.

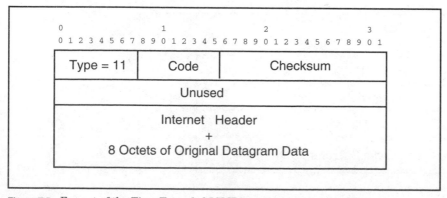

Figure 7.5   Format of the Time Exceeded ICMP message.

TABLE 7.3   Time Exceeded Codes

Code	Meaning
0	Time-to-Live exceeded
1	Fragment Reassembly Time exceeded

The *Pointer* field in the ICMP *Parameter Problem* message identifies the octet at which the error was detected. Figure 7.6 displays the format of the *Parameter Problem* message. Table 7.4 displays the *Parameter Problem* code values.

7.4.4   Congestion Problems

The IP protocol is very simple; a host or router processes a datagram and sends it on as quickly as possible. However, delivery does not always proceed smoothly. A number of things can go wrong.

One or more hosts sending UDP traffic to a slow server may flood the server with UDP traffic. This will cause the server to discard the overflow traffic.

A router may run out of buffer space and be forced to discard some datagrams. A slow wide area connection, such as a 56-kilobit-per-second link between two 10-megabit-per-second LANs, can create a bottleneck. Congestion causes datagrams to be discarded, resulting in retransmissions which produce even more traffic.

7.4.5   Source Quench

The *Source Quench* message shown in Figure 7.7 was intended to relieve this problem but did not succeed. The details of exactly how the congested system should execute a source quench were left to the implementer, leaving open the question:

When—and to whom—should a router or host send a Source Quench message?

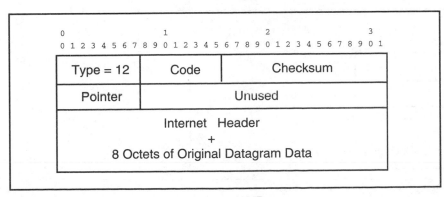

Figure 7.6   Format of the Parameter Problem ICMP message.

TABLE 7.4   Parameter Problem Codes

Code	Meaning
0	The value in the pointer field identifies the octet where an error occurred.
1	A required option is missing (used in the military community to indicate a missing security option).
2	Bad length

Usually, ICMP error messages tell a source host why one of its datagrams has been discarded. However, in a congestion situation, it is possible that the datagrams that happen to be discarded do not come from the hosts that are generating very heavy traffic. Moreover, exactly how a host should handle an incoming *Source Quench* message was left equally fuzzy.

The current *Router Requirements* document (RFC 1812) stipulates that in fact, *Source Quench* messages *should not* be sent. Work is proceeding on more effective congestion control mechanisms.

7.4.6   Redirect

More than one router might be attached to a LAN. If a local host sends a datagram to the wrong router, the router forwards the datagram but sends a *Redirect* message to the source host, as shown in Figure 7.8. The host should switch subsequent traffic to the shorter route.

Redirect messages can be used to cut down on manual network administration. A host can be configured with a single default router and can dynamically learn about routes that pass through other routers.

The format of the redirect message is shown in Figure 7.9. Redirect codes are listed in Table 7.5. Some routing protocols can choose a delivery path based on a datagram's *Type of Service* (TOS) field. Codes 2 and 3 provide advice that reflects these considerations.

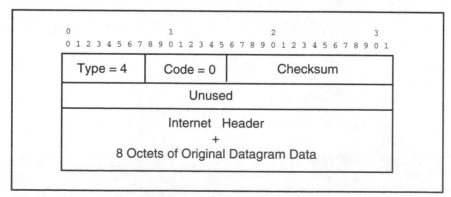

Figure 7.7   Format of the Source Quench ICMP message.

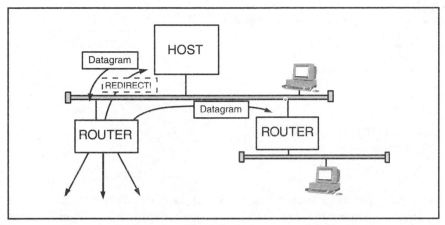

Figure 7.8   Correcting host routing via a redirect message.

7.4.7   Handling of Incoming ICMP Error Messages

When ICMP error messages arrive at a source host, what happens to them?[1] Vendors vary greatly in the way that they implement networking software, and TCP/IP standards try to allow for a lot of leeway. The guidelines given for the various message types are:

Destination Unreachable	Deliver the ICMP message to the transport layer. Action should depend on whether the reason is transient or permanent (e.g., communication administratively forbidden).
Redirect	The host *must* update its routing table.
Source Quench	Deliver to the transport layer or to an ICMP processing module.
Time Exceeded	Deliver to the transport layer.
Parameter Problem	Deliver to the transport layer; optionally, notify the user.

---

[1]There have been host implementations that ignore most or all incoming ICMP messages.

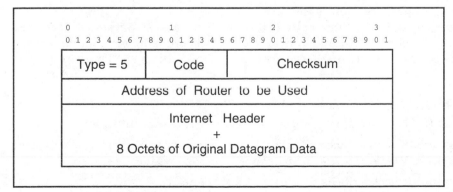

Figure 7.9   Format of the Redirect ICMP message.

TABLE 7.5   Redirect Codes

Code	Meaning
0	Redirect datagrams for the Network.
1	Redirect datagrams for the Host.
2	Redirect datagrams for the Type of Service and Network.
3	Redirect datagrams for the Type of Service and Host.

Sometimes, error conditions must be handled by cooperation between the operating system, the communications software, and the application that is communicating.

7.5 PATH MTU DISCOVERY

When performing a function (such as a file transfer) which carries bulk data from one host to another, the size of the datagrams that are used can have a big impact on performance. IP and TCP headers use up at least 40 bytes of overhead.

- If data is sent in 80-byte datagrams, overhead is 50 percent.
- If data is sent in 400-byte datagrams, overhead is 10 percent.
- If data is sent in 4000-byte datagrams, overhead is 1 percent.

To minimize overhead, we would like to send the biggest datagrams that we can. But recall that there are limits on the biggest datagram size [the Maximum Transmission Unit (MTU)] for each medium. If datagrams are too large, they will be fragmented, and fragmentation slows performance.

For many years, hosts would avoid fragmentation by setting the "effective MTU for sending" to 576 for traffic to any nonlocal destination. This often crippled performance unnecessarily.

Clearly, it would be very helpful to know the biggest datagram size that can be sent along a path and delivered intact. There is a very simple mechanism called *Path MTU discovery* that determines this size. The way that it works is:

- The *Don't Fragment* flag in IP headers is set equal to 1.
- The Path MTU size starts out at the MTU for the local interface.
- If the datagram is too large for some router to forward, the router will send back an ICMP *Destination Unreachable* message with code = 4.
- The sending host reduces the datagram size and tries again.

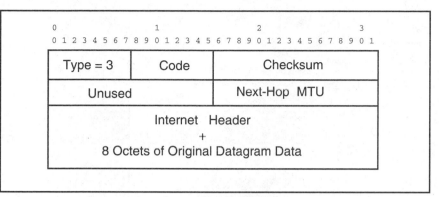

Figure 7.10   Destination Unreachable reporting MTU size.

What size should the host try next?[2] If the router has up-to-date software, its *Destination Unreachable* message will include the MTU size that could have been transmitted (see Figure 7.10).[3]

Because paths can change dynamically, the *Don't Fragment* flag can be left on throughout the communication. Routers will send corrections as needed.

If old router software is used, the router will not provide the next-hop MTU size. A host can make a reasonable guess by picking the next lower level from the list of standard MTU sizes (see Chapter 4). The resizing procedure continues until a feasible value has been found and the destination is reached.

Of course, a path change might provide an opportunity to use a larger MTU. A system that has dropped down to a small MTU can periodically try a larger value to see if improvement is possible.

7.6 ICMP QUERY MESSAGES

Not all ICMP messages signal errors. Some are used to probe the network for useful information. Is host X up? Is router Y running? How long does it take for a round trip to Z and back? What is my address mask?

Specifically, the ICMP query messages include:

- *Echo* request and reply messages that can be exchanged with hosts and routers.

- *Address Mask* request and reply messages that enable a system to discover the address mask that should be assigned to an interface.

- *Timestamp* request and reply messages that retrieve the clock setting at a

---

[2]The specification suggests that IP should store the Path MTU value and should make the value available to the transport protocols.

[3]Sometimes firewalls are configured to discard *all* incoming ICMP messages. Obviously, this mechanism will not work unless these ICMP messages can be delivered.

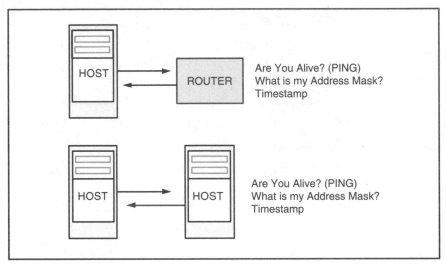

Figure 7.11   ICMP Query messages.

target system. (The response also was intended to give some idea of how long that system typically takes to process a datagram.)

Figure 7.11 summarizes the ICMP query services. The *ping* program, which sends "Are you alive?" echo messages, is used on a daily basis by network managers. *Address Mask* queries are used occasionally, while *Timestamp* messages rarely are seen.

7.6.1   Echo Request and Reply

The *Echo Request* and *Echo Reply* are used to check whether systems are active. Type = 8 is used for the request and Type = 0 for the reply. The number of octets in the data field is variable and can be selected by the sender.

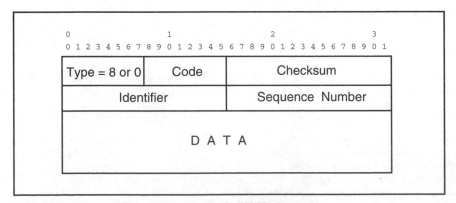

Figure 7.12   Format of Echo Request or Reply ICMP messages.

The responder must send back the same data that it receives. The *Identifier* field is used to match a reply with its original request. A sequence of echo messages can be sent to test whether the network is dropping messages and to estimate the average round-trip time. To do this, the identifier is held fixed while the sequence number (which starts at 0) is incremented for each message. The format of the echo message is shown in Figure 7.12.

The famous *ping* command, available on just about every TCP/IP system, is built on the ICMP echo request and reply messages. In the dialogue below, we first test that host *ring.bell.com* is alive. Then we send it a sequence of 14 messages, each containing 64 octets of data. Note that messages 0, 1, and 2 were dropped. Round-trip times are displayed at the right.

```
> ping ring.bell.com
ring.bell.com is alive

> ping -s ring.bell.com 64 14
64 bytes from ring.bell.com: icmp_seq = 3. time = 21. ms
64 bytes from ring.bell.com: icmp_seq = 4. time = 18. ms
64 bytes from ring.bell.com: icmp_seq = 5. time = 17. ms
64 bytes from ring.bell.com: icmp_seq = 6. time = 19. ms
64 bytes from ring.bell.com: icmp_seq = 7. time = 17. ms
64 bytes from ring.bell.com: icmp_seq = 8. time = 17. ms
64 bytes from ring.bell.com: icmp_seq = 9. time = 17. ms
64 bytes from ring.bell.com: icmp_seq = 10. time = 18. ms
64 bytes from ring.bell.com: icmp_seq = 11. time = 17. ms
64 bytes from ring.bell.com: icmp_seq = 12. time = 17. ms
64 bytes from ring.bell.com: icmp_seq = 13. time = 17. ms

—ring.bell.com PING Statistics—
14 packets transmitted, 11 packets received, 21% packet loss
round trip (ms) min/avg/max = 17/17/21
```

7.6.2   Address Mask

Recall that an organization may choose to break up its local address field into a subnet part and a host part. When a system comes up, it may not have been configured to know how many bits have been assigned to the subnet address field. To find out, the system can broadcast an *Address Mask Request.*

The response should be sent by an authorized address mask server. Normally, we would expect that server to be a router, but a host might occasionally be used. The reply will put 1s into the network and subnet fields of a 32-bit address mask field.

An address mask server can be configured so that if it has been off-line for a while, it will broadcast an *Address Mask Reply* as soon as it becomes active. This is done for the benefit of systems that started up while the server was unavailable.

Figure 7.13 displays the format of the *Address Mask Request* and *Reply.* The type is 17 for the request and 18 for the reply. Generally, the identifier and sequence number can be ignored.

Actually, the preferred method of determining the address mask is to use a boot protocol, such as the *Dynamic Host Configuration Protocol,* or *BOOTP.* These processes are more efficient because they provide a full set of configu-

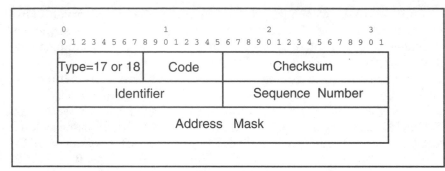

Figure 7.13   Format of an Address Mask message.

ration parameters. They also often are more accurate. For example, a Unix workstation might be misconfigured and respond incorrectly to *Address Mask Request* messages. As a result, your system might receive several answers, and some of them might be wrong.

7.6.3   Timestamp and Timestamp Reply

Timestamp messages report the time at a system and were intended to give a sense of how long the remote system spends buffering and processing a datagram. Note the fields:

Originate timestamp   Time that the sender last touched the message
Receive timestamp   Time that the echoer first touched it
Transmit timestamp   Time that the echoer last touched it

If possible, the time that is returned should be measured in milliseconds since midnight, Universal Time (formerly Greenwich Mean Time). Most implementations actually return the same time in the *Receive timestamp* and *Transmit timestamp* fields.

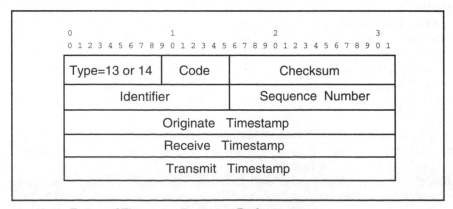

Figure 7.14   Format of Timestamp Request or Reply messages.

This protocol could provide a very simple way for one system to synchronize its clock with another. Of course the synchronization will be a rough one because of possible network delays. There is a far more capable *Network Time Protocol* that has been defined for Internet time synchronization.

Type 13 is used for the *Timestamp Query* and 14 is used for the *Timestamp Reply*. The format of the message is shown in Figure 7.14.

7.7 VIEWING ICMP ACTIVITIES

Below we show the ICMP part of a *netstat* report on protocol statistics. This report shows ICMP activity since the last initialization.

```
> netstat -s

icmp:
  1075 calls to icmp_error

  Output histogram:
    echo reply: 231
    destination unreachable: 1075

  2 messages with bad code fields
  0 messages < minimum length
  21 bad checksums
  0 messages with bad length

  Input histogram:
    echo reply: 26
    destination unreachable: 1269
    source quench: 2
    echo: 231
  231 message responses generated
```

The system has sent 1075 *Destination Unreachable* messages. There were 231 *Echo Requests* received, and all were answered. There were 26 *Echo Replies* received.

The local system has recorded the fact that 21 ICMP messages were received with bad *ICMP* checksums.

There were 1269 *Destination Unreachable* messages and two *Source Quenches* that were received by this system.

The *netstat* report that follows includes information about routing. From this report, we can see that new routers were discovered dynamically via *Redirect* messages. There were 12 unreachable destinations reported by *Destination Unreachable* messages. There were over 349 uses of *wildcard*— that is, default, route selections.

```
> netstat -rs
routing:
  0 bad routing redirects
  0 dynamically created routes
  2 new routers due to redirects
  12 destinations found unreachable
  349 uses of a wildcard route
```

7.8 ROUTER DISCOVERY

Although many LANs have a single default router, there are a substantial number of LANs that have two or more routers.

What happens when a router is added to a LAN? Redirect messages will notify systems of new routes. But suppose that the default router has crashed.

The *Router Discovery* protocol provides a robust method, based on ICMP messages, for keeping track of LAN routers. The basic idea is that routers will periodically advertise their presence. Hosts need to listen for these advertisements.

The preferred way for a router to announce its presence is to send advertisements to the *all-systems multicast address,* 224.0.0.1. But not all host systems support multicast addresses, so sometimes the broadcast address, 255.255.255.255, must be used.

A host that is coming up may not be able to wait around to find out about the routers on its LAN. The host can ask routers to send their advertisements right away. The preferred way to do this is to send a *Router Solicitation* message to the *all-routers multicast address,* 224.0.0.2. Again, since not all systems support multicast addresses, sometimes the broadcast address, 255.255.255.255, must be used instead.

A typical scenario for a router is:

- Each router interface is configured with the *advertisement address*—either 224.0.0.1 or 255.255.255.255—to use on a connected LAN.

- When the router initializes, if multicasting can be used, the router starts listening on the *all-routers* multicast address (224.0.0.2). It also listens for broadcasts.

- The router announces its presence to hosts on a connected LAN by transmitting a *Router Advertisement* to the advertisement address for that LAN. The advertisement lists all of the router's IP addresses for that interface.

- The router shows that it is still alive by periodically multicasting another *Router Advertisement*. (A 7 to 10 minute period is suggested.)

- The router sends an advertisement when a host asks it to do so.

For a host, the scenario is:

- Each host interface is configured with the *Solicitation Address*—either 224.0.0.2 or 255.255.255.255.

- When a host initializes, it starts to listen for *Router Advertisements*.

- At start-up, the host optionally can send a *Router Solicitation* message to the solicitation address. Routers will respond either to the host's IP address or to the advertisement address.

- When a host hears from a new router, the host adds a *default* route via this router to its routing table. The entry is assigned the lifetime timeout value, (typically, 30 minutes), which was announced in the *Router Advertisement.*

- The lifetime timeout is reset whenever a fresh advertisement is received from the router. If the timeout expires, the router's entry is removed from the host's routing table.

- To tell the world that it is about to shut down gracefully, a router can send an advertisement with a lifetime of 0.

If more than one router is present, how does a host choose the router that should forward a particular datagram? Each advertisement includes a *preference level* number. If no specific routing information is included in the host's routing table, the host chooses a router with the *highest* preference level. If that router does not in fact provide the best route, it will send back an ICMP redirect message.

Router Advertisements are Type 9 ICMP messages and *Router Solicitations* are Type 10.

7.8.1   Dead Routers

Router discovery will help a host find out that a local router has crashed, but only after a very long timeout—possibly 30 minutes. Host TCP/IP implementations are supposed to include built-in detection algorithms to detect whether a router is dead. Positive clues are easy. For example:

- The existence of active TCP sessions connected via the router

- Receipt of an ICMP redirect message from the router

Negative clues include:

- Failure to respond to ARP requests

- Many consecutive TCP retransmission timeouts

If there is evidence that a router is down, the final test is to send a *ping* request.

IP version 6 provides a better method of discovering whether a local router—or local host—has ceased to communicate.

7.9   RECOMMENDED READING

ICMP is defined in RFC 792. RFC 1122, the Host Requirements document, and RFC 1812, Router Requirements, contain some extremely useful clarifications. Router discovery is described in RFC 1256.

Path MTU discovery is described in RFC 1191, and additional advice is available in RFC 1435.

8

IP Routing

8.1 INTRODUCTION

Routing is the most important function that IP does. In large networks, IP routers exchange information that keeps their routing tables up to date. How is this done?

There is no single required protocol for updating IP routing table information.

Instead, network administrators always have been free to select any routing information protocol that meets their own internal requirements. Over the years, numerous protocols have been designed—and improved—by standards groups and by vendors. By long tradition, these protocols are called *Interior*[1] *Gateway Protocols,* or IGPs.

Separating the way that routing tables get updated from the rest of IP was a very good idea. Routing has become more and more sophisticated and efficient, while the basic IP protocol has remained unchanged.

Today, there are several IGPs in common use. Currently, the venerable *Routing Information Protocol* (RIP) is still popular. RIP chooses routes based on simple hop count estimates.

Sites that have selected Cisco routers often opt for Cisco's proprietary *Internet Gateway Routing Protocol* (IGRP) or *Enhanced IGRP* (EIGRP). IGRP uses sophisticated cost measurements that take many factors—including current load and reliability—into consideration.

The more complex *Open Shortest Path First* (OSPF) is implemented for some large networks. And there are organizations that use OSI's *Intermediate System to Intermediate System* (IS-IS) protocol, which can route both OSI and IP traffic. Both OSPF and IS-IS build detailed maps for at least part of a network and generate paths before choosing a route.

---

[1]Also called *Internal* Gateway Protocols.

Allowing free choice of protocols within an end-user organization has worked quite well. However, a standard *is* needed when routing across the chains of Service Provider networks that connect end-user sites together. Although there still is occasional use of the older *Exterior Gateway Protocol* (EGP), many Service Providers have converted to the *Border Gateway Protocol* (BGP).

In this chapter, we will take a look at how each of these protocols work and will examine their distinguishing features.

8.2 AUTONOMOUS SYSTEMS

How can so many different choices of routing protocol be allowed? The Internet model partitioned the world into units called *Autonomous Systems* (ASs). Simply stated, an Autonomous System was "somebody's network." Formally, an Autonomous System was:

> *A connected segment of a network topology that consists of a collection of subnetworks (with hosts attached) interconnected by a set of routes.*[2]

Most importantly, the subnetworks that make up an Autonomous System were under a common administration.

A typical Autonomous System was a network owned by a company or an Internet Service Provider (ISP). Actually, nobody was interested in your Autonomous System unless you wished to communicate with other Autonomous Systems. In this case, you needed to register with the InterNIC to get your own unique *Autonomous System number*. Figure 8.1 shows companies, Service Providers, Autonomous Systems, and their use of Interior Gateway Protocols and BGP.[3]

An organization's network administrators make their own decisions about the types of routers that they will use inside their network and the routing protocol or protocols that they will employ.

How were separate Autonomous Systems tied together? The Internet has been finding ways to do this for many years. As you might guess, the unique Autonomous System numbers that we mentioned earlier played a role. The *Exterior*[4] *Gateway Protocols* that make it all work are discussed later in this chapter.

The definition of Autonomous System and use of Autonomous System numbers changed in 1996. Recall that many Internet Service Providers have been delegated big address blocks, and they assign subblocks to their customers. Traffic can be routed to the provider using its short prefix. The provider can then use longer prefixes to identify customer sites.

[2]From RFC 1812, Requirements for IP version 4 Routers.

[3]Often, there is no need for a steady exchange of routing information across a link between an organization and a Service Provider. The information needed is just entered manually.

[4]Also called *External* Gateway Protocols.

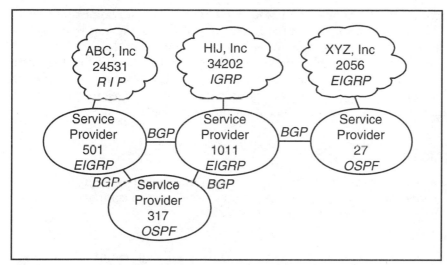

Figure 8.1 Autonomous Systems and routing protocols.

For routing purposes, a single AS number is sufficient to identify the entire cluster of networks made up of the provider's network and all of its customer networks. As stated in RFC 1930, the new definition is:

An AS is a connected group of one or more IP prefixes run by one or more network operators which has a SINGLE and CLEARLY DEFINED routing policy.

Most networks that connect to the Internet today have a very simple routing policy, namely, they have a single service provider and they are willing to exchange data with any other network connected to the Internet. Such a network does not need a separate AS number.

However, a business might have multiple service providers—or it might use the Internet as an inexpensive way to access its clients or suppliers and limit communications to these sites. This business would need its own AS number. This number would be used as an index to information that is used to define and enforce its routing policy.

Recall that the Internet Assigned Numbers Authority (IANA) has set aside blocks of IP addresses for private use. Just in case anyone needs them, the IANA also has reserved AS numbers 64,512 through 65,535 for private use.

8.3   IP ROUTING

An IP datagram follows a path made up of a sequence of hops. A node is one hop away—or adjacent—if there is a direct Local Area Network (LAN) or Wide Area Network (WAN) connection to the node. Routers that are separated by one hop are called *neighbors*.

A source route that contains a preselected list of hops can be put into an IP header. However, the use of source routes is rare.[5] A datagram is routed by choosing its *next-hop* destination at each router along its path.

Next-hop routing is flexible and robust. A change in network topology usually can be configured by updating one, or just a few, routers. Routers can inform one another of both temporary and permanent changes in the network and dynamically can switch traffic onto alternative routes as needed.

8.4   ROUTING METRICS

How do we decide that one route is better than another? There has to be some kind of *metric* (distance measurement) that can be used to compare routes.

8.4.1   Distance Vector Protocols

Very simple routing protocols just use end-to-end hop counts to compare routes. Some improvement is gained by using *weighted* hop counts; for example, a hop across a high-speed LAN could be given weight 1, while a hop across a slow medium (such as a 19.2-kilobit-per-second point-to-point line) could be given a weight of 10. This assures that a path across fast links will be preferred to a path across slow links. *RIP* chooses routes based on hop counts.

More sophisticated protocols combine elements such as bandwidth, delay, reliability, current load, or dollar cost into the calculation of a routing metric. *IGRP* and *EIGRP* use fine-tuned metrics.

Algorithms that base routing decisions entirely on metric values are called *distance vector* algorithms.

8.4.2   Link State Protocols

Recently, a lot of attention has been focused on *link state* routing algorithms. A link state router builds a map of the network and discovers paths from the router to any destination.

A cost metric is assigned to each link in the map. A total cost is computed for each path starting at the router by combining the link costs. Then the best path or paths for carrying traffic are selected.

The routers send updates to one another whenever the topology changes. After a change, all paths are recalculated. *OSPF* and *IS-IS* are link state protocols.

Link state algorithms are sometimes called *Shortest Path First,* or SPF. This refers to the name of the computer science algorithm that is used to calculate the shortest paths from one node to all other nodes in a network.

---

[5]In fact, since source routes sometimes are used as part of a cracker's break-in strategy, many routers are configured to discard any datagrams that contain source routes.

8.5   ROUTING TABLES

Recall that a host or a router consults its routing table in order to forward datagrams toward a remote destination. This table matches each destination with the address of the router to be used as the next hop.

Destinations listed in a routing table can include supernets,[6] networks, subnets, and individual systems.

A default destination entry also can be included. It is represented as destination 0.0.0.0.

There is no standard format for routing tables, but a simple table entry would contain items such as:

- Address of a destination network, subnet, or system
- IP address of the next-hop router to be used
- Network interface to be used to reach the next-hop router
- Mask for this destination
- Distance to the destination (e.g., number of hops to reach the destination)
- Number of seconds since the route was last updated

In order to keep routing tables small, most or all of the entries identify destination supernets, networks, or subnets. The idea is that if you can get to a router on a host's network, and then to a router on the host's subnet, you will be able to get to that host.

Sometimes a few entries that contain the complete IP addresses of some systems are included in a routing table.

Let's look at a couple of sample routing tables to get a feel for how they work.

8.6   A RIP ROUTING TABLE

The routing entries in Table 8.1 come from a RIP router owned by a university. The table lists destinations and tells you the next-hop router to which you should forward datagrams heading for that destination. The table also contains the metric (distance) information that helped the router to choose the next hop.

The routing table includes entries for many different subnets of network 128.36.0.0, along with routes to the three networks 130.132.0.0, 192.31.2.0, and 192.31.235.0. (These entries were retrieved using *HP OpenView for Windows Workgroup Node Manager.*) The table is split into two parts because of the large number of columns (although four of the columns in the second part are unused by RIP).

---

[6]Recall that a supernet is a classless block of IP addresses which have a common prefix.

TABLE 8.1   Routing Table in a RIP Router

ip Route Dest	ip Route Mask	ip Route Next Hop	ip Route Type	ip Route Protocol
0.0.0.0	0.0.0.0	128.36.0.2	indirect	rip
128.36.0.0	255.255.255.0	128.36.0.62	direct	local
128.36.2.0	255.255.255.0	128.36.0.7	indirect	rip
128.36.11.0	255.255.255.0	128.36.0.12	indirect	rip
128.36.12.0	255.255.255.0	128.36.0.21	indirect	rip
128.36.13.0	255.255.255.0	128.36.0.12	indirect	rip
128.36.14.0	255.255.255.0	128.36.0.21	indirect	rip
128.36.15.0	255.255.255.0	128.36.0.21	indirect	rip
128.36.16.0	255.255.255.0	128.36.0.36	indirect	rip
128.36.17.0	255.255.255.0	128.36.0.12	indirect	rip
128.36.19.0	255.255.255.0	128.36.0.10	indirect	rip
128.36.20.0	255.255.255.0	128.36.0.10	indirect	rip
128.36.21.0	255.255.255.0	128.36.0.5	indirect	rip
128.36.22.0	255.255.255.0	128.36.0.5	indirect	rip
128.36.126.0	255.255.255.0	128.36.0.41	indirect	rip
130.132.0.0	255.255.0.0	128.36.0.2	indirect	rip
192.31.2.0	255.255.255.0	128.36.0.1	indirect	rip
192.31.235.0	255.255.255.0	128.36.0.41	indirect	rip

8.6.1   Using the Route Mask

To find a match for a destination address such as 128.36.2.25, you would compare 128.36.2.25 with each entry's *Route Destination*. The entry's *Route Mask* tells you how many bits of 128.36.2.25 must match corresponding bits in the Route Destination. For example, the third entry in Table 8.1 has Route Mask 255.255.255.0, which means that the first 3 bytes, 128.36.2, need to match—and they do.[7]

Suppose that we got a match for two different entries. The one with the longer mask is preferred.

8.6.2   Default Route

The first entry in Table 8.1 is the *default* route. It announces that if no match is found, traffic should be forwarded to the neighbor router with address 128.36.0.2.

---

[7]More formally, the result of performing a logical AND of the destination address and entry's Route Mask is compared with the entry's Route Destination.

TABLE 8.1   Routing Table in a RIP Router (Continued)

ip Route Dest	ip Route Metric 1	ip Route Metric 2	ip Route Metric 3	ip Route Metric 4	ip Route Metric 5	ip Route if Index	ip Route Age (seconds)
0.0.0.0	2	−1	−1	−1	−1	1	153,836
128.36.0.0	0	−1	−1	−1	−1	1	0
128.36.2.0	1	−1	−1	−1	−1	1	30
128.36.11.0	1	−1	−1	−1	−1	1	13
128.36.12.0	1	−1	−1	−1	−1	1	15
128.36.13.0	1	−1	−1	−1	−1	1	14
128.36.14.0	1	−1	−1	−1	−1	1	16
128.36.15.0	1	−1	−1	−1	−1	1	17
128.36.16.0	12	−1	−1	−1	−1	1	24
128.36.17.0	1	−1	−1	−1	−1	1	16
128.36.19.0	14	−1	−1	−1	−1	1	27
128.36.20.0	1	−1	−1	−1	−1	1	28
128.36.21.0	1	−1	−1	−1	−1	1	5
128.36.22.0	1	−1	−1	−1	−1	1	5
128.36.126.0	1	−1	−1	−1	−1	1	23
130.132.0.0	2	−1	−1	−1	−1	1	25
192.31.2.0	3	−1	−1	−1	−1	1	10
192.31.235.0	1	−1	−1	−1	−1	1	25

8.6.3 Using Subnet 0

By the way, the administrators of this network are doing something that is frowned upon by the standards writers. They have assigned subnet number 0 to the LAN that this router lives on. Recall that you are not supposed to use 0 as a subnet number. However, some facilities need to use every number that they can get their hands on. Router vendors understand this, and the router can handle these addresses.

8.6.4 Direct and Indirect Destinations

Note that the second table entry says that LAN 128.36.0 is of type *direct,* which means that this subnet is directly connected to this router. The protocol is *local,* which indicates that this route was learned by looking at the configuration of the router.

The remainder of the entries list remote subnets and networks that are reached indirectly, by forwarding traffic to other routers. These routes were learned via RIP.

8.6.5 Routing Metrics

Looking at the second part of the table, we see that it has space for several metrics. RIP only uses one metric, a simple hop count reporting the distance to a destination. The unused columns are set to -1. Note that a 0 metric is assigned to subnet 128.36.0, which is directly connected to the router. Many of the other destinations are one hop away. However, subnet 128.36.19.0 is 14 hops away.

This router actually is a Shiva *Lanrover,* and it has a large number of phone lines attached to interface 1.

8.6.6 Route Age

The *Route Age* column tracks the number of seconds since each route was updated or validated. Entries that have been learned via RIP will age out and be invalidated if they are not reconfirmed within 3 minutes.

8.7 AN IGRP/BGP ROUTING TABLE

The routing entries in Table 8.2 come from an Internet router owned by an Internet Service Provider (ISP). The table lists destinations and identifies the next-hop router to be used for datagrams heading for each destination. The table also contains information that helps the router to recalculate the next hop if the network topology changes.

The routing table includes entries for many different networks and subnets. (These entries also were retrieved using an HP *OpenView* management system.)

8.7.1 Using the Route Mask

To find a match for destination 128.121.54.101, we apply the *route mask* for each entry to 128.121.54.101 and compare the result with the Route Destination. Applying the mask 255.255.255.0 in the fourth entry, we get 128.121.54.0, which matches the entry's destination.

IGRP can produce ties—there might be multiple entries with the same destination and mask. In this case, the route with the best metric is used. Or, if the metrics are close, IGRP can split traffic across two or more paths.

8.7.2 Default Route

The first entry in the table is the *default* route. If no match is found, traffic is forwarded to the neighbor router with address 130.94.40.250.

TABLE 8.2   IGRP and BGP Routing Entries

ip Route Dest	ip Route Mask	ip Route Next Hop	ip Route Type	ip Route Protocol
0.0.0.0	0.0.0.0	130.94.40.250	indirect	ciscoIgrp
128.121.50.0	255.255.255.0	128.121.50.50	direct	local
128.121.52.0	255.255.255.0	128.121.50.55	direct	local
128.121.54.0	255.255.255.0	128.121.50.50	direct	local
128.6.0.0	255.255.0.0	130.94.0.49	indirect	ciscoIgrp
128.96.0.0	255.255.0.0	130.94.40.250	indirect	ciscoIgrp
130.33.0.0	255.255.0.0	130.94.16.2	indirect	ciscoIgrp
130.44.0.0	255.255.0.0	130.94.0.49	indirect	ciscoIgrp
130.68.0.0	255.255.0.0	130.94.0.49	indirect	ciscoIgrp
130.94.1.24	255.255.255.248	130.94.0.49	indirect	ciscoIgrp
130.94.1.32	255.255.255.248	130.94.0.49	indirect	ciscoIgrp
130.94.2.8	255.255.255.248	130.94.0.49	indirect	ciscoIgrp
130.94.2.16	255.255.255.248	130.94.0.49	indirect	ciscoIgrp
130.94.7.0	255.255.255.248	130.94.0.49	indirect	ciscoIgrp
130.94.7.8	255.255.255.248	130.94.0.49	indirect	ciscoIgrp
44.0.0.0	255.0.0.0	130.94.15.201	indirect	bgp
128.3.0.0	255.255.0.0	130.94.40.201	indirect	bgp
129.210.0.0	255.255.0.0	130.94.15.201	indirect	bgp
13.0.0.0	255.0.0.0	130.94.15.201	indirect	bgp

8.7.3 Direct and Indirect Destinations

The next three destinations are of type *direct,* which means that they are subnets that are directly connected to the router. Their protocol is *local,* which indicates that the routes were learned via manually entered configuration information.

Next, there are several entries for remote (indirect) destinations whose routes were learned via Cisco's proprietary IGRP protocol.

8.7.4 Small Subnets

The set of destinations starting with entry 130.94.1.24 look like host addresses at first glance, but the Route Mask shows that all of these entries actually represent small subnets. The host part occupies only the last 3 bits of the address. For example, the binary representation of 24 is 00011000, and so all

TABLE 8.2   IGRP and BGP Routing Entries (Continued)

ip Route Dest	ip Route Metric 1	ip Route Metric 2	ip Route Metric 3	ip Route Metric 4	ip Route Metric 5	ip Route if Index	ip Route Age (seconds)
0.0.0.0	10647	1170	21000	0	255	6	12
128.121.50.0	0	−1	−1	−1	−1	1	0
128.121.52.0	0	−1	−1	−1	−1	1	35
128.121.54.0	0	−1	−1	−1	−1	1	0
128.6.0.0	12610	1536	61000	2	255	3	11
128.96.0.0	14647	1170	61000	2	255	6	16
130.33.0.0	8710	1536	22000	1	255	2	18
130.44.0.0	16610	1536	101000	4	255	3	37
130.68.0.0	12710	1536	62000	3	255	3	39
130.94.1.24	82125	128	40000	0	255	3	41
130.94.1.32	182571	56	40000	0	255	3	42
130.94.2.8	10510	1536	40000	0	255	3	42
130.94.2.16	10510	1536	40000	0	255	3	43
130.94.7.0	10610	1536	41000	1	255	3	2
130.94.7.8	12510	1536	60000	1	255	3	3
44.0.0.0	0	−1	−1	−1	−1	6	51766
128.3.0.0	0	−1	−1	−1	−1	6	42049
129.210.0.0	0	−1	−1	−1	−1	6	586765
13.0.0.0	0	−1	−1	−1	−1	6	224463

of the bits for 24 actually lie within the subnet part of the address. The hosts on this subnet would have addresses ranging from 130.94.1.25 to 130.94.1.30.

8.7.5   Border Gateway Protocol Entries

Finally, the table ends with a list of remote destinations that were learned via the *Border Gateway Protocol,* which provides information used to route between different Autonomous Systems on the Internet.

8.7.6   Routing Metrics

If we examine the second part of Table 8.2, we see that a 0 metric is assigned to destinations that are reached on three directly connected subnets. As before, unused metric values in the table are set equal to −1.

Values have been assigned to all five metrics for Cisco's IGRP. However, there is no attempt to display meaningful metrics for Internet destinations in remote Autonomous Systems, which were learned via the Border Gateway Protocol.

All of the interfaces on the router are numbered, and the datagram is sent via the interface identified in the *IfIndex* column.

8.7.7 Route Age

For IGRP, the *Route Age* column tracks the number of seconds since the route was updated or validated. Entries that have been learned via a protocol will age out if they are not reconfirmed from time to time. For BGP, the Route Age is used to track the stability of long distance routes.

8.8 PROTOCOLS THAT MAINTAIN ROUTING TABLES

How did these routers learn their table entries? How are entries kept up to date? How is the best choice for the next-hop router discovered? These chores are the job of the routing protocol. The simplest routing protocol is:

- Study diagrams of your network to find the best paths. Choose the next hop so that these paths are followed.
- Enter data into your routing tables manually.
- Update your routing tables manually.

This is exactly what should be done for a simple branch office router like the one in Figure 8.2. The small branch office router in the figure needs only two entries—one for the local network 192.101.64.0 and a default entry for the "cloud."

Manual routing can be used throughout a very small network. But for a complex network that is growing and changing, and has multiple potential routes to a destination, manual routing will turn into a nightmare.

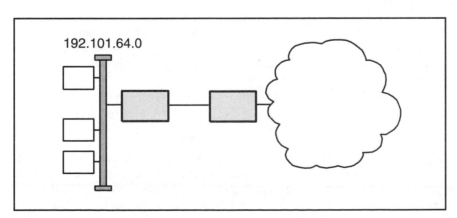

Figure 8.2 Branch office routing.

At some point it becomes impossible for humans to analyze and respond to network conditions. A routing protocol is then needed to automate:

- Exchange of information between routers about the current state of the network

- Recomputation of the best next-hop selection as changes occur

Over the years, there has been substantial research into routing protocols. Many have been implemented and their merits have been hotly debated. What are the characteristics of a winning protocol?

- When the network changes, it should respond quickly.
- It should compute optimal routes.
- It should scale up well as your network grows.
- It should be frugal in its use of computer resources.
- It should be frugal in its use of transmission resources.

But computing optimal routes in a large network may require considerable CPU and memory, and quick response may require immediate transmission of a mass of information. A good protocol design has to balance these requirements.

We'll start off our study of routing protocols by looking at a very simple one—RIP.

8.9   ROUTING INFORMATION PROTOCOL

The Interior Gateway Protocol in widest use today is RIP, which is derived from the Xerox Network Systems (XNS) routing protocol. RIP's popularity is based on its simplicity and availability.

RIP was included in the Berkeley Software Distribution TCP/IP implementation and still is distributed with Unix systems as program *routed*.

In fact, a RIP *routed* program is a standard part of most vendors' host and router TCP/IP packages. RIP is included in a free bundle of software from Cornell University called *gated*. RIP was in widespread use for several years before being standardized in RFC 1058. A second version was proposed in 1993 and improved in 1994. (The original version now has been declared "history," which means obsolete.)

RIP computes routes using a simple *distance vector* routing algorithm. Every hop in the network is assigned a cost (usually 1). The total metric for a path is the sum of the hop costs. RIP chooses the next hop so that datagrams will follow a least-cost path.

Figure 8.3 shows how distance estimates propagate across the network. The router at the upper left can tell that datagrams forwarded via Router A will reach Network N in fewer hops than datagrams forwarded via Router B.

RIP's strong points are its simplicity and availability. Often there is no reason to use more functional—and complicated—methods for a small network

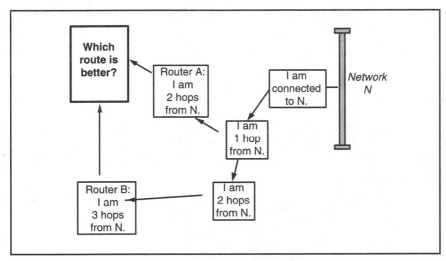

Figure 8.3 Discovering hop-counts to a destination.

or a network with a simple topology. However, for large, complex networks, RIP has some serious shortcomings. For example:

- The maximum metric for any path is 15. Sixteen means *"I can't get there!"* Because it is easy to run out of hops on a big network, RIP usually is configured with a cost of 1 for each hop, whether that hop is a slow dialup or a high-speed fiber link.
- After a disruption in the network, RIP often is slow to reestablish optimal routes. In fact, after a disruption, datagram traffic may run around in circles for a while.
- RIP cannot respond to changes in delay or load across links. It cannot split traffic to balance the load.

8.9.1 Initializing RIP

To get started, each router only needs to know the networks to which it is connected. A RIP router broadcasts this information to each of its LAN neighbor routers. It also sends the routing information to neighbors at the other end of a point-to-point line or virtual circuit.

As shown in Figure 8.4, the news spreads like gossip—each router passes the information to its own neighbors. For example, Router C very quickly learns that it is two hops away from subnet 130.34.2.0.

Like all automated routing protocols, RIP has to send routing updates, receive routing updates, and recompute routes. A RIP router sends information to its neighbor routers every 30 seconds. Sending out routing information is called "advertising" routes.

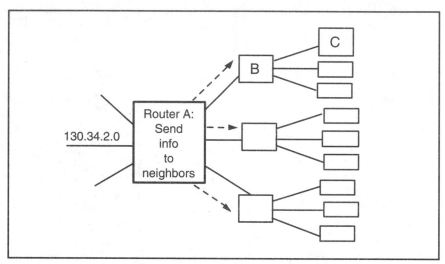

Figure 8.4 Propagating routing information.

A host on a LAN could eavesdrop on RIP broadcast advertisements and use them to update its own routing table—or at least assure itself that its routers are alive.

8.9.2 Updating a RIP Table

As shown in Figure 8.5, Router A has been sending traffic to network 136.10.0.0 through Router B. Router A receives an update from Neighbor D that announces a shorter route and therefore changes its routing table. Note

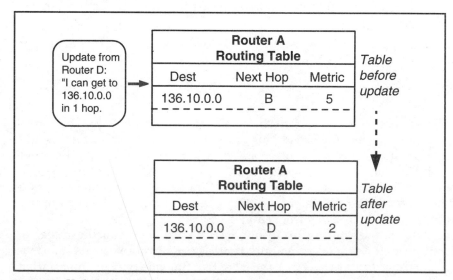

Figure 8.5 Updating routing tables with RIP.

that the hop from Router A to Router D is added to D's metric to compute the distance (2) from A to 136.10.0.0.

8.9.3 RIP Version 1 Mechanisms

Now let's walk through the formal steps for RIP version 1 routing. We start our routing table off with distances that we know about. Then, whenever an update arrives from a neighbor, we recheck our table and see if any entries can be added or improved:

1. A cost is assigned to traversing each attached subnetwork (usually 1).
2. A router sends its current routing table to its neighbors every 30 seconds.
3. When a router receives a neighbor's table, it checks each incoming entry. The cost assigned to the subnetwork on which the update arrived is added to each metric.
4. If a destination is new, it is added to the local routing table.
5. If a destination already is in the table, but the update provides a shorter route, the entry is replaced.

It would be nice if routes always got better and better, but sometimes a link or a router will go down, and our traffic will have to take a longer path. We find out about bad news two ways:

1. Router A has been sending traffic to a destination via Router X, and X sends an update that announces that the number of hops to that destination has increased (or perhaps that the destination cannot be reached). Router A changes its entry accordingly.
2. Router A has been sending traffic to a destination via Router X but has not received any updates for 3 minutes. Router A has to assume that Router X has crashed and mark all routes through X as unreachable (by giving them a metric value of 16). If no new route to an unreachable destination is discovered within 2 minutes, its entry is deleted.[8] In the meantime, Router A's updates tell other systems that Router A cannot reach the destination.

8.9.4 RIP Version 1 Update Messages

As we mentioned earlier, update messages are sent between RIP routers at regular intervals. In addition, request messages may be sent to a neighbor to ask for routing information. Typically, a system would send out a request:

- During system initialization
- When performing a network monitoring function

---

[8]This is called *garbage collection.*

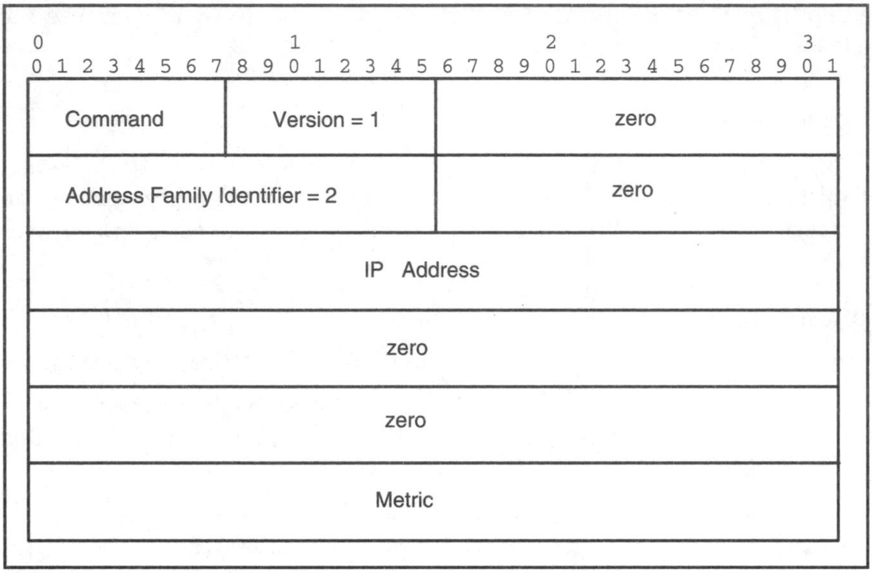

```
0                    1                    2                    3
0 1 2 3 4 5 6 7 8 9 0 1 2 3 4 5 6 7 8 9 0 1 2 3 4 5 6 7 8 9 0 1
```

Command	Version = 1	zero
Address Family Identifier = 2		zero
IP Address		
zero		
zero		
Metric		

Figure 8.6   Format of a RIP version 1 message.

The format of the RIP version 1 request or response/update message is shown in Figure 8.6. A command field of 1 indicates a request and 2 indicates a response or spontaneous update.

8.9.5   RIP Version 1 Message Details

When the original RIP RFC was written, it was anticipated that these routing messages would be used for other network protocols besides IP, and so an *address family identifier* field and space for up to 14 octets per address were included.

The address family, IP address, and metric fields can be repeated, and the message can contain up to 25 address entries. The maximum message size is 512 octets. If more than 25 entries need to be sent, multiple messages are used.

An update contains all destinations and metrics in the sender's routing table. A request contains an entry for each address for which a metric is desired. A single entry with address 0 and a metric of 16 asks for a complete routing table update.

Regular RIP updates are sent via the User Datagram Protocol (UDP) from source port 520 to port 520 at destination routers. However, requests can be sent from any port, and the response would then be sent back to the requesting port.

8.9.6   Fine-Tuning RIP

The preceding sections have described the basic RIP protocol. However, RIP implementations need some additional features in order to solve some problems:

- With a 30-second interval between updates, it could take a long time for changes to percolate through a big network.

- After a change—especially if some connectivity has been lost—there is a tendency for traffic to run around in circles.

The next sections will describe how these problems can be handled.

8.9.7 Triggered Updates and Hold Down

Triggered updates speed up the process of discovering changes. Whenever a router changes its metric for a route, it sends updates announcing the change.

Note that one new update will trigger others that may trigger others. However, this spurt of messages will prevent a lot of user traffic from wandering along bad routes.

Since there will be a tendency for a lot of updates to be sent at the same time, each system waits a random amount of time before sending. Also, the bandwidth used by triggered updates can be reduced by sending only those entries that actually have changed, rather than the entire routing table.

While adjustments are going on, a router that has discovered that a destination is unreachable may receive an obsolete update that indicates that a defunct route is available. If this update is accepted, not only would the router replace good information with bad, it would trigger updates that spread the misinformation.

For this reason, some vendors implement a *hold down* rule that sets a period of time during which updates are ignored for a destination that has just been marked unreachable.

8.9.8 Split Horizon and Poisoned Reverse

Why does RIP traffic sometimes run around in circles? The reason is that after a change, it can take a while for all of the routers to get updated to accurate information. Figure 8.7 shows a very simple example.[9] Router D has two paths to *Network N*. One is a direct 1-hop link, and the other is a long 10-hop path. When the short link is disrupted, Router D replaces the route with the alternate, which has a metric of 10.

But the routing entries for *Network N* in the RIP messages sent by Routers A, B, and C would have the form:

Network N   Metric = 2

These messages would *not* have any way to say that the route is through Router D. What would happen if Router D got a scheduled update from A before it has had a chance to tell Router A about its own change?

---

[9]The example is taken from RFC 1058.

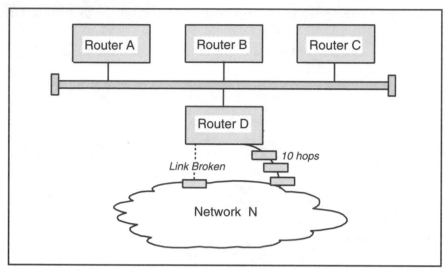

Figure 8.7   Routing after a network disruption.

- D would update its routing table entry to:

Destination	Next hop	Metric
Network N	A	3

- D would start sending its traffic to A (who would send it back).
- D also would send updates to A, B, and C announcing that it can get to *Network N* in three hops.
- A would reply that it now can get to *N* in four hops. Routers B and C would be equally confused and, depending on the timing of updates, might send traffic for *N* to each other, to A, or to D.
- RIP updates would bounce back and forth.

The good news is that the metrics at A, B, and C would be increased by each update message until they finally reached 11 and the correct route would be discovered. A couple of simple mechanisms can prevent the period of confusion from occurring.

Split Horizon means that a router should not report a route to the next-hop system for that route. For the example in Figure 8.7, this says that Routers A, B, and C must not tell Router D that they have a route to *Network N,* because that route actually passes through D.

Poisoned Reverse goes further. With Poisoned Reverse, Routers A, B, and C in Figure 8.7 prevent misunderstanding by sending updates across the LAN that say "Don't try to get there through me!" Specifically, the updates would include the entry:

Network N   Metric = 16

This cures the problem for the configuration shown above, but for networks containing big loops, traffic may still run around in circles for a while, even if the destination actually can't be reached at all. Fortunately, the metrics will eventually increase to 16, and correct routing will be restored. The process is called *counting to infinity*.

Incidentally, don't count on finding triggered updates, hold down, Split Horizon, and Poisoned Reverse in every RIP router that you encounter. There are dozens of versions of RIP, written for every sort of box.[10] When in doubt about what a particular router does, move that router into a branch office and configure it manually.

There are several obvious deficiencies in the version 1 RIP message protocol. These will be described in the sections that follow.

8.9.9 No Subnet Mask

Note that masks are not included in RIP version 1 messages (see Figure 8.6), so a RIP version 1 router cannot tell whether an address represents a subnet or a host address.

For a long time, router vendors solved this problem by requiring users to choose one subnet mask and use it for an entire network. A router connected to that network would learn the subnet mask by checking the configuration of its interface to the network.

A router that was not directly connected to a network would have no way to learn the network's subnet mask. If it received a routing entry for a subnet of the remote network, the entry would be useless. For this reason, version 1 routers do not send subnet or host entries for a network to routers that are not connected to that network. An external router is sent a single entry for an entire network (e.g., 145.102.0.0).

Note that this can result in *extra* entries as well as too few entries. If a network has Classless Inter-Domain Routing (CIDR) addressing, a separate entry must be provided for each of the class C addresses in the bunch. In contrast, one entry that included a mask could have represented the entire CIDR network.

8.9.10 LAN Broadcasts

Version 1 messages are broadcast onto LANs. Every network interface must therefore absorb and examine the message. The use of LAN *multicasts* would make a lot more sense.

---

[10]Someone probably has implemented RIP in a toaster.

8.9.11   Lack of Authentication

Another worrying problem in version 1 is the lack of authentication for RIP messages. Suppose that someone with access to a network injected false routing messages (using forged source addresses) that reported that most destinations were unreachable? This would cause serious disruption of service.

8.9.12   Can't Distinguish Fast from Slow Links

A network administrator is allowed to manually assign a hop count to a link. Thus a 9600-bit-per-second point-to-point link could be assigned a hop count of, say, 5 to indicate that it has less capacity than a 10,000,000-bit-per-second LAN link.

Unfortunately, when hops add up to more than 15, destinations become unreachable. Therefore, administrators usually must assign the same hop count, 1, to slow and fast links.

There is one advantage to having a small maximum hop count. Recall that an unreachable destination sometimes causes a temporary routing loop to form. The metrics in update messages for the loop increase up to 16 fairly quickly, and this breaks the loop. A bigger limit would slow down recovery from looping.

8.9.13   Excessive Traffic

Routing tables grow to a substantial size on large networks. Sending entire routing tables in updates can impose a heavy overhead on a network. Routers also are slowed down by having to process dozens or hundreds of entries in update messages, most of which have not changed.

Even small periodic updates are a problem for switched long distance connections. Dialups or X.25 circuits may be used for links with only occasional, sporadic traffic. To save money, these circuits are closed when there is no traffic to send. Whenever possible, manual configuration is used to describe the remote networks, but sometimes this is not feasible.

Newer routing protocols solve these problems by sending updates only after a change and reporting only the changed routes. Routers exchange periodic (Hello) messages to show that they are alive, except across switched links, where the neighbor is assumed to be alive until an attempted transmission fails.

8.10   RIP VERSION 2

Although RFC 1058, which defined version 1, was published in 1983, version 2 of RIP did not appear until 1993. In that time, a lot of work had been accomplished in designing complex new protocols that solved all of the problems. But many organizations liked the simplicity, ease of installation, and ease of use of RIP.

Version 1 of RIP has been declared "historic," and users should upgrade to version 2. Version 2 offers simple solutions for a few of RIP's problems.

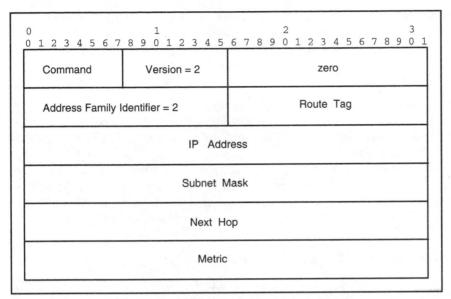

Figure 8.8   Format of a RIP version 2 message.

However, to preserve interworking with RIP version 1 routers, the changes are modest. The maximum hop count remains at 15, and complete routing tables still are exchanged every 30 seconds. But tables are *multicast*, rather than broadcast.

Most version 2 changes are based on packing more information into the update messages. Version 2 routing updates have the format shown in Figure 8.8.

Subnet Mask	Put it into the message.
Next Hop	Used to report routes through *other* routers. For example "go to *Network N* via Router B." Figure 8.9 shows how a single "bilingual" router (Router A) translates between RIP and IGRP and passes next-hop information between the two sets of routers.

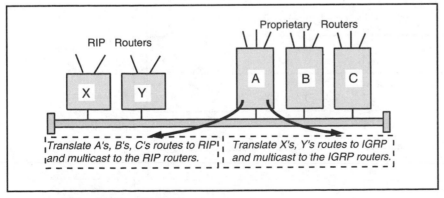

Figure 8.9   Using the *Next-Hop* field to report routes.

External Routes   The new *Route Tag* field includes information learned by an external protocol, such as BGP. A popular use for the tag is to state the Autonomous System number of an external network.

8.10.1 RIP Version 2 Authentication

Optionally, the space for the first entry in an update can be used for authentication. The space is identified as an authentication field by using X'FFFF as the Address Family Identifier. The type of authentication being used is named in the next field.

The remaining 16 bytes contain the authentication information. Although the only type[11] defined in the version 2 RFC was a simple password sent in the clear, router vendors are moving to authentication based on MD5. Figure 8.10 shows the format of a message that starts with an authenticator.

8.11 MOVING TO MORE SOPHISTICATED PROTOCOLS

Two approaches have been taken in developing more advanced protocols. Like RIP, Cisco's proprietary IGRP is a distance vector protocol, but IGRP fixes RIP's shortcomings. OSPF and OSI's IS-IS are Link State protocols. They create network maps, discover all routes to a destination, and then com-

---

[11]Although there is only one type, its identifier is 2.

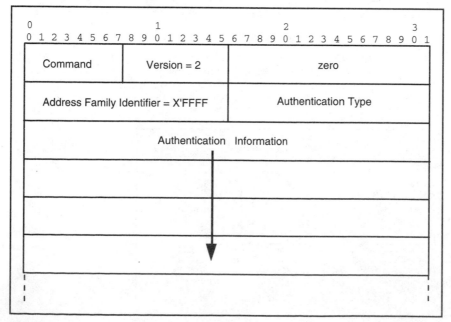

Figure 8.10 Version 2 message starting with an authenticator.

pare their metrics to choose the best paths.

These protocols all support advanced features, such as the ability to split traffic among several roughly equivalent routes.

In addition, there is a move toward supporting routing based on *Type of Service* (TOS). For example, one low-delay route might be reserved for interactive traffic, while another route that could provide good throughput, but not low delay, would be used for bulk data transfers.

8.12   IGRP AND EIGRP

Although IGRP is a distance vector protocol, IGRP's metric is computed using a formula that takes many factors into account, including the network's delay and bandwidth. In addition, IGRP can base routes on the current load level ("occupancy") of each link, as well as the current end-to-end error rate.

IGRP can split traffic across paths that are equally—or almost equally—good. When there are several paths to a destination, more of the traffic is sent along the paths that have higher bandwidth.

A Service Provider's boundary IGRP router can gather information from several external Autonomous Systems. Routes to external networks are reported to internal routers. Thus, IGRP supports routing between different Autonomous Systems.

EIGRP uses the same metrics and routing formulas as IGRP, but it introduces some important improvements. EIGRP cuts down greatly on routing traffic by sending updates only after a change and sending only the altered entries. EIGRP also includes an algorithm that prevents loops from forming.

In the sections that follow, we'll first describe IGRP and then discuss the EIGRP enhancements.

8.12.1   IGRP Routing

Like RIP, an IGRP router would broadcast its routing table to its neighbors at periodic intervals. And like RIP, an IGRP router starts up with routing table entries for its directly connected subnetworks. Its table is enlarged by updates sent by neighboring routers. IGRP update messages do not include subnet masks. In place of the simple hop counts used in RIP, IGRP routing updates carry several types of metric information, namely:

Delay   The delay is the time (in units of 10 microseconds) that it takes a bit to reach a destination when there is no load on the network.

Bandwidth   The metric is 10,000,000 divided by the smallest bandwidth on the path (measured in kilobits per second). For example, the smallest bandwidth of 10 kilobits per second yields a metric of 1,000,000.

Load   The load is the proportion of the path's bandwidth that currently is in use. It is coded on a scale from 0 to 255; 255 means 100 percent loaded.

Reliability   This is the fraction of the datagrams that are arriving undamaged. It is coded on a scale from 0 to 255; 255 means that 100 percent are arriving undamaged.

Hop count   The number of hops to the destination.

Path MTU   The biggest Maximum Transmission Unit (MTU) that can be carried across every link on the path.

The delay, bandwidth, and MTU are derived from router configuration information. The load and reliability are computed dynamically, from information exchanged by the routers. Table 8.3 shows some examples of coded delay and bandwidth values.

Recall that Table 8.2 displayed metric values returned by a Simple Network Management Protocol (SNMP) poll of a Cisco router. For example:

ip Route Dest	ip Route Metric 1	ip Route Metric 2	ip Route Metric 3	ip Route Metric 4	ip Route Metric 5	ip Route If Index	ip Route Age (seconds)
128.6.0.0	12610	1536	61000	2	255	3	11
128.96.0.0	14647	1170	61000	2	255	6	16
128.112.0.0	10667	1170	21200	1	255	6	23

For IGRP/EIGRP the metric values have the following meanings:

Metric1   Overall route metric

Metric2   Bandwidth metric

Metric3   Sum of interface delays

Metric4   Route hop count

Metric5   Interface reliability (255 means 100 percent)

TABLE 8.3   IGRP Delay and Bandwidth Measurements

Medium	Default delay values (10-microsecond units)	Bandwidth metric (10,000,000 ÷ bandwidth in kilobits per second)
Satellite (500-megabit per second)	200,000 (2 seconds)	20
Ethernet (10 megabit)	100 (1 millisecond)	1,000
1.544 megabit	2000 (20 milliseconds)	6,480
64 kilobit	2000	156,250
56 kilobit	2000	178,570
10 kilobit	2000	1,000,000
1 kilobit	2000	10,000,000

8.12.2   Other Configured IGRP Values

Configuring an IGRP router is fairly straightforward. In addition to the IP addresses, subnet masks, MTUs, bandwidths, and link delays, you can specify:

- A variance factor V. If M is the smallest path metric, paths with metric less than M × V will be used.

- Whether hold downs are enabled or disabled.

Timers also are configurable, although the defaults frequently are used. The default values are:

- Updates are broadcast every 90 seconds.

- If no update is received from a next-hop router within 270 seconds, its entries are timed out. If there is no alternative route, the destination is marked unreachable.

- A hold down, during which no new path for an unreachable destination will be accepted, will last 280 seconds.

- If no update for a destination has been received for 540 seconds (the *flush* time), it is removed from the routing table.

8.12.3   IGRP Protocol Mechanisms

Like RIP, an IGRP router sends periodic updates to its neighbors. The updates include the entire current routing table, with all metrics.

A hold down prevents a defunct route from being reinstalled by a stale message. No new route to the destination is accepted during the hold down period. (Hold downs can be disabled if desired.)

Split horizon is used to prevent advertising a path to a router that is upstream for the path. IGRP also provides its own version of route poisoning. If the metric for a route increases by more than a factor of 1.1, it is very likely that a loop is building, and so the route is removed.

Triggered updates are sent after a change—such as removal of a route. A route is removed when:

- Communication to a neighbor times out. Routes via the neighbor are removed.

- A next-hop router sends a notification that a route is unusable.

- The metric has increased sufficiently so that route poisoning is invoked.

8.12.4   Exterior Routing

One reason that IGRP is popular with Internet Service Providers is that it handles routing between Autonomous Systems. IGRP updates can include multiple routes to outside networks. A default route to the outside world can be selected from several proposed external routes.

8.12.5   Enhanced IGRP Features

Enhanced IGRP uses the same metrics and distance calculations found in IGRP. However, it improves IGRP significantly by supporting subnet masks and doing away with periodic updates. Only changes are sent, and EIGRP makes sure that these changes have been received by requiring them to be acknowledged. Simple periodic *Hello* messages are used to discover neighbors and check that they are still active. Another important enhancement is use of the *Diffusing Update ALgorithm* (DUAL), which guarantees loop-free routing.

8.12.6   EIGRP DUAL

The basic idea behind DUAL is simple. It is based on the observation:

> *If a path consistently takes you closer to a destination, then the path cannot be a loop.*

Turning the statement around, if a path is a loop, it will contain a router whose distance to the destination is bigger than the previous router's distance (as illustrated in Figure 8.11).

DUAL is designed to find paths that have the property that each router along the way is closer to the destination. Router E in Figure 8.11 will view with great suspicion an update from its *downstream* neighbor Z that reports a much bigger metric than is in E's own current table.

8.12.7   DUAL Topology Table

To implement DUAL, EIGRP saves information that IGRP would have thrown away. EIGRP saves the route information that each neighbor has sent in its updates. (IGRP would have discarded information about any

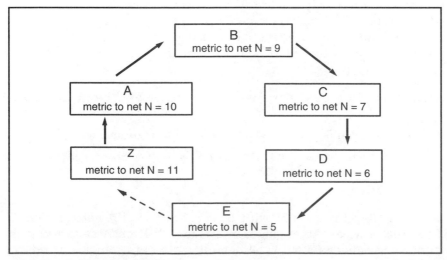

Figure 8.11 Routing around a loop.

routes that were not optimal.) The information is stored in a side table called the *Topology Table*. Topology Table data includes:

Destination	Neighbor	Neighbor's metric	My current best metric

8.12.8 DUAL Feasible Successors

The interesting entries in the Topology Table are the ones that describe *feasible successors*. My *feasible successors* are neighbors who are closer to the destination than I am (currently).

As long as there is at least one feasible successor, I have a route to the destination and am in *Passive* state for DUAL. However, when an update changes the picture so that I have no feasible successor, I start to query my neighbors to find out whether traffic has simply been switched to a longer route or a loop is forming.

Let's walk through the process a little more formally:

1. Suppose that I have reached the point where I have only one feasible route to a destination, via Router Z.

2. An update arrives from Z that increases Z's metric. Furthermore, the new distance from Z to the destination is *greater* than my current distance. This could indicate that a loop is forming.

3. I go into *Active* state and start a *route recomputation*.

4. But while the recomputation is in progress, I will continue to route through Z.

5. I send an update message (which is called a "query") to all neighbors except Z. The message announces my new, bigger metric distance for the destination.

6. If a neighbor has one or more feasible routes, it sends a reply and announces its good routes to the destination.

7. A neighbor that does not have a feasible route will enter *Active* state (if it is not already in it) and send queries to *its* neighbors. (Optionally, it can report back immediately that it already is in *Active* state and is recomputing.)

8. Queries will "diffuse" through the network until either a feasible route is found or a router that knows that the destination is unreachable is contacted.

9. When a router has found a feasible route or established unreachability, it sends back replies to the queries that it has received.

10. When all of its own queries have been answered, a router returns to passive state.

EIGRP has proven that distance vector routing is far from dead. In the next sections, we will look at its competitor, link state routing.

8.13   OPEN SHORTEST PATH FIRST

In 1988, the Internet Engineering Task Force (IETF) started work on a new standard protocol to replace RIP. The result was the *Open Shortest Path First* (OSPF) Interior Gateway Protocol, a routing protocol intended for use within Autonomous Systems of all sizes. In 1990, OSPF was recommended as a proposed standard. The protocol is a nonproprietary public technology.

Recall that a link state router discovers paths by building a map of the network and using the map to build a tree of paths with the router as its root. A metric value is computed for each path, and then the optimal path or paths are selected for each IP Type of Service.

OSPF uses both link state paths and distance metrics. OSPF is designed to scale well and to spread accurate routing information quickly. In addition, OSPF supports:

- Quick detection of topology changes and very rapid reestablishment of loop-free routes

- Low overhead, using updates that report changes rather than all routes

- Traffic splitting across multiple equivalent paths

- Routing based on Type of Service

- Use of multicast on LANs

- Masks for subnetting and supernetting

- Authentication

In April of 1990, when the very large NASA Science internet converted to OSPF, NASA routing traffic decreased dramatically. After a change or disruption in the network, globally correct routing information was reestablished very quickly—typically within a few seconds, as compared to minutes for some older protocols.

Version 2 of OSPF was published in mid-1991, and a revised version 2 came out in March of 1994. This was a 216-page document, so the description below should be considered a rough outline of the protocol.

8.13.1   Autonomous Systems, Areas, and Networks

In the OSPF standard, the term *network* means an IP network, a subnet, or CIDR supernet. Similarly, a *network mask* identifies a network, subnet, or CIDR supernet. An *area* is a set of contiguous networks and hosts, along with any routers having interfaces to the networks.

An Autonomous System that uses OSPF is made up of one or more areas. Each area is assigned a number. Area 0 is a connected *backbone* that links to all of the other areas and glues the Autonomous System together. Figure 8.12 illustrates this topology.

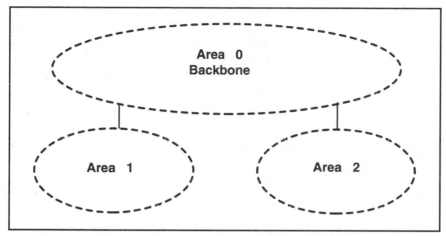

Figure 8.12   OSPF backbone and areas.

8.13.2 OSPF Area Routing

Routing within an area is based on a complete link state map for the area. OSPF scales well because a router needs to know detailed topology and metric information *only* about an area that it belongs to.

Every OSPF router in an area keeps an identical *routing database* describing the topology and status of all of the elements in the area. The database is used to construct the area map. This database includes the state of every router, each router's usable interfaces, its connected networks, and its adjacent routers.

Whenever a change occurs (such as a link going down), this information is propagated through the area. This promotes accurate routing and quick response to trouble. For example, if OSPF routing were used in the network in Figure 8.13, Router A would quickly be informed that the link to B was

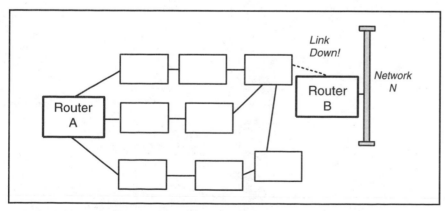

Figure 8.13   Using complete route information.

down and would realize immediately that there was no usable route to *Network N.*

A router that is initializing obtains a copy of the current routing database from an adjacent neighbor. After that, only changes need to be communicated. Changes get known quickly because OSPF uses an efficient flooding algorithm to spread update information through the area.

8.13.3   OSPF Area Shortest Paths

A router uses its area database to construct a tree of shortest paths with itself as the root. This tree is used to build the routing table. If Type of Service routing is supported in the area, a separate tree and set of routes is built for each Type of Service value.

8.13.4   The OSPF Backbone, Borders, and Boundaries

Areas are glued together by the backbone. The backbone contains all routers that belong to multiple areas, as well as any networks and routers not assigned to any other area. Recall that areas are numbered, and the backbone is area 0.

A *border* router belongs to one or more areas and to the backbone. If the Autonomous System is connected to the outside world, *boundary* routers learn routes to networks that are external to the Autonomous System.

In Figure 8.14, backbone area 0 includes Routers A, B, C, F, and G. Area 1 includes Routers B and D. Area 2 includes Routers C, E, and F. Routers B, C,

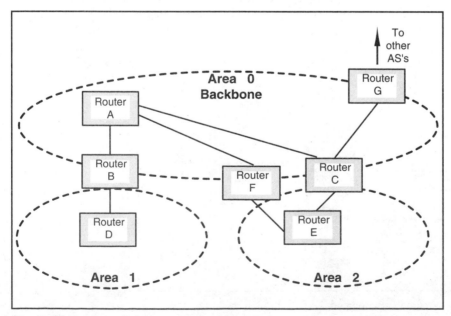

Figure 8.14   Router and areas in an Autonomous System.

and F are border routers. Router G is a boundary router. Router B knows the full topologies of area 1 and of the backbone. Similarly, Routers C and F know the full topologies of area 2 and of the backbone.

The backbone must be contiguous. What happens if a restructuring of the network or an equipment failure causes the backbone to be broken? Sometimes, *virtual links* can be used to tie together the pieces of the backbone.

A virtual link can be defined between two backbone routers that interface to the same area. The virtual link is treated like an unnumbered point-to-point link. The cost of the virtual link is the total path cost of the route between the two routers.

As shown in Figure 8.15, if the link from A to F were lost, Router F would no longer be connected to the other backbone routers via a backbone link. The virtual link F-E-C could be used to restore the connectedness of the backbone.

8.13.5 Routing Across an OSPF Area Border

A border router knows the complete topology of each area to which it connects. Recall that every border router belongs to the backbone, and so it also knows the full backbone topology.

8.13.6 Using Summarized Information Inside an OSPF Area

Each border router summarizes area information and tells the other backbone routers how far it is from networks within its own area(s). This enables every border router to calculate distances to destinations outside its own areas and then pass that information into its own areas.

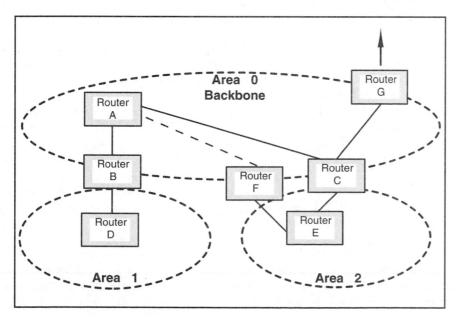

Figure 8.15 Defining a virtual link.

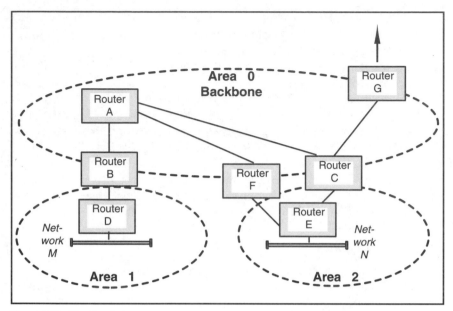

Figure 8.16 Routing between areas.

The summarized information includes a network, subnet, or supernet identifier; a network mask; and the distance from the router to the external network.

For example, in Figure 8.16, suppose that Router E wants to choose a path to *Network M*. Router E uses its area database to find the distances d_c and d_f to the border Routers C and F. Each of these has reported its distance m_c and m_f to *Network M*. Router E can compare $d_c + m_c$ with $d_f + m_f$, and pick the shortest route.

Note that Router B should not bother to pass summarized distance information into area 1. There is only one path out of this area, and so one simple default route entry suffices for all external destinations. If an area has a single border router, or if it just doesn't matter which border router is used, it is called a *stub* area, and one or more default routes can be provided to reach external destinations.

8.13.7   Destinations Outside of the OSPF AS

Many Autonomous Systems are connected to the Internet or to other Autonomous Systems. OSPF boundary routers provide information about distances to networks outside of the Autonomous System.

There are two types of OSPF external distance metrics. Type 1 is equivalent to the local link state metric. Type 2 metrics are long distance metrics—they are measured at a greater order of magnitude. If an analogy is made with driving distances, think of Type 2 routes as being derived from a national road map and measured in hundreds of miles, while Type 1 metrics use a local distance measurement in miles.

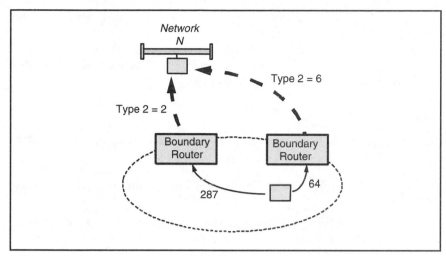

Figure 8.17 Choosing routes using Type 2 metrics.

Figure 8.17 shows two routes to an external network, N. The Type 1 metrics would be ignored in the distance calculation, and the route with Type 2 metric equal to 2 would be chosen.

Another feature of OSPF (especially convenient for Service Providers) is that a router on the boundary of an Autonomous System can act as a *route server* and can advertise entries that identify routes through other boundary routers. Information would include:

Destination, Metric, Boundary Router to be used

8.13.8 OSPF Protocol

Now we are ready to describe some of the internals of the OSPF protocol. Every OSPF router maintains a detailed database of information needed to build its area routing tree, such as descriptions of:

- Each router's interfaces, connections, and associated metrics
- Each multiaccess network and a list of all routers on the network

How does a router get its information? It starts by finding out who its neighbors are by means of Hello messages.

8.13.9 Hello Messages

Each OSPF router is configured with a unique identifier that is used in messages. Usually, the router's smallest IP address is used as its unique ID.

A router periodically multicasts Hello messages on a multiaccess network [such as an Ethernet, Token-Ring, or Fiber Distributed Data Interface (FDDI) LAN] to let other routers know that it is active. It also sends Hello messages to peers attached by point-to-point links or virtual circuits to let these neighbors know that it is awake.

One reason that Hello works well is that a message contains a list of IDs of neighbors whose Hellos the sender already has heard. That way, every router knows whether its messages are getting through.

8.13.10 Designated Router

On a multiaccess network, Hello messages also are used to select and identify a *Designated Router*. The Designated Router has two jobs:

- It is responsible for reliably updating its adjacent neighbors with the latest network topology news.

- It originates *networks links advertisements,* which list all routers connected to the multiaccess network.

In Figure 8.18, Designated Router A exchanges information with Routers B, C, and D on its LAN as well as with Router E, which is connected by a point-to-point link.

8.13.11 Adjacencies

Designated Router A acts as the local expert and keeps up to date on the complete local topology. It then communicates this topology to adjacent routers.

B, C, and D keep their databases synchronized by talking to A. They do not have to talk to one another. Two routers that synchronize databases with one another are called *adjacent.* B and C are *neighbors,* but they are not adjacent to one another.

Clearly, this is an efficient method of keeping the LAN router databases synchronized. It can also be used on frame relay or X.25 networks. Routers can exchange Hellos across virtual circuits, choose a designated router, and

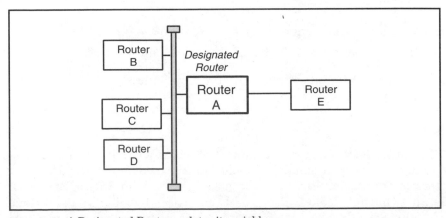

Figure 8.18 A Designated Router updates its neighbors.

synchronize their databases with the designated router. This speeds up synchronization and cuts down on network traffic.

The loss of a designated router would be a pretty disruptive event. For this reason, a backup designated router always is selected and is ready to take over immediately.

8.13.12   Initializing a Routing Database

Now suppose that Router B has just been restarted after being off-line for maintenance. First B listens to Hellos, discovers its neighbors, and finds out that Router A is the Designated Router. Next, B brings itself up to date by talking to A.

More specifically, A and B will exchange *Database Description* messages. These messages contain a list of what each has in its database. Each item has a sequence number that is used to establish which router has the freshest information for the item. (The sequence number of a routing entry is increased[12] whenever the entry is updated.)

After this exchange of information is complete, each knows:

- Which items are not yet in its local database
- Which items are present but out of date

Link State Request messages are used to ask for entries that are in need of an update. *Link State Update* messages respond to these requests. After a full (and acknowledged) exchange of information, the databases are synchronized. Link State Update messages also are used to report changes in the area topology. Topology updates are flooded through an area so that all databases are kept in synchronization.

8.13.13   OSPF Message Types

The five message types used in the OSPF protocol exchanges that we have described are:

Hello	Used to identify neighbors, to elect a Designated Router for a multiaccess network, to find out about an existing Designated Router, and as an "I am alive" signal.
Database Description	During initialization, used to exchange information so that a router can find out what data is missing from its database.
Link State Request	Used to ask for data that a router has discovered is missing from its database or is out of date.
Link State Update	Used to reply to a Link State Request and also to dynamically report changes in network topology.
Link State ACK	Used to confirm receipt of a Link State Update. The sender will retransmit until an update is ACKed.

---

[12]Numbers eventually roll around to 0.

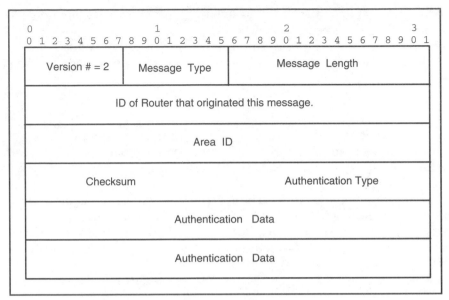

Figure 8.19   The standard 24-octet OSPF message header.

8.13.14   OSPF Messages

OSPF messages are carried directly in IP datagrams, with protocol type 89.

All OSPF messages start with the 24-octet header that is shown in Figure 8.19. The current version is 2. The *type field* contains a number corresponding to the message type. The length is the total length, including the header.

Authentication types are registered with the IANA. Secure and authenticated transmission of routing information is especially important to the robustness of networks.

8.13.15   Contents of an OSPF Link State Update

The critical OSPF routing information is transmitted in Link State Update messages. Updates are sent between adjacent routers. When the designated router on a broadcast network receives an update, it multicasts it to the other routers on the network. Updates propagate through an area quite efficiently. Every newly received link state advertisement must be acknowledged.

A Link State Update message contains items that are called *Advertisements*. Each message can include the following types of Advertisements:

Router Links	The state of each of the interfaces on a router.
Network Links	The list of routers connected to a multiaccess net. This is provided by the Designated Router on the net.
Summary Link to a Network	A route to a network outside the local area but in the Autonomous System. This is provided by a border router.

Summary Link to a Boundary Router	A route through the Autonomous System to the boundary of the Autonomous System. This is provided by a Border router.
AS External Link	A route to a destination in another Autonomous System. This information is provided by a boundary router.

A Link State Update message begins with the standard 24-octet header. The remainder of the message is made up of Advertisements of the various kinds listed above.

8.13.16 OSPF Enhancements

There has been a steady flow of enhancements for OSPF. For example, to decrease cost, it is desirable to disconnect dialup lines and virtual circuits when there is no traffic to send. The protocol has been modified so that periodic Hellos are not sent, so these lines do not need to be kept up all of the time. OSPF also has been extended to support IP multicasting. OSPF is in active use, and further enhancement and refinement can be expected to occur.

8.14 OSI ROUTING

OSI uses the term *Intermediate System* rather than router or gateway. The OSI routing protocol, *IS-IS,* originally was defined to support OSI routers but then was extended to IP.

Like OSPF, IS-IS is a link state protocol and supports hierarchical routing, Type of Service routing, splitting of traffic on multiple paths, and authentication.

IS-IS has two types of routers: level 1 for routing within an area and level 2 for routing to destinations outside an area. (Level 2 routers could be viewed as analogous to routers in the OSPF backbone.) A level 1 Intermediate System router forwards traffic bound for destinations outside of its area to its nearest level 2 router. Traffic is then routed to a level 2 router that is connected to the destination area.

Many of the mechanisms used in OSPF were based on similar (but not identical) mechanisms in IS-IS, for example, the use of link state advertisements, flooding, and sequence numbers.

Some IS-IS proponents believe that it is better to route IP and OSI traffic using this single integrated protocol instead of using separate router-to-router protocols.

8.15 EXTERIOR GATEWAY PROTOCOLS

Recall that by definition, an Interior Gateway Protocol is used within an Autonomous System. Different Autonomous Systems are free to choose the protocol and metric that suits them best. But how can we make reasonable routing decisions about traffic that travels between two different Autonomous Systems?

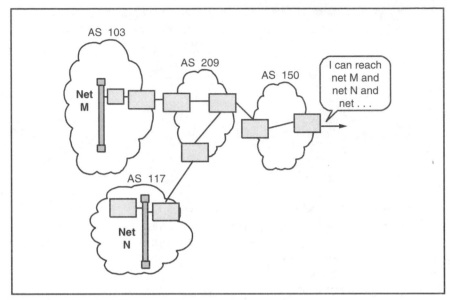

Figure 8.20 Simple EGP messages in a complex network.

8.16   EGP

For many years, the simple Exterior Gateway Protocol (EGP) was used on the Internet to enable an Autonomous System to route information to external networks. EGP is widely available. A very simple design is used. EGP routers in neighboring Autonomous Systems tell one another what networks they can reach.

EGP was designed in the early 1980s when the Internet was still fairly small and had a simple topology consisting of a backbone and a set of networks directly connected to the backbone. As the Internet evolved into its present, mesh-like topology, EGP began to pass reachability information through a chain of Autonomous Systems, as shown in Figure 8.20.

EGP does not disclose the routes that are followed by datagrams that are traveling to external locations. It even hides which Autonomous Systems are traversed along the way. EGP's simple reachability information has proved to be inadequate in the modern environment. Use of EGP is diminishing rapidly. The protocol will be described very briefly here.

8.16.1   EGP Model

An EGP router is configured with the IP addresses of one or more exterior neighbor EGP routers. Usually, exterior neighbors are connected to a common multiaccess network or are joined by a point-to-point link.

EGP enables a router to find out which networks can be reached through its exterior neighbors. EGP has the following ingredients:

Neighbor Acquisition	A router sends a Neighbor Acquisition Request. The receiver sends back both a Neighbor Acquisition Response and a Neighbor Acquisition Request message.
Neighbor Release	To terminate being a neighbor, a router sends a Neighbor Cease message. The receiver sends back a Neighbor Cease message.
Neighbor Reachability	The relationship between acquired neighbors is kept alive by periodic exchanges of Hello and I Heard You messages.
Network Reachability	A router sends a poll to the exterior neighbor, requesting information on reachable networks. The neighbor responds with a Network Reachability message.

The content of Network Reachability messages needs a little more explanation. If the exterior neighbors are connected by a point-to-point link, the message will identify networks that can be reached via the sender. A hop count to each destination also is provided. Figure 8.21 illustrates this configuration—Router A reports reachable networks to Router X.

As shown in Figure 8.22, sometimes several routers from different Autonomous Systems share a common multiaccess network. In this case, EGP Router A will inform EGP Router X of the networks reached via A, B, and C, respectively (and the hop counts for each, if known). Similarly, EGP Router X will inform EGP Router A of the networks reached via X, Y, and Z, respectively.

Routers A and X are *direct* neighbors, but Routers B and C are *indirect* neighbors of Router X.

If Router A were to crash, Router X could try to acquire one of its indirect neighbors, B or C, as a direct EGP neighbor.

EGP messages are carried directly inside IP datagrams whose protocol field is set to 8.

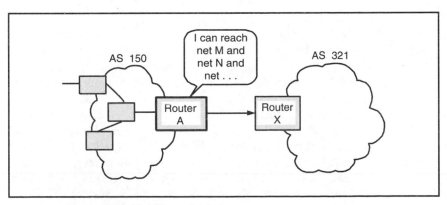

Figure 8.21 Network Reachability messages.

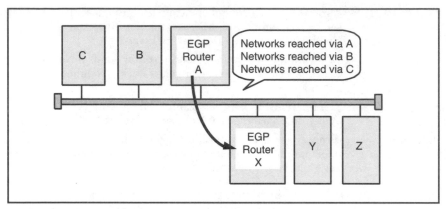

Figure 8.22   Efficient exchanges of EGP information.

8.17   BGP

The Border Gateway Protocol (BGP) is in wide use on the Internet. The current version is BGP-4.

In today's Internet, there are many Service Providers, and they are joined together by a mesh of interconnections. Traffic often transits the networks of several different Service Providers on the way to its destination. For example, the route below starts within JVNC and then traverses MCI, SPRINT, and NYSERNET routers before reaching its destination.

```
> traceroute nyu.edu
traceroute to CMCL2.NYU.EDU      (128.122.128.2), 30 hops max, 40 byte packets
1 nomad-gateway.jvnc.net         (128.121.50.50)   3 ms   3 ms   2 ms
2 liberty-gateway.jvnc.net       (130.94.40.250)  49 ms  10 ms  21 ms
3 border2-hssi2-0.NewYork.mci.net (204.70.45.9)   13 ms  12 ms  19 ms
4 sprint-nap.NewYork.mci.net     (204.70.45.6)    33 ms  25 ms  19 ms
5 sl-pen-2-F4/0.sprintlink.net   (192.157.69.9)   24 ms  21 ms  21 ms
6 ny-nyc-2-H1/0-T3.nysernet.net  (144.228.62.6)   31 ms  29 ms  24 ms
7 ny-nyc-3-F0/0.nysernet.net     (169.130.10.3)   31 ms  23 ms  20 ms
8 ny-nyu-1-h1/0-T3.nysernet.net  (169.130.13.18)  21 ms  34 ms  19 ms
9 NYU.EDU                        (128.122.128.2)  19 ms  22 ms  21 ms
```

The purpose of BGP is to support routing across a chain of Autonomous Systems while preventing loops from forming. To do this, BGP systems exchange information that describes paths to the networks that they can reach. Unlike EGP, BGP reveals the entire chain of Autonomous Systems that must be traversed to reach a network.

For example, (as shown in Figure 8.23), a BGP system in Autonomous System 34 reports to AS 205 that *Networks M* and *N* are in the AS. AS 205 reports a path to *M* and *N* through itself *and* AS 34. Then AS 654 reports a path to *M* and *N* through itself *and* AS 205 *and* AS 34. The path grows along the way, and the full path is reported to the next system. Thus, BGP reachability information includes the entire chain of Autonomous Systems that are crossed on the way to a destination network.

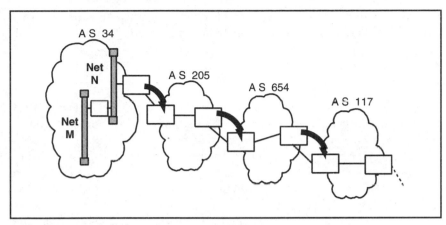

Figure 8.23   A BGP chain of Autonomous Systems.

The path is reported in the order that would be used to reach the destination, that is:

AS 654, AS 205, AS 34

If AS 117 passes the information on, it would add itself in front:

AS 117, AS 654, AS 205, AS 34

Note how easy it is to detect and prevent loops. If an AS receives an advertisement and sees its own ID on a path, it simply throws the advertisement away.

In addition to reporting paths to individual networks, BGP can identify aggregate sets of networks using CIDR prefixes.

8.17.1   BGP Route Aggregation

An Internet *route* consists of a destination network and instructions on how to get there. There has been explosive growth in the number of routes because of the rapid increase in the number of networks.

An interim solution has gotten routing under control. The current method of route reduction is to assign a block of addresses with a common prefix to a Service Provider. The provider assigns subblocks to its customer networks.

The size of the Service Provider prefix is identified by a number that indicates the length in bits of the IP address prefix. Traffic can be sent from external ASs to the provider and its customers by means of a single route that corresponds to the prefix. The provider can then use longer prefixes to forward the traffic to each of its customer's ASs.

This is simple for incoming traffic, but we have to reverse the scenario to see what a Service Provider must do on an ongoing basis with outgoing advertisements. Client ASs will inform their Service Provider of routes to their internal networks. The Service Provider will *aggregate* routes with a common prefix into single routing entries before advertising them to the outside world.

8.17.2 BGP Mechanisms

A BGP system opens a TCP connection to well-known port 179 at a BGP neighbor. Each opening message identifies the sender's Autonomous System and BGP identifier and may include authentication information.

Once the connection is open, the peers exchange their route information. The connection remains open, and updates are sent as needed. To make sure that they still are in contact, the systems periodically exchange Keep-alive messages (usually every 30 seconds).

A Service Provider network carries traffic between Autonomous Systems, and is very likely to include multiple systems that speak BGP. These systems will communicate with one another via *internal* BGP connections. An *external* BGP connection is used to communicate to a BGP peer system in a different Autonomous System. (These connections are referred to as *links,* even though they are TCP connections that might possibly pass through intermediate routers.)

One big difference between BGP and other routing protocols is that the systems that exchange BGP routing information can be hosts—they do not have to be routers. One possible configuration is to give a host the job of talking to all external BGP systems in neighboring Autonomous Systems. The host could then be used as a route server, passing information to boundary routers in its own AS.

8.17.3 Contents of BGP Update Messages

A BGP update message can contain at most one feasible route. But it also can include a list of one or more *withdrawn* routes, which are routes that should no longer be used.

A route description is made up of a sequence of *path attributes,* which include:

Origin of Path Information	One of: Source was the IGP of the original AS. Source was EGP. Other.
AS Path	The path along which this update was carried.
Next Hop	IP address of the boundary router that should be used as the next hop to the destination. This might be a router belonging to the local Autonomous System or an external router that is directly connected to both the sender and to the recipient for this update.
Multi-exit Discriminator	If I have multiple exit points that connect me to my neighbor's Autonomous System, my neighbor can assign numbers to indicate which exit is better. The neighbor's *smallest* number indicates a better route.

Local Preference	Purely internal information, used when sending BGP updates to systems in the local AS. When there are multiple BGP routes to a destination, a *bigger* number is preferred.
Atomic Aggregate	Indicates that an Autonomous System has aggregated several destinations into a single route entry.
Aggregator	The IP address and AS number of the last system that aggregated several routes into this one.
Reachable Nets	A list of prefixes for networks that can be reached via this route.

8.17.4 Making Choices

Figure 8.24 illustrates the difference between the Multi-exit Discriminator and Local Preference. Systems in AS 117 want to reach *Network N* in AS 433. AS 654 has two routes to that destination, and AS 654 announces that the one through Router E is better. However, AS 117 has an internally assigned local preference for the route to *Network N* via AS 119.

8.17.5 Using Aggregation

The purpose of route aggregation is to avoid including a lot of unnecessary information in remote routing tables. A Service Provider can aggregate the routes reported by its client Autonomous Systems.

As shown in Figure 8.25, BGP routers in AS 650, 651, and 652 can report their routes, but the provider in AS 117 aggregates them into one entry. The fact that this has been done is signaled by the *Atomic Aggregate* attribute.

Note that AS 652 might be a local Service Provider and might already be aggregating its customer's routes, so it is possible that more of the route is

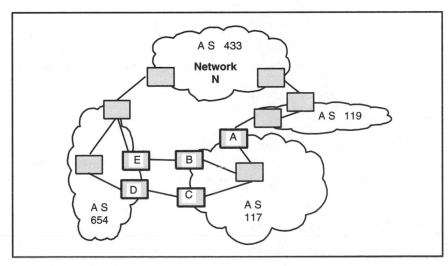

Figure 8.24   Preferred routes.

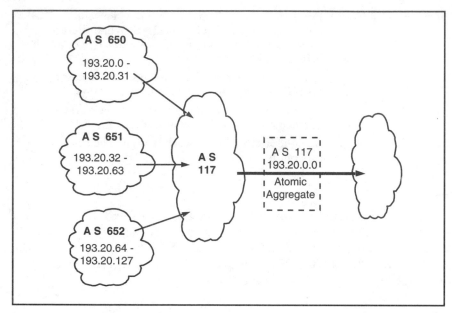

Figure 8.25 Aggregating routes.

hidden from remote systems. Each of the aggregating Autonomous System routers will forward traffic to the customer destinations based on their own routing tables.

8.17.6 Withdrawing BGP Routes

A route is terminated when:

- It is on a list of withdrawn routes sent in an update.
- An update provides a replacement route.
- The BGP system closes its connection. All routes learned via that system are voided.

8.18 RECOMMENDED READING

Routing is a very important function, and many RFCs have been written on the subject. Some of the most important and useful ones are listed below. Check the RFC index for the most current versions.

RIP:

RFC 1058 *Routing Information Protocol*

RFC 1723 RIP *Version 2 Carrying Additional Information*

RFC 1582 *Extensions to RIP to Support Demand Circuits*

OSPF:

RFC 1583 *OSPF Version 2*

RFC 1793 *Extending OSPF to Support Demand Circuits*

RFC 1586 *Guidelines for Running OSPF Over Frame Relay Networks*

RFC 1584 *Multicast Extensions to OSPF*

RFC 1403 *BGP OSPF Interaction*

BGP:[13]

RFC 1771 *A Border Gateway Protocol 4 (BGP-4)*

RFC 1773 *Experience with the BGP-4 Protocol*

RFC 1772 *Application of the Border Gateway Protocol in the Internet*

See Cisco's on-line documentation at *www.cisco.com* for technical information on IGRP and EIGRP.

---

[13]The OSI Inter-Domain Routing Protocol (IDRP) is expected to supplant BGP some time in the future.

User Datagram Protocol

9.1 INTRODUCTION

Now that we have dealt with the physical movement of bits across media and the routing of datagrams across an internet, we are ready to turn to the services that applications will use directly for the transfer of data. The first of these, *User Datagram Protocol* (UDP), is very straightforward. UDP enables applications to send individual messages to one another.

Why define this kind of service? There are many applications that can be built on top of User Datagrams in a very natural way. For example, a simple exchange of User Datagrams can be used to execute a quick database lookup. We already have encountered an important service that is based on UDP, namely the *Domain Name System* (DNS) (see Figure 9.1).

The overhead of sending and receiving the many messages required to set up and take down a connection is avoided by simply sending a query and a response. UDP also is a perfect building block for constructing monitoring, debugging, management, and testing functions.

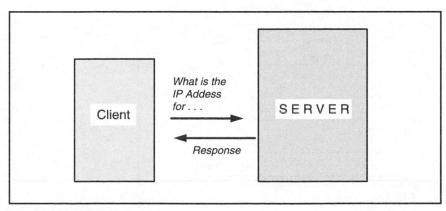

Figure 9.1 A DNS request and response.

UDP is a very basic service, simply passing individual messages to IP for transmission. Since IP is unreliable, there is no guarantee of delivery. If an application sends a query in a UDP Datagram and a response does not come back within a reasonable amount of time, it is up to the application to retransmit the query.

Sometimes this results in duplicate queries showing up at a server. If the application includes a transaction identifier with its query message, the server will be able to recognize duplicates and discard them. This mechanism is the application's responsibility, not UDP's.

9.1.1   Broadcast and Multicast

Another advantage of UDP is that it can be used by applications that need to broadcast or multicast messages. For example, a BOOTP client broadcasts a request for initialization parameters.

9.2   APPLICATION PORTS

What happens to data when it arrives at a destination host? How does it get delivered to the appropriate process?

As shown in Figure 9.2, for each layer, there is a protocol identifier that indicates what should be done with incoming data. At layer 2, an Ethernet type of X'08-00 in a frame header indicates that the frame should be passed to IP. At layer 3, the *Protocol* field in the IP header identifies the layer 4 protocol to which datagrams should be passed (e.g., 6 for TCP, 17 for UDP).

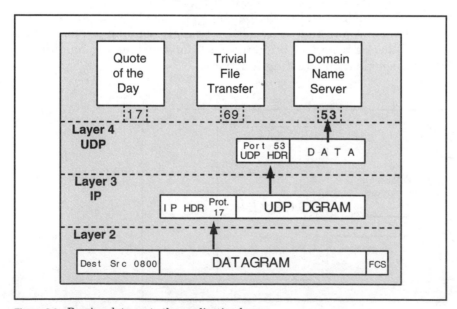

Figure 9.2   Passing data up to the application layer.

TABLE 9.1   Examples of UDP Well-Known Ports

Service	Port/protocol	Description
Echo	7/udp	Echo User Datagram back to sender.
Discard	9/udp	Discard User Datagram.
Daytime	13/udp	Report time in a user-friendly way.
Quote	17/udp	Return a "quote of the day."
Chargen	19/udp	Character generator.
Nameserver	53/udp	Domain name server.
Bootps	67/udp	Server port used to download configuration info.
Bootpc	68/udp	Client port used to receive configuration info.
TFTP	69/udp	Trivial File Transfer Protocol port.
SunRPC	111/udp	Sun Remote Procedure Call.
NTP	123/udp	Network Time Protocol.
SNMP	161/udp	Used to receive net management queries.
SNMP-trap	162/udp	Used to receive network problem reports.

A host can be expected to participate in many simultaneous communications at any time. How do UDP datagrams get sorted out and delivered to appropriate application layer processes?[1] The answer is that every UDP communication endpoint is assigned a 16-bit identifier called a *port* number.[2]

Port numbers from 0 to 1023 are reserved for standard services. Standard ports are called *well-known* ports. The use of a well-known port enables a client to identify the service that it wishes to access. For example, the UDP-based Domain Name Service is accessed at well-known port 53.

How do well-known ports get assigned? As you might guess, the Internet Assigned Numbers Authority (IANA) is in charge of this function. Port numbers for specific applications are registered with the IANA and published in its *Assigned Numbers* Request For Comments document. A partial list of UDP ports taken from the current *Assigned Numbers* RFC document is displayed in Table 9.1.

---

[1]This process of passing data to the correct process is sometimes referred to as *demultiplexing*.

[2]The term *port* was a poor choice for this identifier. Application port numbers identify clients and servers and have *nothing* to do with systems hardware or the physical path that data follows.

Several of these well-known services provide building blocks for testing, debugging, and measurement. For example, the *echo* service at port 7 does what its name implies—it returns any datagram that is sent to it. *Discard* at port 9, on the other hand, just throws datagrams away. A *character generator* responds to any message with a datagram containing between 0 and 512 bytes. The number of bytes is randomly chosen.

The *quote of the day* service responds to any datagram by sending back a message—for example, some fortune cookie wisdom to brighten your day when you log off. On many systems, you can run a *fortune* program that requests a quote:

```
> fortune
Churchill's Commentary on Man:
 Man will occasionally stumble over the truth, but most of the
 time he will pick himself up and continue on.
```

A *daytime* server responds to any datagram with a message containing the current date and time in a readable ASCII format. In contrast, the *Network Time Protocol* (NTP) provides a robust method for synchronizing the clocks at computers across a network.

A BOOTP server and client are used to initialize an unconfigured device. A workstation can find out its IP address, its address mask, the location of its default router, the addresses of its important servers, and, if needed, the name and location of a software download file from a boot server. The workstation's software is downloaded using the *Trivial File Transfer Protocol*.[3]

We already have talked to the *Nameserver* at port 53 via the *nslookup* command. Ports 161 and 162 are used by the *Simple Network Management Protocol*.

Apart from the official number assignments, any system running TCP/IP may reserve a range of numbers for important network services and applications.

The remaining port numbers (above 1023) are allotted to clients by a host's networking software on an as-needed basis. The scenario that follows indicates how this happens:

1. A user invokes a client program (such as *nslookup*).

2. The client process executes a system routine that says "I want to perform UDP communication. Give me a port."

3. The system routine selects an unused port from the pool of available ports and gives it to the client process.

We shall see that TCP also identifies its sources and destinations with 16-bit port identifiers. For example, port 21 is used to reach a *File Transfer* service and port 23 is used to reach a *telnet* login service.

---

[3]The Trivial File Transport Protocol will be described in Chapter 14.

TCP and UDP numbers are independent of each other. One process may be sending messages from UDP port 1700 while another is engaged in a session at TCP port 1700. There are some services that can be accessed via both TCP and UDP. In this case, the IANA makes an effort to assign the same numbers to the UDP and the TCP ports assigned to a service. However, as endpoints of communication, these still are different "places."

9.3 SOCKET ADDRESSES

The combination of the IP address and port used for a communication is called a *socket address*. Note that a socket address provides all of the information a client or server needs in order to identify its communicating partner.

The IP header contains the source and destination IP addresses. The UDP or TCP header contains the source and destination port numbers. Thus every UDP or TCP message carries the socket addresses for its source and destination.

Below, we show part of the display that results from using the *netstat -na* command to display local and remote socket addresses for current active communications at *tigger*. The socket addresses are written in the form *IP-Address.port-number*.

```
> netstat -na
Active Internet connections (including servers)
Proto  Recv-Q  Send-Q  Local Address         Foreign Address       (state)
tcp         0       0  127.0.0.1.1340        127.0.0.1.111         TIME_WAIT
tcp         0       0  128.121.50.145.25     128.252.223.5.1526    SYN_RCVD
tcp         0       0  128.121.50.145.25     148.79.160.65.3368    ESTABLISHED
tcp         0     438  128.121.50.145.23     130.132.57.246.2219   ESTABLISHED
tcp         0       0  128.121.50.145.25     192.5.5.1.4022        TIME_WAIT
tcp         0       0  128.121.50.145.25     141.218.1.100.3968    TIME_WAIT
tcp         0       0  128.121.50.145.25     35.8.2.2.3722         TIME_WAIT
tcp         0       0  128.121.50.145.1338   165.247.48.4.25       ESTABLISHED
tcp         0       0  128.121.50.145.25     128.173.4.8.3826      ESTABLISHED
tcp         0       0  128.121.50.145.25     192.48.96.14.3270     ESTABLISHED
.  .  .
udp         0       0  *.7                   *.*
udp         0       0  *.9                   *.*
udp         0       0  *.37                  *.*
udp         0       0  *.19                  *.*
udp         0       0  *.111                 *.*
.  .  .
```

For example, the boxed entry shows a TCP login session from client port 2219 at IP address 130.132.57.246 to the standard *telnet* port, 23, at 128.121.50.145. Entries such as *.7 and *.9 represent UDP services at *tigger* waiting for clients to make requests.

9.4 UDP PROTOCOL MECHANISMS

What mechanisms are needed to make the User Datagram Protocol work? First of all, UDP has been assigned its unique protocol identifier, 17. This number is placed in the IP *Protocol* field of outgoing UDP messages. Incoming messages with 17 in the IP *Protocol* field are delivered to UDP.

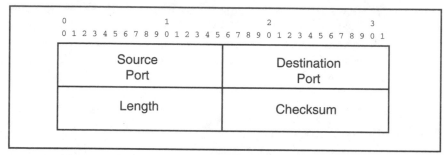

Figure 9.3   The UDP Header.

UDP forms a message by adding a simple header to application data. This header contains the source and destination port numbers.

9.4.1   UDP Header

Figure 9.3 displays the UDP header format. The header contains the 16-bit source and destination port numbers that identify the endpoints of the communication. A length field indicates the total number of octets in the UDP header and data part of the message. A checksum field is provided to validate the contents of the message.

9.4.2   Checksum

Recall that the IP header contained a checksum field used to validate the fields in the IP header. The purpose of the UDP checksum is to validate the *contents* of a UDP message.

The UDP checksum is computed on the combination of a specially constructed *pseudo header* containing some IP information, the UDP header, and the message data.

The format of the pseudo header that is put together by the checksum function is shown in Figure 9.4. Note that the source address, destination address, and protocol field are taken from the IP header.

Use of the UDP checksum in a particular communication is optional. If unused, the field is 0. If a checksum has been computed and the value turns out to be 0, this is represented as a field of 1s.

9.4.3   Other UDP Functions

Apart from submitting and accepting datagrams, UDP must obey the common sense rules of passing options down from an application to IP and passing error notifications up from IP to the application.

9.4.4   Sample UDP Messages

Figure 9.5 contains side-by-side displays of the IP and UDP portions of a query and its corresponding response. The displays were generated by a

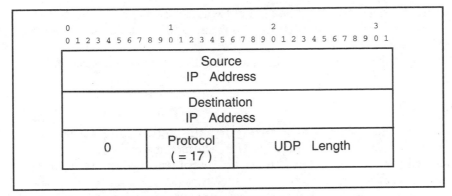

Figure 9.4 Pseudo header fields included in the UDP checksum.

Network General *Sniffer* LAN monitor. The query contains a request for status information and was sent to a host by a network management station. The data portions of the query and response messages are not displayed here.

The request was sent from source IP address 128.1.1.1 and UDP port 1227 to destination IP address 128.1.1.10 and UDP port 161. (Network management queries always are sent to UDP port 161.)

```
IP:    ----- IP Header -----          IP:    ----- IP Header -----
IP:                                    IP:
IP:    Version = 4, header length - 20 bytes   IP:    Version = 4, header length = 20 bytes
IP:    Type of service = 00            IP:    Type of service = 00
IP:         000. .... = routine        IP:         000. .... = routine
IP:         ...0 .... = normal delay   IP:         ...0 .... = normal delay
IP:         .... 0... = normal throughput   IP:         .... 0... = normal throughput
IP:         .... .0.. = normal reliability   IP:         .... .0.. = normal reliability
IP:    Total length = 131 bytes        IP:    Total length = 160 bytes
IP:    Identification = 21066          IP:    Identification = 2015
IP:    Flags = 0X                      IP:    Flags = 0X
IP:    .0.. .... = may fragment        IP:    .0.. .... = may fragment
IP:    ..0. .... = last fragment       IP:    ..0. .... = last fragment
IP:    Fragment offset = 0 bytes       IP:    Fragment offset = 0 bytes
IP:    Time to live = 60 seconds/hops  IP:    Time to live = 64 seconds/hops
IP:    Protocol = 17 (UDP)             IP:    Protocol = 17 (UDP)
IP:    Header checksum = 2A13 (correct)   IP:    Header checksum = 7061 (correct)
IP:    Source address = [128.1.1.1]    IP:    Source address = [128.1.1.10]
IP:    Destination address = [128.1.1.10]   IP:    Destination address = [128.1.1.1]
IP:    No options                      IP:    No options
IP:                                    IP:
UDP:   ----- UDP Header -----          UDP:   ----- UDP Header -----
UDP:                                    UDP:
UDP:   Source port = 1227 (SNMP)       UDP:   Source port = 161 (SNMP)
UDP:   Destination port = 161          UDP:   Destination port = 1227
UDP:   Length = 111                    UDP:   Length = 140
UDP:   No checksum                     UDP:   Checksum = 4D4F (correct)
UDP:                                    UDP:
```

Figure 9.5 IP and UDP headers for a request and response.

In both IP headers, the IP *Protocol* field was set to 17, meaning "UDP." A UDP checksum was not computed for the request, but a UDP checksum was included in the response.

The *Sniffer* analyzer recognizes that port 161 is a network management port.

9.5 UDP OVERFLOWS

When an application acquires a UDP port, the networking software will reserve some buffers to hold a queue of User Datagrams arriving at that port. A UDP-based service has no way to predict or control how many datagrams will be sent to it at any time.

If the service is bombarded with more datagrams than it can handle, the overflow simply will be discarded. The fact that this has happened will show up in networking statistics reports under a heading such as "UDP Socket Overflows." For example, the report below was produced by the *netstat* command.

```
> netstat -s
 . . .
udp:
  0 incomplete headers
  0 bad data length fields
  0 bad checksums
  17 socket overflows
```

9.6 RECOMMENDED READING

User Datagram Protocol is defined in RFC 768. RFCs 862 to 865 discuss the *echo, discard, character generator,* and *quote of the day* UDP services. RFC 867 describes the *daytime* utility and RFC 1119 presents version 2 of the *network time service.* The BOOTP protocol will be described in Chapter 11. Additional UDP services are examined in other chapters.

10

Transmission Control Protocol

10.1 INTRODUCTION

IP was kept simple so that the network layer could focus on performing one important function—routing data from its source to its destination. The job of turning an exchange of datagram traffic into a solid, reliable application-to-application data connection is carried out by TCP, which is implemented in the end hosts. Services such as the World Wide Web (WWW), terminal login, file transfer, and mail transfer run on top of TCP connections.

10.1.1 Major TCP Services

We may view TCP as providing *data calls,* analogous to voice telephone calls. A caller identifies the destination. At the other end, a listening application is alerted that there is an incoming call and picks up the connection. The two ends exchange information for a while. When they are finished, both parties say "good-bye" and hang up.

IP makes a best-effort attempt to deliver datagrams, but some may be destroyed along the way, while others can arrive out of order. A datagram may wander around the network for a fairly long time and turn up unexpectedly. It is up to TCP to assure that data is delivered *reliably, in sequence, and without confusion or error.*

An application in a fast, powerful host could swamp a slow recipient with data. TCP provides the *flow control* that enables the *receiver* to regulate the rate at which the sender may transmit data. TCP also has mechanisms that let it respond to network conditions, adjusting its own behavior to optimize performance.

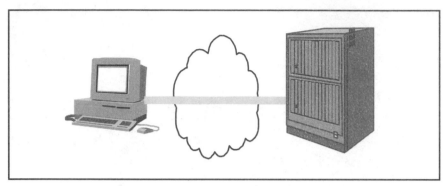

Figure 10.1 A client calling a server.

10.1.2 TCP and the Client/Server Model

TCP operates very naturally in a client/server environment (see Figure 10.1). A server application *listens* for incoming connection requests. For example, World Wide Web, file transfer, and terminal access servers listen for incoming clients. A client application initiates TCP communication by invoking communications routines that establish a connection with a server.[1]

The "client" may actually be another server; for example, a mail server connects to a peer mail server in order to transfer mail between computers.

10.2 TCP CONCEPTS

In what form does an application pass data to TCP? In what form does TCP pass data to IP? How do sending and receiving TCPs identify the application-to-application connection that a unit of data belongs to? These questions will be answered in the sections that follow, which deal with TCP's conceptual design.

10.2.1 Outgoing and Incoming Data Streams

The *conceptual* model for a connection is that an application sends a stream of data to a peer application. At the same time, it receives a stream of data from its peer. TCP provides a *full duplex* service that simultaneously manages *two streams* of data, as shown in Figure 10.2.

10.2.2 Segments

TCP must convert an application's outgoing stream[2] of data into a form that can be delivered in datagrams. How is this done?

---

[1]Chapter 21 describes the socket programming interface.

[2]This is just a model. In reality, applications pass pieces of data of varying size to TCP for transmission.

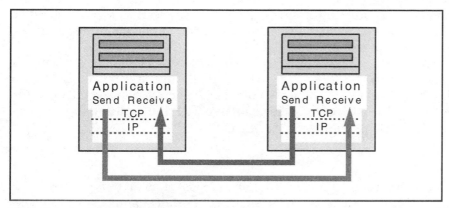

Figure 10.2 Applications exchanging streams of data.

The application passes data to TCP and TCP places this data into a *send* buffer. TCP slices off a chunk of data and adds a header, forming a *segment*. Figure 10.3 shows how data in a TCP *send* buffer is packaged into a segment. TCP passes the segment to IP for delivery in a single datagram. Packaging data in good-sized chunks makes efficient use of transmission facilities, so TCP would like to wait until a reasonable amount of data has collected before creating a segment.

10.2.3 Push

But sometimes big, efficient data chunks are not appropriate for a particular application. For example, suppose that an end user's client program has initiated an interactive session with a remote server, and the user has just typed a command followed by *return*.

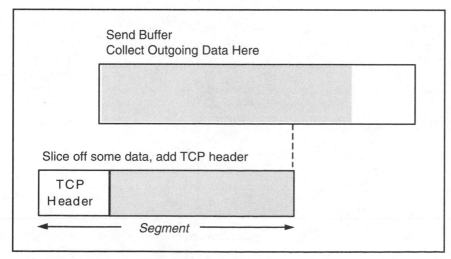

Figure 10.3 Creation of a TCP segment.

The user's client program wants to let TCP know that this data should be sent to the remote host and delivered to the server application promptly. The *push* feature makes this happen

If you view a trace of an interactive session, you will see many segments that contain very little data, and you are likely to see a push signaled in every data segment. On the other hand, push would not be used during a file transfer (except for the very last segment), so TCP could pack data into segments as efficiently as possible.

10.2.4 Urgent Data

Recall that application data transmission is modeled as an ordered stream of bytes flowing to its destination. But using the example of an interactive session again, suppose that a user has pressed an *attention* or *break* key. The remote application should be able to jump over intervening bytes and notice this as soon as possible.

There is an *Urgent Data* mechanism that marks specific information in a segment as *urgent*. TCP can signal its peer that a segment contains urgent data and can point to where that data is. The peer TCP can pass this information to the destination application.

10.2.5 Application Ports

A client must identify the service that it wants to reach. This is done by specifying the service host's IP address and its TCP port number. Just as for the User Datagram Protocol (UDP), TCP port numbers range from 0 to 65,535. Recall that ports in the range 0 to 1023 are well-known ports, used to access standard services.

TABLE 10.1   Well-Known TCP Ports and Their Applications

Port	Application	Description
9	Discard	Discard all incoming data
19	Chargen	Exchange streams of characters
20	FTP-Data	File Transfer data transfer port
21	FTP	File Transfer dialogue port
23	TELNET	*Telnet* remote login port
25	SMTP	Simple Mail Transfer Protocol port
110	POP3	PC mail retrieval service
119	NNTP	Network news access

Some sample well-known TCP ports and their applications are listed in Table 10.1. *Discard* at port 9 and *chargen* at port 19 are TCP versions of the utility services already described for UDP. Keep in mind that traffic sent to TCP port 9 will be totally separate from traffic sent to UDP port 9.

What about the ports used by clients? There are a few instances in which a client will operate out of a well-known port, but most of the time, a client that wants to open a connection just asks the operating system to assign it an unused, unreserved port number. At the end of the connection, the client will relinquish the port back to the system, and it can be reused by another client. Since there are more than 63,000 TCP ports in the pool of unreserved numbers, there is no shortage of ports for clients.

10.2.6 Socket Addresses

Recall that the combination of the IP address and the port used for communication is called a *socket address*. A TCP connection is completely identified by the socket addresses at its two ends. Figure 10.4 shows a connection between a client with socket address (128.36.1.24, port = 3358) and a server with socket address (130.42.88.22, port = 21).

Every datagram header contains its source and destination IP addresses. As we shall see later, the source and destination port numbers are carried in the TCP segment header.

Usually a server is capable of handling many clients at the same time. The server's unique socket address would be accessed simultaneously by all of its clients, as shown in Figure 10.5.

Since a datagram containing a segment for a particular TCP connection identifies both of the IP addresses and both of the ports, it is easy for a server to keep track of multiple client connections.

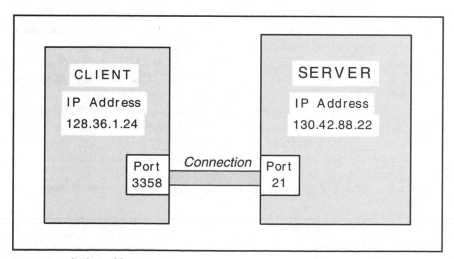

Figure 10.4 Socket addresses.

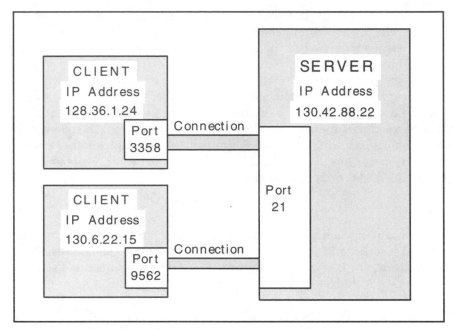

Figure 10.5 Multiple clients connecting to a server socket address.

10.3   TCP RELIABILITY MECHANISMS

The sections that follow describe the mechanisms that TCP uses to deliver data reliably, in order, and without loss or duplication.

10.3.1   Numbering and Acknowledgment

TCP employs numbering and acknowledgment (ACK) to transfer data reliably. The TCP numbering scheme is unusual: *every octet* sent on a TCP connection is viewed as having a sequence number. A segment's TCP header contains the sequence number *of the first octet of data in the segment.*

The receiver is expected to ACK received data. If an ACK does not arrive within a timeout interval, the data is retransmitted. This strategy is called *positive acknowledgment with retransmission.*

The receiving TCP keeps a close watch on incoming sequence numbers to keep arriving data in order and to make sure that no data is missing. Since ACKs occasionally are lost or late, duplicate segments may arrive at the receiver. The sequence numbers pinpoint which data has been duplicated so that it can be discarded.

Figure 10.6 shows a simplified view of TCP timeout and retransmission.

10.3.2   TCP Header Fields for Ports, Sequencing, and ACKs

As shown in Figure 10.7, the first few fields of the TCP header provide space for the source and destination ports, the sequence number of the first byte of

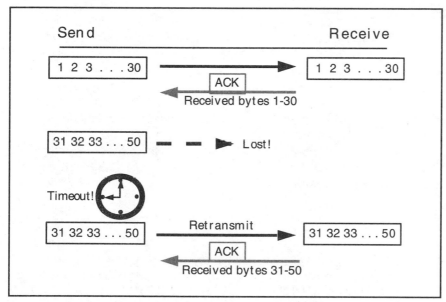

Figure 10.6 TCP timeout and retransmission.

enclosed data, and the acknowledgment value, which contains the sequence number of the *next* byte expected from the other end. In other words, if TCP has received all bytes up to 30 from its partner, it would write 31 into the acknowledgment field of the segment that it is about to transmit.

One small detail should be noted. Suppose that TCP has transmitted bytes 1 through 50 and currently has no more data that needs to be sent. If some data arrives from the partner, TCP needs to acknowledge it. To do this, TCP will send a header without any data attached. Naturally, the header would contain the acknowledgment value. The sequence number field would contain 51, the number of the next byte that TCP *intends* to send. When TCP sends data later on, the new TCP header also will have 51 in the sequence number field.

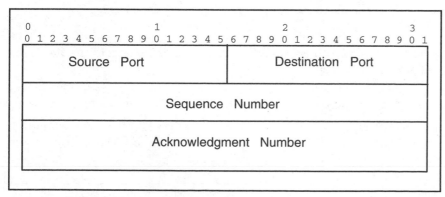

Figure 10.7 Initial fields of the TCP header.

10.4 ESTABLISHING A CONNECTION

How do two applications start a connection? Before communicating, each calls a subroutine that creates a block of memory that will be used to store the TCP and IP parameters for the connection—such as the socket addresses, current sequence numbers, initial IP Time-To-Live value, and so forth.

A server application waits for incoming clients. A client that wishes to access the server issues a *connect* request, identifying the server's IP address and port.

There is one technical point. Rather than starting their byte numbering at 1, each side generates a random[3] *initial sequence number*. We will explain why this is a good idea later in this chapter.

10.4.1 A Connection Scenario

The connection procedure is called a three-way handshake because three messages (called SYN, SYN, and ACK) are exchanged to set up the connection.

Three important pieces of information are exchanged during connection setup. Each side notifies its partner of:

1. How much buffer space it has available to receive data

2. The maximum amount of data that an incoming segment may carry

3. The initial sequence number it will use for numbering outgoing data

Note that each side uses items 1 and 2 to *set limits on what the other side can do*. A PC might have a small receive buffer, while a supercomputer might have a very large buffer. A PC's memory structure might limit incoming chunks of data to 1K, while a supercomputer might be able to handle larger segments.

The ability to control how the other side sends data is an important feature in the scalability of TCP/IP.

Figure 10.8 shows a sample connection scenario. Very simple initial sequence numbers are shown, to promote easy readability. (Note that in this example, the client is able to receive bigger segments than the server can.)

The steps are:

1. The server initializes and is ready to accept connections from clients. (This is called a passive open.)

2. The client asks TCP to open a connection to a server at a given IP address and port. (This is called an active open.)

3. The client TCP picks an initial sequence number (1000 in the example). The client TCP sends a *synchronize segment* (called a SYN) carrying this

---

[3]The specification originally advised that an initial sequence number should be generated from a 32-bit internal clock that is supposed to be incremented roughly every 4 microseconds.

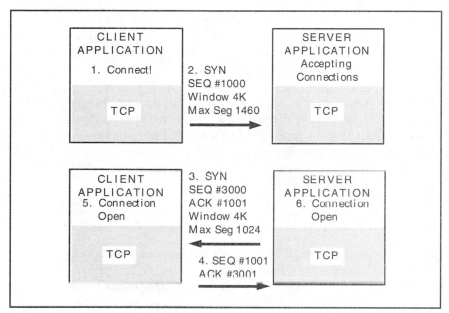

Figure 10.8 Setting up a connection.

sequence number, the size of the receive window (4K), and the size of the biggest segment that the client can receive (1460 bytes).

4. When the SYN arrives, the server TCP picks *its* initial sequence number (3000). The server TCP sends a SYN segment containing its initial sequence number, (3000), an ACK 1001, meaning that the first data byte sent by the client should be numbered 1001, the size of its receive window (4K), and the size of the biggest segment that it can receive (1024 bytes).

5. When the client TCP receives the server's SYN/ACK message, the client TCP sends back an ACK 3001, meaning that the first data byte sent by the server should be numbered 3001.

6. The client TCP notifies its application that the connection is open.

7. When the server TCP receives the client TCP's ACK, the server TCP notifies its application that the connection is open.

The client and server have announced their rules for receiving data, have synchronized their sequence numbers, and are ready to exchange data.[4]

---

[4]The TCP specification also allows for an (unlikely) scenario in which peer applications perform an active open to each other at the same time.

10.4.2 Setting Up IP Parameters

The programming calls that set up a connection can set parameter values for the IP datagrams that will carry the data for the connection. When specific values are not given, system defaults are used.

For example, an application can choose a specific IP Precedence and Type of Service. Since each end of a connection chooses its own Precedence and Type of Service independently, theoretically, these could be different for the two directions of data flow. In practice, they usually are the same.

When applications use the IP government/military security option, both ends of the connection must operate at the same security level, or the connection will be terminated.

10.5 DATA TRANSFER

Data transfer begins after completion of the three-way handshake.[5] Figure 10.9 shows a straightforward exchange of data. In order to keep the numbering simple, 1000-byte messages are used. *Each segment's TCP header includes an ACK field which identifies the sequence number of the next byte expected from the partner.*

The first segment sent by the client contains bytes 1001 to 2000. Its ACK field announces that 3001 is the sequence number of the next byte expected from the server.

The server responds with a segment that contains 1000 bytes of data (starting at 3001). The ACK field in the TCP header indicates that bytes 1001 to 2000 have been received in perfect condition, so the sequence number of the next byte expected from the client is 2001.

---

[5]The TCP standard actually allows inclusion of data in the handshake segments. This data would not be delivered to an application until the handshake was complete.

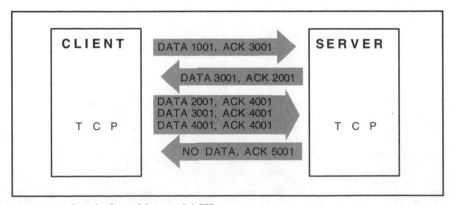

Figure 10.9 Simple flow of data and ACKs.

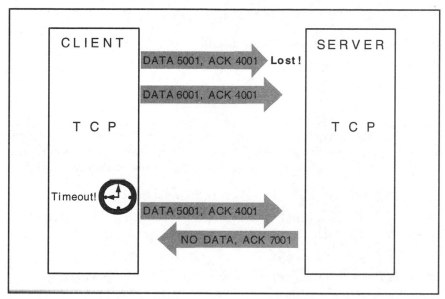

Figure 10.10   Data loss and retransmission.

Next, the client sends segments starting at bytes 2001, 3001, and 4001 in quick succession. Note that the client did not have to wait for an ACK after each segment. Data can be sent to a partner as long as the partner has some unused buffer space available.[6]

The server saves bandwidth by using a single ACK to indicate that all of the segments were received in perfect condition.

Figure 10.10 shows a transfer in which the first segment is lost. After a timeout period, the segment is retransmitted. Note that once the missing segment arrives, the receiver can send a single ACK that confirms that both segments have arrived safely.

10.6   CLOSING A CONNECTION

The normal termination of a connection is carried out by means of a three-way handshake similar to the connection opening. Either side can launch the close, which usually follows the pattern:

A: "I'm finished. I have no more data to send."

B: "OK."

B: "I'm finished too."

A: "OK."

---

[6]Later we shall see that the receiver controls the amount of data that can be sent at any time very precisely.

The following pattern also is valid but is rarely seen:

A: "I'm finished. I have no more data to send."

B: "OK. But here is some data..."

B: "I'm finished too."

A: "OK."

In the example, the server initiates the close, as is often the case in real client/server interactions. For example, after a *telnet* user types "logout," the server will invoke a call to close the connection. The steps in Figure 10.11 are:

1. The server application tells TCP to close the connection.
2. The server TCP sends a Final Segment (FIN), informing the partner that it will send no more data.
3. The client TCP sends an ACK of the FIN segment.
4. The client TCP notifies its application that the server wishes to close.
5. The client application tells its TCP to close.
6. The client TCP sends a FIN message.
7. The server TCP receives the client's FIN, and responds with an ACK.
8. The server TCP notifies its application that the connection is closed.

Both sides may initiate a close simultaneously. In this case the normal close is completed when each partner has sent an ACK.

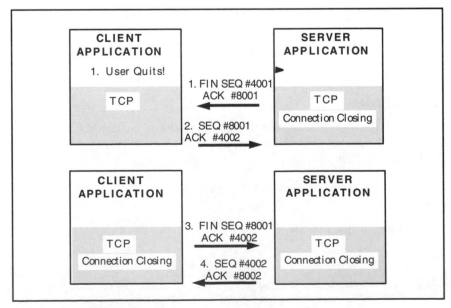

Figure 10.11   Closing a connection.

10.6.1 Abrupt Close

Either side can call for an abrupt close. This may be done because the application wishes to abort the connection or because TCP has detected a serious communication problem that cannot be resolved. An abrupt close is requested by sending one or more "resets" to the partner. A reset is signaled by means of a flag in the TCP header.

10.7 FLOW CONTROL

The TCP data receiver is in charge of its incoming flow of data. The receiver decides how much data it is willing to accept, and the sender must stay within this limit. The discussion that follows describes at a conceptual level the way that this is done. Vendors can implement these mechanisms in any way that is convenient for them.

During connection setup, each partner assigns receive buffer space to the connection and announces, "Here's how many bytes you can send me." This number is usually an integer multiple of the maximum segment size.

Incoming data flows into the receive buffer and stays there until it is absorbed by the application associated with that TCP port. Figure 10.12 shows a receive buffer that can hold 4K bytes.

Buffer space is used up as data arrives. When the receiving application removes data, space is cleared for more incoming data.

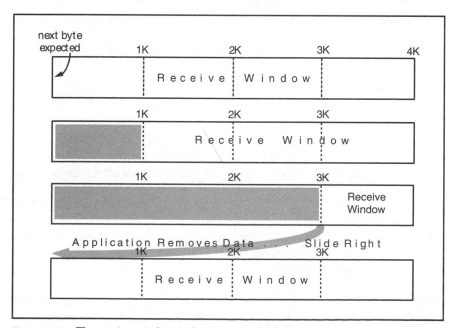

Figure 10.12 The receive window within a receive buffer.

10.7.1   Receive Window

The *receive window* consists of any space in the receive buffer that is not occupied by data. Data will remain in a receive buffer until the targeted application accepts it. Why wouldn't an application remove its data immediately?

A simple scenario should help to explain this. Suppose that a client is sending a file to a file transfer server that is running at a very busy multiuser computer. The file transfer server program will read data from the receive buffer and write it out to disk. When the server performs its disk Input/Output (I/O), the program will have to wait for the I/O to complete. In the meantime, other programs will be scheduled and run by the operating system. More data may arrive while the file transfer server process is waiting to run again.

The receive window extends from the last acknowledged byte to the end of the buffer. In Figure 10.12, initially the whole buffer is available, and so there is a 4K-byte receive window. 1K bytes arrive, and the receive window shrinks to 3K.[7] Two more segments containing 1K bytes each arrive, causing the receive window to shrink to 1K bytes.

Finally, the application absorbs the 3K bytes of data in the buffer, making space available for more incoming data. This can be visualized as *sliding* the window to the right. Now all 4K buffer bytes are free.

Every ACK sent by the receiver contains an update on the current state of its receive window. The flow of data from the sender is regulated according to these window updates.

Most of the time, the receive buffer size set at start-up is maintained throughout the connection. However, the TCP standard does not restrict the way that an implementation manages its buffers. The receive buffer size can grow or shrink, as long as the receiver never "takes back" an allowance that it has granted the sender.

What happens when segments arrive which are in the window but are out of order? Virtually all implementations hold onto any data falling within the window and ACK the entire block of contiguous data when the missing bytes arrive. This is fortunate, since throwing away data would lead to poor performance.

10.7.2   Send Window

The data transmitter keeps track of two things; how much data has been sent and acknowledged and the current size of the partner's receive window. The active *send space* extends from the first unacknowledged octet to the right edge of the current receive window. The *send window* part of this space indicates how much additional data may be sent to the partner.

The initial sequence number and initial receive window size are announced during connection setup. An example, (see Figure 10.13), illustrates some of the data transmission mechanisms.

---

[7]For simplicity, all of the segments in this example contain 1K bytes of data. For real sessions, actual amounts will, of course, vary, depending on the needs of the application.

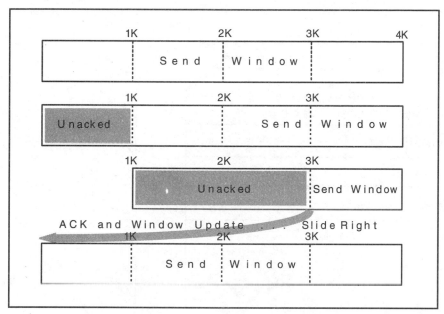

Figure 10.13 A send window.

1. The sender starts out with a 4K-byte send window.

2. The sender transmits 1K bytes. A copy of these bytes must be kept until the bytes have been acknowledged, since they may need to be retransmitted.

3. An ACK arrives for the first 1K bytes, and another 2K bytes are sent. The result is shown in the third part of Figure 10.13. 2K bytes still have to be kept.

4. Finally, an ACK arrives that reports that all of the transmitted bytes have been received. The ACK also updates the receiver's window size to 4K.

There are some interesting features to be noted:

- The sender does not have to wait for an ACK for each segment of data that is transmitted. The only limitation on transmission is the size of the receive window. (For example, the sender could transmit 4K 1-byte segments.)

- Suppose that a sender has to retransmit data that was sent in several very short (e.g., 80-byte) segments? The data can be repackaged in the most efficient way for retransmission—for example, into a single segment.

10.8 TCP HEADER

Figure 10.14 displays the format of a segment—that is, a TCP header and data. The TCP header starts with its source and destination port identifiers. The *Sequence Number* field identifies the position in the outgoing data

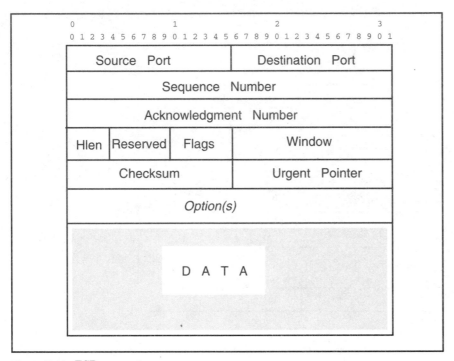

Figure 10.14 TCP segment.

stream held by the data in this segment. The *ACK* field identifies where we expect the next incoming segment to be located in the incoming data stream.

There are six flags:

URG.   1 if urgent data is included

ACK.   1 for all but the initial SYN segment

PSH.   Indicates that data should be delivered promptly

RST.   Indicates an error; also used to abort a session

SYN.   Set to 1 during connection setup

FIN.   Set to 1 during graceful close

The *Data Offset* field contains the TCP header length, measured in 32-bit words. The TCP header must end on a 32-bit boundary.

10.8.1   Maximum Segment Size Option

The *maximum segment size* (MSS) option is used to announce the size of the biggest chunk of data that can be received (and reassembled) by the system. The name is quite misleading. Normally, a *segment* is defined as a TCP header plus data. In spite of its title, a system's *maximum segment size* is defined as:

The size of the biggest datagram that can be received − *40*

In other words, the MSS reports the receiver's biggest *data payload* size when the TCP and IP headers each are 20 bytes long. If there are any header options, their lengths must be subtracted. Thus, to find out how much data can be stuffed into a segment, TCP needs to calculate:

$$Announced\ MSS + 40 - (size\ of\ TCP\ and\ IP\ headers)$$

Usually, partners exchange their MSS values in the initial "SYN" messages that start up a connection. If a system does not announce its maximum segment size, a default MSS of 536 bytes will be assumed.

The maximum segment size is encoded with a 2-byte introducer followed by a 2-byte value, so the biggest size possible is 2^{16} 1, or 65,535 bytes.

The MSS imposes a strict top limit on the size that TCP can send—the recipient cannot handle anything bigger. But the sending TCP might possibly have to send *smaller* segments because of the Path MTU size for the connection.

10.8.2   Use of Header Fields in the Connection Request

The first segment sent to start a connection has the SYN flag set to 1 and the ACK flag set to 0. The initial SYN is the *only* segment that will have an ACK field of 0. Note that firewalls use this fact to screen out incoming TCP session requests.

The *Sequence Number* field contains the *initial sequence number*. The *Window* field contains the initial *receive window* size. The only option currently defined for TCP is the maximum segment size that this TCP is willing to receive.[8] The maximum segment size occupies 32 bits and usually is included in the connection request (in the *Options* field). The length of a TCP SYN header that contains the MSS option is 24 bytes.

10.8.3   Use of Header Fields in the Connection Response

In a response accepting the connection, both the SYN and ACK flags are 1. The responder's initial sequence number is in the *Sequence Number* field, and the receive window size is in the *Window* field. The maximum segment size that the responder is willing to receive usually is included in the connection response (in the *Options* field). This can be different from the initiator's size—they do not have to be the same.

A connection request can be rejected by sending a response whose reset flag, RST, is equal to 1.

10.8.4   Choosing the Initial Sequence Number

During connection startup, the TCP specification states that each end of the connection picks an *initial sequence number* from some 32-bit internal clock. Why bother to do this?

---

[8] If no MSS is provided, the default is 536.

Imagine what happens after a system has crashed. Suppose that a user has opened a connection just before the crash and has sent a small amount of data. After it recovers, the system does not remember anything that it was doing before the crash—including the connections that were running and the port numbers that were assigned. Users will have to start their connections again. Port numbers will be handed out on a first-come first-served basis, and some of these may be ports that were in use for other connections just a few seconds earlier.

In the meantime, a system at the far end of a connection may be totally unaware that its partner has crashed and restarted. There could be a lot of confusion as old data that took a long time coming through the network gets intermingled with data from fresh connections. Hitching fresh starts to some clock value helps to avoid this problem. Old data is likely to be numbered with values that are outside of the new sequence number range.[9]

10.8.5 General Field Usage

To prepare a TCP header for transmission, the sequence number of the first octet of enclosed data is entered into the *Sequence Number* field.

The number of the next octet expected from the partner is filled into the *Acknowledgment Number* field, and the ACK bit is set to 1. The *Window* field contains the current size of the receive window, that is: *The number of bytes, starting from the acknowledgment number, that can be received.* Note that this provides extremely precise flow control. The partner is told the exact status of the receive window throughout the session.

If the application has signaled a push to TCP, the PUSH flag is set to 1. The receiving TCP is supposed to react to the PUSH flag by delivering the data to its application promptly when the application is willing to receive it.

An URGENT flag set to 1 indicates that urgent data is pending, and the Urgent Pointer points to the last octet of urgent data. Recall that a typical use of urgent data is to send a break or interrupt signal from a terminal.

Urgent data is sometimes referred to as *out-of-band* data. This term is misleading. Urgent data is sent within the TCP data stream. It is up to the local TCP implementation to provide some mechanism that warns an application that urgent data is waiting and lets an application examine urgent data before it has read all of the bytes that lead up to this data.

The RESET flag is set to 1 to abort a connection. It also can be set in a response to a segment that does not make sense for any current connection that TCP is managing.

The FIN flag is set to 1 in the messages that are used to close a connection.

---

[9]Crackers have accessed machines that used predictable initial sequence numbers by forging the source IP address of a trusted host in their messages. A cryptographic hash function based on an internal secret is a better way to pick an initial number.

10.8.6 Checksum

The IP checksum was applied only to the IP header. The checksum in the TCP header is applied to the entire TCP segment,[10] as well as to a pseudo header made up of information extracted from the IP header. The pseudo header shown in Figure 10.15 is similar to that used for the UDP checksum.

The TCP length is computed by adding the length of the TCP header to the length of the data. The TCP checksum is *required,* not optional as in UDP. The checksum for an incoming segment is computed and compared to the content of the checksum field in the TCP header. If the values do not match, the segment is discarded.

10.9 SAMPLE TCP SEGMENTS

The display in Figure 10.16 shows a Network General *Sniffer* analysis of a series of TCP segments. The first three segments set up a connection between a *telnet* client and *telnet* server. The last segment carries 12 bytes of data.

The *Sniffer* analyzer translates most numbers to decimal form. However, the flags values are reported in hex. Flags = 12 means flag pattern 010010. The checksum also is reported in hex.

10.10 KEEPING A SESSION ALIVE

10.10.1 Window Probes

An active sender and a sluggish receiver can lead to a 0-byte receive window. This is also called a *closed window.* When space opens up, an ACK will be sent updating the window size. But what if the ACK is lost? Both sides could wait forever.

---

[10]The checksum field in the TCP header is set to 0 during the checksum computation.

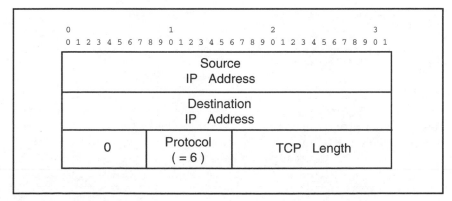

Figure 10.15 Pseudo header fields included in the TCP checksum.

```
1. SYN                              2. SYN
TCP:  --- TCP header ---            TCP:  --- TCP header ---
TCP:                                TCP:
TCP:  Source port = 2345            TCP:  Source port = 23 (Telnet)
TCP:  Dest port = 23 (Telnet)       TCP:  Dest port = 2345
TCP:  Initial sequence number =     TCP:  Initial sequence number =
      16421121                            390272001
                                    TCP:  Acknowledgment number =
                                          16421122
TCP:  Data offset = 24 bytes        TCP:  Data offset = 24 bytes
TCP:  Flags = 02                    TCP:  Flags = 12
TCP:  ..0. .... = (No urg ptr)      TCP:  ..0. .... = (No urg ptr)
TCP:  ...0 .... = (No ack)          TCP:  ...1 .... = Acknowledgment
TCP:  .... 0... = (No push)         TCP:  .... 0... = (No push)
TCP:  .... .0.. = (No reset)        TCP:  .... .0.. = (No reset)
TCP:  .... ..1. = SYN               TCP:  .... ..1. = SYN
TCP:  .... ...0 = (No FIN)          TCP:  .... ...0 = (No FIN)
TCP:  Window = 2048                 TCP:  Window = 4096
TCP:  Checksum = F2DA (correct)     TCP:  Checksum = C13A (correct)
TCP:                                TCP:
TCP:  Options follow                TCP:  Options follow
TCP:  Max segment size = 1460       TCP:  Max segment size = 1024

3. ACK                              4. DATA
TCP:  --- TCP header ---            TCP:  --- TCP header ---
TCP:                                TCP:
TCP:  Source port = 2345            TCP:  Source port = 23 (Telnet)
TCP:  Dest port = 23 (Telnet)       TCP:  Dest port = 2345
TCP:  Sequence number =             TCP:  Sequence number =
      16421122                            390272002
TCP:  Acknowledgment number =       TCP:  Acknowledgment number =
      390272002                           16421122
TCP:  Data offset = 20 bytes        TCP:  Data offset = 20 bytes
TCP:  Flags = 10                    TCP:  Flags = 18
TCP:  ..0. .... = (No urg ptr)      TCP:  ..0. .... = (No urg ptr)
TCP:  ...1 .... = Acknowledgment    TCP:  ...1 .... = Acknowledgment
TCP:  .... 0... = (No push)         TCP:  .... 1... = Push
TCP:  .... .0.. = (No reset)        TCP:  .... .0.. = (No reset)
TCP:  .... ..0. = (No SYN)          TCP:  .... ..0. = (No SYN)
TCP:  .... ...0 = (No FIN)          TCP:  .... ...0 = (No FIN)
TCP:  Window = 2048                 TCP:  Window = 4096
TCP:  Checksum = DF43 (correct)     TCP:  Checksum = 9FEF (correct)
TCP:  No TCP options                TCP:  No TCP options
                                    TCP:  [12 byte(s) of data]
```

Figure 10.16 "Sniffer" analyzer display of TCP headers.

To avoid this, the sender sets a *persist timer*[11] when the receive window closes. When the timer expires, a segment called a *window probe* is sent to the partner. (Some implementations include data in the probe.) The probe causes the partner to send back an ACK that reports the current window status.

If the window is still 0, the persist timer is doubled. This repeats until the timer value reaches a maximum of 60 seconds. TCP will continue to send

---

[11]Its value is set equal to the retransmission timeout.

probes every 60 seconds forever—or until the window opens or a user inter-rupts the process or a timeout imposed by the application expires.

10.11   KILLING A SESSION

10.11.1   Timing Out

A partner's system may crash, or the route to the partner may be totally dis-rupted by loss of a gateway or link. What prevents TCP from retransmitting the same data forever? There are several mechanisms.

After reaching a first threshold number of retransmissions, TCP advises IP to check whether a dead router is the problem and notifies the application that there is a problem. TCP proceeds to transmit until a second threshold is reached; then it breaks the connection.

Of course, before all this happens, an ICMP message may arrive stating that the destination was unreachable for some reason. In some implementa-tions, TCP will still keep trying to reach the destination until it times out. (After all, the problem might get fixed.) Then it reports the Destination Unreachable status to its application.

An application also can place its own time limit on data delivery, along with an action to be taken when the time expires. Typically, the action is to break the connection.

10.11.2   Keep-Alives

What happens to a connection when neither end has any data to send for a long time? The connection is maintained in an idle state. During the period that it is idle, the network could crash or wires could be cut and sewn togeth-er again. As long as the network is up when the partners start to exchange data again, they will not lose their session. This design suited the Department of Defense (DOD) sponsors very well.

But any connection—even an idle one—occupies quite a lot of memory in a computer. Some administrators would like to reclaim resources that are unused. Hence, many TCP implementations send *keep-alive* messages that test an idle connection. TCP sends keep-alives to the partner periodically, to check that it is still there. A keep-alive triggers an ACK in response. The use of keep-alives is an optional feature. If it is available at a system, an applica-tion can disable it for its own connections. The suggested *default* value of the timeout is 2 hours!

Recall that applications can set their own application-layer timers at levels that make sense for the application and can abort sessions that are idle.

10.12   PERFORMANCE

How well can TCP perform? There are many factors that affect performance. Most basic are resources such as memory and bandwidth. Figure 10.17 sum-marizes the elements that have an impact on performance.

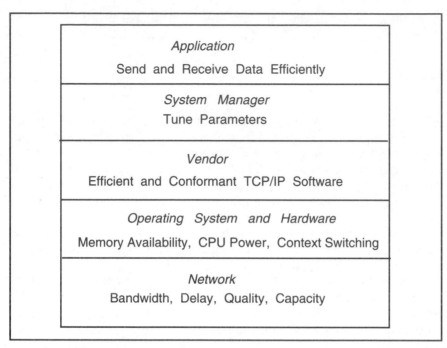

Figure 10.17 TCP performance factors

The bandwidth and delay of the underlying network impose limits on throughput. Poor transmission quality causes lots of discarded datagrams. Discards provoke retransmissions, with the result that effective bandwidth is cut.

A receiver that provides lots of input buffer space allows a sender to keep transmitting without pause. This is especially important for networks with large delays, where a long time elapses between sending data and receiving an ACK and window update. To support a steady flow of data from a source, the destination needs a receive window whose size is at least (bandwidth × delay).

For example, if you can send data at 10,000 bytes per second, and it takes 2 seconds for an ACK to arrive, the receiver must provide a receive buffer of at least 20,000 bytes in order to maintain a steady flow of data. A receive buffer that could hold only 10,000 bytes would cut throughput in half.

Another factor that has an impact on performance is a host's ability to react to high-priority events and rapidly *switch context*—that is, stop doing one thing and take care of another. A host may be supporting many local interactive users, background batch processes, and dozens of communications connections. Switching context to take care of communications housekeeping is a hidden overhead. An implementation that integrates TCP/IP with the operating system kernel can cut back significantly on context switching overhead.

Capable CPU resources are needed to quickly carry out the steps required to process the TCP header. A CPU that cannot compute checksum values rapidly can slow down data transmission.

Finally, vendors should make it easy to set configuration parameters so that network managers can tune TCP to local conditions. For example, the ability to match buffer sizes to bandwidth and delay can improve throughput substantially. Unfortunately, there are many implementations that hard code fixed configuration choices into the software.

Now suppose that your environment is perfect—lots of resources and the host switches contexts faster than Billy the Kid ever drew his gun. Will you get wonderful performance?

Maybe. The quality of the TCP software is important. Over the many years of TCP experience, many performance problems have been diagnosed and solved. Software that conforms to RFC 1122, which defines requirements for the communications layers in Internet hosts, encompasses these solutions.

Silly Window Syndrome (SWS) avoidance and the algorithms of Jacobson, Karn, and Partridge are especially important. These important algorithms are discussed in the sections that follow.

And finally, software developers can make a big difference by writing programs that do not perform unnecessarily small sends and receives and have built-in timeouts that free network resources when no useful work is being accomplished.

10.13 PERFORMANCE ALGORITHMS

Now we come to the most complex part of TCP. Many mechanisms that improve performance or solve bottleneck problems have been discovered. We'll introduce several of them in this chapter.

- The *slow start* prevents a new session from piling large amounts of traffic onto a network that might already be congested.

- Curing the *Silly Window Syndrome* prevents poorly designed applications from loading the network with overhead messages.

- The *delayed ACK* cuts down overhead by reducing the number of stand-alone acknowledgment messages.

- *Computing retransmission timeouts* based on a session's real round-trip times reduces unnecessary retransmissions but does not delay too long when a retransmission really is needed.

- Throttling back TCP transmissions when the network is *congested* allows routers to get back to normal and provides a fair share of the network resources to all of its sessions.

- Sending *duplicate ACKs* when out-of-order segments arrive enables the partner to retransmit before the timeout expires.

10.13.1 Slow Start

What would happen if you turned on every appliance in your house at the same time? This is not very different from what happens every morning when everyone arrives at the office and races for their electronic mail. However, the *slow start* prevents the network from blowing its fuse.

A new connection that immediately starts to transfer bulk data across a congested network could cause a severe problem. The idea behind slow start is that a new connection starts out carefully, increasing its rate of transfer gradually according to network conditions. Here is how it is done.

At connection setup, the partner announces its receive window. A second *congestion window* size is calculated for the purpose of throttling transmissions to match network conditions. The sender is limited by the congestion window rather than the (larger) receive window.

The congestion window starts off at one segment. For each segment that is ACKed successfully, the congestion window is increased by one segment—as long as it still is smaller than the receive window. If the network is not overloaded, the congestion window soon will reach the size of the receive window. During normal steady state transmission, the size of the congestion window is identical to the size of the receive window.

Note that Slow Start is not really slow. After one ACK, the congestion window equals two segments. If ACKs for two segments arrive successfully, the window will increase to four segments. If these are ACKed successfully, the window could increase to eight segments. In other words, the window grows to its natural size exponentially.

Suppose a timeout occurs instead of an ACK. We'll discuss what TCP does after a timeout a little later.

10.13.2 Silly Window Syndrome

In the earliest implementations of TCP/IP, a phenomenon called *Silly Window Syndrome* (SWS) was observed to happen fairly frequently. To understand SWS, let's look at a worst-case (and, one hopes, unlikely) scenario:

1. A sending application is transmitting data quickly.

2. The receive application is reading 1 byte of data at a time out of the receive buffer—slowly.

3. The receive buffer fills up.

4. The receive application reads 1 byte, and TCP sends an ACK saying "I have room for 1 byte of data."

5. The sending TCP packages 1 byte and transmits it.

6. The receive TCP sends an ACK saying "Thanks. I got it, and I have no more room."

7. The receive application reads 1 byte, sends an ACK, and the process repeats.

The slow receive application already has plenty of data waiting for it, and hustling those bytes into the right edge of the window performs no useful function—but it does add a lot of unnecessary traffic to the network.

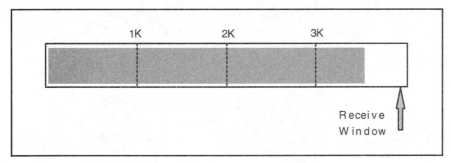

Figure 10.18 Buffer with a very small receive window.

The situation does not have to be this extreme to cause trouble. A fast sender, a slow receive application that reads chunks of data that are small relative to the maximum segment size, and an almost full receive buffer can trigger it. Figure 10.18 shows a buffer whose condition could trigger Silly Window Syndrome.

The solution is simple. Once the receive window has shrunk to less than a given target size, TCP starts to lie. TCP must not tell the sending partner about *additional* amounts of window space that open up when the receive application performs small reads. Instead, TCP must keep the extra resources a secret until the target amount of buffer space is available. The recommended amount is one segment—except for the case where the whole receive buffer holds only one segment, in which case one-half of the buffer will do. The target size at which TCP tells the truth is:

$$minimum \left(\frac{1}{2} \ Receive \ Buffer \ Size, \ Maximum \ Segment \ Size \right)$$

TCP starts to lie when the window size is less than this amount and tells the truth when the window size is at least this amount. Notice that no harm is being done by holding back the sender, since the receive application has not absorbed most of the data that already is waiting for it.

The solution is easiest to understand for the byte-at-a-time worst-case scenario above. We'll assume that the receive buffer can hold several segments, which usually is the case. The fast sender will fill up the receive buffer, and then the sender will be told that no room is available until enough space opens up to receive a complete segment.

10.13.3 Nagle Algorithm

The sender should independently avoid transmitting short segments by accumulating data for a while before dispatching it. The Nagle algorithm introduces a very simple idea that reduces the number of tiny datagrams presented to a network.

Nagle recommends holding back transmission (even of pushed data) if the sender is waiting for an ACK of some previously transferred data.

Accumulated data will be sent if the previous data is ACKed, if a full-sized segment can be sent, or if a preset timeout expires. The algorithm can be disabled for real-time applications which must send data as soon as possible.

10.13.4 Delayed ACK

The delayed ACK is another mechanism that can improve performance. Cutting down on the number of ACKs that are sent saves bandwidth that can be used for other traffic. If the partner TCP delays a little before sending an ACK, there is a chance that:

- The partner can acknowledge multiple segments with a single ACK message.

- The partner's application may get some data to send within the timeout period, in which case the ACK will be contained in the outgoing header and a separate message won't be needed.

To avoid delaying a stream of full-sized segments (e.g., while performing a file transfer), an ACK should be sent for at least every second full segment.

Many implementations use a 200-millisecond timeout. But note that delaying the ACK will not slow down transmission. If tiny segments are arriving, there will be plenty of buffer space, and the transmitter can keep sending (although retransmissions will be slowed). If full segments are arriving, every second one will immediately enable an ACK.

10.13.5 Retransmission Timeout

After sending a segment, TCP sets a timer and listens for an ACK. If the ACK does not arrive within the timeout period, TCP retransmits the segment. But how long should the timeout be?

If the retransmission timeout is too short, the sender will clutter the network with unnecessary segments and burden the receiver with extraneous duplicates. But timeouts that are too long prevent brisk recovery when a segment really has been destroyed and will decrease throughput.

How do you choose an ideal timeout? A value that works well on a high-speed Local Area Network (LAN) would be disastrous for a multihop long-distance connection, so it is clear that "one size fits all" will not work here. Moreover, even during a single connection, network conditions may change and delays may increase or decrease.

Algorithms of Jacobson, Karn, and Partridge[12] enable TCP to adapt to changing conditions and are now mandated for TCP implementations. These algorithms are sketched below.

Common sense tells us that the best basis for estimating a good retransmission timeout for a particular connection is to keep a watch on the connec-

---

[12]See *Congestion Avoidance and Control,* Van Jacobson, and *Improving Round-Trip Time Estimates in Reliable Transport Protocols,* Karn and Partridge.

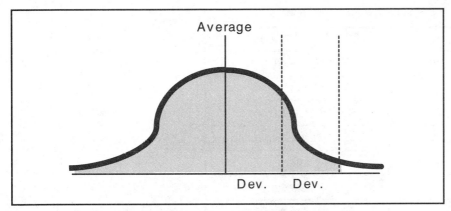

Figure 10.19   Looking at a distribution of round-trip times.

tion, recording the *round-trip times* that elapse between the transmission of data and the arrival the of acknowledgment of that data.

Elementary statistics can suggest a good idea of what to do next (see Figure 10.19.) Obviously, it makes sense to compute the average round-trip time. But the average would be a bad choice for a timeout, because half of the round-trip values can be expected to be bigger than the average. However, if we add on a couple of deviations, we get an estimate that allows for normal variability without causing retransmission waits that are too long.

It is not necessary to perform the heavy-duty calculation involved in finding the formal statistical standard deviation. A simple rough estimate will do—use the absolute value of the difference between the latest value and the average:

$$\text{Latest deviation} = |\text{latest round trip} - \text{average}|$$

Another factor needs to be considered in order to calculate a good retransmission timeout. That is the fact that round-trip times will change depending on network conditions. What has happened in the last minute is much more important than what happened an hour ago.

For example, suppose that we kept a running average of round-trip times for a very long session. We'll assume that at the start the network was lightly loaded and there were 1000 small values, but then traffic built up and there was a significant increase in delays.

For example, if 1000 values with average 170 were followed by 50 values with average 282, the current average would be

$$170 \times 1000/1050 + 282 \times 50/1050 = 175$$

A more responsive measure, the *Smoothed Round-Trip Time* (SRTT), puts a heavier weight on recent values:

$$\text{New SRTT} = (1 - \alpha) \times (\text{old SRTT}) + \alpha \times \text{Latest Round-Trip Time}$$

TABLE 10.2   Computing the Smoothed Round Trip Time

Old SRTT	Latest RTT	(7/8)(old SRTT) + (1/8)(RTT)
230.00	294	238.00
238.00	264	241.25
241.25	340	253.59
253.59	246	252.64
252.64	201	246.19
246.19	340	257.92
257.92	272	259.68
259.68	311	266.10
266.10	282	268.09
268.09	246	265.33
265.33	304	270.16
270.16	308	274.89
274.89	230	269.28
269.28	328	276.62
276.62	266	275.29
275.29	257	273.00
273.00	305	277.00

The value of α lies between 0 and 1. *Increasing* α causes the current round-trip time to have a greater effect on the smoothed average. Since computers can divide by powers of 2 very quickly by shifting binary numbers to the right, a value of $(1/2)^n$ is always chosen for α, typically 1/8, so that we have:

$$\text{New SRTT} = \frac{7}{8} \times \text{old SRTT} + \frac{1}{8} \times \text{latest Round-Trip Time}$$

Table 10.2 shows how quickly the SRTT formula adjusts if the current SRTT is 230, and a change in network conditions causes a sequence of longer Round-Trip Times.[13] The value calculated in column 3 is fed back into the next line of column 1 as the old SRTT.

Now what value should be chosen for the retransmission timeout? A look at the sample Round-Trip Times shows that there are fairly large deviations

---

[13] We're assuming that no timeouts occur.

between individual times and the current average. It makes sense to allow a good-sized margin for deviations. If we can come up with a reasonable estimate for a smoothed deviation (SDEV), the formula below would provide a good retransmission timeout[14] value:

$$T = \text{RETRANSMISSION TIMEOUT} = \text{SRTT} + 2 \times \text{SDEV}$$

This is the formula referenced in RFC 1122. However, some implementations use:

$$T = \text{SRTT} + 4 \times \text{SDEV}$$

To compute SDEV, first calculate the absolute value of the current deviation:

$$\text{DEV} = |\text{ latest Round-Trip Time} - \text{old SRTT}|$$

Next, use a smoothing formula to fold in this latest value:

$$\text{New SDEV} = \frac{3}{4} \times \text{old SDEV} + \frac{1}{4} \times \text{DEV}$$

One question remains—how do we get started? Set:

Initial TIMEOUT = 3 seconds

Initial SRTT = 0

Initial SDEV = 1.5 seconds

Van Jacobson has defined a fast algorithm that computes the retransmission timeout very efficiently.

10.13.6　Sample Statistics

How well does this timeout work? Quite remarkable performance improvements were observed when this timeout was implemented. We can get a good indication of a worst-case level of retransmission overhead by looking at some simple *netstat* statistics from *tigger*. *Tigger* is an Internet server, and it communicates with hosts all over the world:

```
tcp:
1301644 packets sent
        879137 data packets (226966295 bytes)
        21815 data packets (8100927 bytes) retransmitted

2012869 packets received
        762469 acks (for 226904227 bytes)
        35803 duplicate acks
        0 acks for unsent data
        1510769 packets (314955304 bytes) received in-sequence
        9006 completely duplicate packets (867042 bytes)
        74 packets with some dup. data (12193 bytes duped)
        13452 out-of-order packets (2515087 bytes)
```

---

[14]Standards documents refer to this as the RTO (for Retransmission TimeOut).

Less than 2.5 percent of the TCP data segments were retransmitted by *tigger*. For 1.5 million incoming data segments (the rest were pure ACKs), 0.6 percent were duplicates. It seems safe to assume that incoming percentages of data loss will be about the same as outgoing percentages. Hence "unnecessary" retransmissions made up about 0.6 percent of the traffic.

10.13.7 Calculations after a Retransmission

The Round-Trip Time used in the calculations above is the time between transmission of a segment and receipt of its acknowledgment. But suppose that the acknowledgment does not arrive within the timeout period and the data must be retransmitted?

Karn observed that the Round-Trip Time cannot be updated when this happens. The current Smoothed Round-Trip Time and *Smoothed Deviation* have to be saved until some segment is acknowledged without an intervening retransmission. At that point, the calculations resume, using the saved values and the new measurements.

10.13.8 Actions after a Retransmission

But what should be done in the meantime? In fact, TCP's behavior should change quite drastically after a retransmission. The predominant cause of data loss is congestion. Hence, the reaction to a retransmission should be to:

- Slow down retransmissions
- Fight congestion by throttling back on the total traffic burden for the network

10.13.9 Exponential Backoff

After a retransmission, the timeout is doubled. What happens if the timer expires again? Data is retransmitted and the retransmission timeout is doubled again. This is called *exponential backoff*.

If failures continue, the timeout will continue to double until it latches at a prespecified maximum (e.g., 1 minute). Only one segment will be sent whenever the timeout expires. After a preconfigured number of transmissions without an ACK, the connection will time out.

10.13.10 Curing Congestion by Reducing Data on the Network

The reduction in data level is a little complicated. It begins just like the Slow Start described earlier. But as we approach the threshold level of traffic that initially got us into trouble, we will really slow down, gradually adding one segment at a time to the congestion window. We'll define a threshold value for slowing down. First compute the danger threshold:

$$Threshold = \frac{1}{2}\ minimum(current\ congestion\ window,\ partner's\ receive\ window)$$

If this is bigger than two segments, use it as the threshold. Otherwise, set the threshold to two segments. The complete recovery algorithm is:

- Initially, set the congestion window to one segment.

- For each ACK received, increase the congestion window by one segment until the threshold is reached. (This is like the usual Slow Start.)
- From now on, a small increment will be added to the congestion window each time an ACK arrives. The increment is chosen so that the rate of increase is at most one segment added per Round-Trip Time.[15]

The idealized scenario below gives a rough idea of how recovery proceeds. We assume that the partner's receive window (and our current congestion window) equaled eight segments when the timeout occurred, so the threshold is four segments. Also suppose that the receiving application reads data immediately, so the receive window stays at eight segments.

1. Send one segment. (Congestion window = one segment.)
2. ACK received, send two segments.
3. ACK for two segments received. Send four segments. (We have reached the threshold.)
4. ACK for four segments received. Send five segments.
5. ACK for five segments received. Send six segments.
6. ACK for six segments received. Send seven segments.
7. ACK for seven segments received. Send eight segments. (The congestion window now equals the receive window.)

As long as all data is ACKed within the retransmission timeout, this process will continue until the congestion window reaches the receive window size. Figure 10.20 illustrates what happens. The window size grows exponentially (doubling) during the initial "slow" start and then levels off to linear growth.

10.13.11 Duplicate ACKs

Some implementations support an optional feature called *fast retransmit*. The purpose is to speed up retransmission under suitable conditions. The basic idea is that the receiver will send duplicate ACKs that indicate that there is a "hole" in its data.

Whenever the receiver gets an out-of-order segment, it sends an ACK that identifies the first byte of the *missing* data (see Figure 10.21).

The sender does not retransmit right away, because IP might deliver data out of order under normal circumstances. However, after receiving a preset number of duplicate ACKs (e.g., three), the missing segment is retransmitted without waiting for a timeout.

Note that each duplicate ACK signals that the partner has just received a segment. Several duplicate ACKs indicate that the network is still delivering reasonable amounts of data, so we assume that it is not badly congested.[16]

---

[15]The increment is (MSS/N) where N is the size of the congestion window in segments.

[16]As part of this algorithm, the congestion window is reduced somewhat, in case there actually is some congestion. The drastic recovery process described in the previous section is not used.

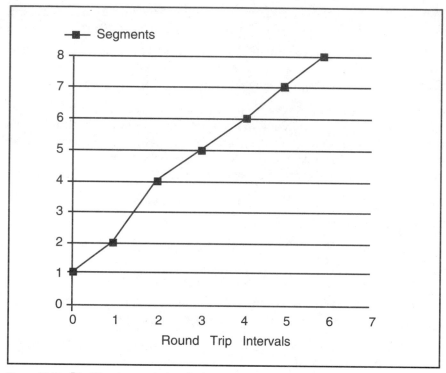

Figure 10.20 Limiting transmission during congestion.

10.13.12 What to Do after a Source Quench

According to the *Host Requirements* standard, TCP must perform the same slow start described in the previous section when a source quench arrives. However, source quench messages are not particularly well targeted or effective—the connection that receives the source quench may not be sending very much traffic. The current *Router Requirements* specification states that routers should *not* send source quench messages.

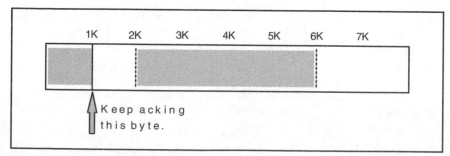

Figure 10.21 Duplicate ACKs.

10.13.13   TCP Statistics

Finally, let's look at a *netstat* statistics report to see many of these mechanisms at work.

```
tcp:
1301644 packets sent
879137 data packets (226966295 bytes)
21815 data packets (8100927 bytes)
  retransmitted
132957 ack-only packets (104216 delayed)
4 URG only packets
1038 window probe packets
248582 window update packets
22413 control packets
2012869 packets received
762469 acks (for 226904227 bytes)
35803 duplicate acks
0 acks for unsent data
1510769 packets (314955304 bytes)
  received in-sequence
9006 completely duplicate packets
  (867042 bytes)
74 packets with some dup. data
  (12193 bytes duped)
13452 out-of-order packets
  (2515087 bytes)
530 packets (8551 bytes) of data after
  window
526 window probes
14158 window update packets
402 packets received after close
108 discarded for bad checksums
0 discarded for bad header offset fields
7 discarded because packet too short
6378 connection requests
9539 connection accepts
14677 connections established
  (including accepts)
18929 connections closed
  (including 643 drops)
4100 embryonic connections dropped
572187 segments updated rtt
  (of 587397 attempts)

11014 retransmit timeouts
26 connections dropped by rexmit timeout

1048 persist timeouts
535 keepalive timeouts
472 connections dropped by keepalive
```

"Packets" are segments.

Retransmissions.

Note the large number of delayed ACKs.

Probes check if zero windows have opened.

These are SYNs and FINs.

These signal that segments arrived out of order.

These are the result of timeouts when the data really was delivered.

For efficiency, some retransmitted data was repackaged to contain additional bytes.bytes

Probably due to data included in probe messages.

These are late retransmissions.
Bad TCP checksums.

Failed to update the Round-Trip Time because an ACK did not arrive before the timeout expired.

Repeated failed retransmissions showed that the connection was gone.

Timeouts to probe a zero window.
Timeouts to check an idle connection.

10.14 VENDOR CONFORMANCE

The current TCP standard requires conformant implementations to adhere to Slow Start when initiating a connection and to use the algorithms of Karn and Jacobson to estimate retransmission timeouts and control congestion. Tests have shown that these mechanisms produce significant improvement in performance.

What happens when you install a system that does not adhere to these standards? Not only does it provide poorer performance for its own users, but it will be a bad neighbor for the other systems on the network, hampering recovery from temporary congestion and producing excessive traffic that causes datagram discards.

10.15 PERFORMANCE BARRIERS

TCP has proved to be very flexible, running over networks that carry hundreds or millions of bits per second. The protocol has achieved good results on modern Ethernet, Token-Ring, and Fiber Distributed Data Interface (FDDI) LANs, as well as low bandwidth links and links with long delays (such as satellite links).

TCP has been tuned so that it is responsive to abnormal conditions such as congestion. However, there are features in the current protocol that limit TCP throughput over emerging technologies that offer bandwidth in the hundreds and thousands of megabytes. To understand the problem, let's look at a simple (although unrealistic) example.

Suppose that we are performing a file transfer between two systems and want to send the data out in a steady stream as efficiently as possible. Suppose that:

- The receiver's maximum segment size is 1 kilobyte.
- The receive window is 4 kilobytes.
- *There is enough bandwidth to send two segments per second.*
- The receive application is absorbing the data as soon as it arrives.
- ACKs arrive after 2 seconds.

The sender will be able to send data steadily, because just as the window allocation becomes exhausted, an ACK arrives that allows another segment to be sent:

```
SEND SEGMENT 1.
SEND SEGMENT 2.
SEND SEGMENT 3.
SEND SEGMENT 4.
```

Two seconds have elapsed.

```
RECEIVE ACK OF SEGMENT 1, CAN SEND SEGMENT 5.
RECEIVE ACK OF SEGMENT 2, CAN SEND SEGMENT 6.
RECEIVE ACK OF SEGMENT 3, CAN SEND SEGMENT 7.
RECEIVE ACK OF SEGMENT 4, CAN SEND SEGMENT 8.
```

Two more seconds have elapsed.

```
RECEIVE ACK OF SEGMENT 5, CAN SEND SEGMENT 9.
. . .
```

If the receive window had been only 2 kilobytes, the sender would have been forced to wait 1 second out of every 2 before sending more data. In fact, to keep a steady flow going, the receive window must have a size that is at least:

$$\text{Window} = \text{bandwidth} \times \text{Round-Trip Time}$$

Although the example was exaggerated (in order to provide very simple numbers), a small window can cause trouble for high-delay satellite links.

Now let's see what happens for very high-bandwidth connections. For example, if the bandwidth and transmission rate are 10 million bits per second but the Round-Trip Time is 100 milliseconds (1/10 second), a steady-flow receive window would have to hold at least 1,000,000 bits, that is, 125,000 bytes. But the largest number that can be written in the TCP receive window header field is 65,536.

Another problem could arise at very high transmission rates because sequence numbers get used up very quickly. If we could transmit data at 4 gigabytes per second on a connection, the sequence numbers would wrap around in 1 second. It would not be possible to distinguish old duplicate datagrams that had been delayed for more than 1 second while crossing an internet from fresh new data.

There is active ongoing research into enhancing TCP/IP to remove these barriers.

10.16   TCP FUNCTIONS

TCP is a protocol of considerable size. This chapter has dealt with the many jobs that TCP has to perform. The list below summarizes TCP's functions:

- Associating ports with connections
- Establishing connections by means of a three-way handshake
- Performing a Slow Start to avoid overloading the network
- Segmenting data for transmission
- Numbering data
- Handling incoming duplicate segments
- Computing checksums
- Regulating the flow of data with receive and send windows
- Terminating connections in an orderly fashion
- Aborting connections
- Signaling urgent data

- Positive acknowledgment with retransmission
- Calculating retransmission timeouts
- Throttling back traffic when the network is congested
- Signaling that segments have arrived out of order
- Probing closed receive windows

10.17 TCP STATES

A TCP connection passes through a number of stages. First the connection is set up by means of an exchange of messages, then data is transmitted, and then it is closed by means of an exchange of messages. Each step in the progress of a connection corresponds to a connection *state*. The TCP software at each end of a connection keeps track of the state at its end of the connection at all times.

Below, we sketch a typical progression of states at the server and client ends of a connection. This is not intended to represent an exhaustive investigation of all possible state transitions. See RFC 793 and the *Host Requirements* document for a complete discussion of TCP states.

During connection setup, the sequence of states is slightly different at the server and client ends. The server states during setup are shown in Table 10.3. The client states during setup are shown in Table 10.4.

In the unlikely case that partners simultaneously try to initiate a connection with each other, each would pass through the states CLOSED, SYN-SENT, SYN-RECEIVED, and ESTABLISHED.

TABLE 10.3   Server State Transitions

Server state	Event	Description
CLOSED		A fictitious state prior to starting the connection.
	Passive Open by Server application.	
LISTEN		The server waits for a connection from a client.
	Server TCP receives SYN, sends SYN/ACK.	The server has received a SYN and has sent a SYN/ACK. It is waiting for an ACK.
SYN-RECEIVED		
	Server TCP receives ACK.	
ESTABLISHED		The ACK has been received and the connection is open.

TABLE 10.4   Client State Transitions

Client state	Event	Description
CLOSED		A fictitious state prior to starting the connection.
	Client application requests a connection. Client TCP sends SYN	
SYN-SENT		The client TCP has sent a SYN to the server.
	Client TCP receives SYN/ACK and sends ACK.	The client has received a SYN/ACK from the server, and has sent back an ACK.
ESTABLISHED		Data transfer can proceed.

The ends of a connection remain in ESTABLISHED state until one end initiates a *close* by sending a FIN segment. During a normal termination, the closer passes through the states shown in Table 10.5. The closer's partner passes through the states shown in Table 10.6.

10.17.1   Viewing the States of TCP Connections

The *netstat -an* command can be used to examine the current state of connections. The display below shows connections in *listen, startup, established, closing,* and *time-wait* states.

Note that the connection port number is tacked onto the end of each local and foreign address. We see that there is TCP traffic in both send and receive queues.

```
> netstat -an
Active Internet connections
Pro    Recv-Q  Send-Q  Local Address        Foreign Address       (state)
t
tcp    0       0       128.121.50.145.25    128.252.223.5.1526     SYN_RCVD
tcp    0       0       128.121.50.145.25    148.79.160.65.3368     ESTABLISHED
tcp    0       0       127.0.0.1.1339       127.0.0.1.111          TIME_WAIT
tcp    0       438     128.121.50.145.23    130.132.57.246.2219    ESTABLISHED
tcp    0       0       128.121.50.145.25    192.5.5.1.4022         TIME_WAIT
tcp    0       0       128.121.50.145.25    141.218.1.100.3968     TIME_WAIT
tcp    0       848     128.121.50.145.23    192.67.236.10.1050     ESTABLISHED
tcp    0       0       128.121.50.145.1082  128.121.50.141.6000    ESTABLISHED
tcp    0       0       128.121.50.145.1022  128.121.50.141.1017    ESTABLISHED
tcp    0       0       128.121.50.145.514   128.121.50.141.1020    CLOSE_WAIT
tcp    0       1152    128.121.50.145.119   192.67.239.23.3572     ESTABLISHED
tcp    0       0       128.121.50.145.1070  192.41.171.5.119       TIME_WAIT
tcp    579     4096    128.121.50.145.119   204.143.19.30.1884     ESTABLISHED
tcp    0       0       128.121.50.145.119   192.67.243.13.3704     ESTABLISHED
tcp    0       53      128.121.50.145.119   192.67.236.218.2018    FIN_WAIT_1
tcp    0       0       128.121.50.145.119   192.67.239.14.1545     ESTABLISHED
tcp    0       0       *.19                 *.*                    LISTEN
tcp    0       0       *.13                 *.*                    LISTEN
tcp    0       0       *.9                  *.*                    LISTEN
tcp    0       0       *.7                  *.*                    LISTEN
tcp    0       0       *.37                 *.*                    LISTEN
```

TABLE 10.5   Closer's State Transitions

Closer state	Event	Description
ESTABLISHED	Local application requests a close.	
	TCP sends FIN/ACK.	
FIN-WAIT-1		The closer is waiting for the partner's response. Recall that fresh data may still arrive from the partner at this stage.
	TCP receives ACK.	
FIN-WAIT-2		The closer is has received an ACK from the partner, but not a FIN. The closer waits for the FIN, accepting incoming data in the meantime.
	TCP receives FIN/ACK . Sends ACK.	
TIME-WAIT		The connection is held in limbo to allow duplicate data or a duplicate FIN that may still exist out in the network to arrive or be discarded. The time-wait period is twice the estimated maximum segment lifetime.
CLOSED		All information about the connection is deleted.

10.18   IMPLEMENTATION ISSUES

From the beginning, TCP was intended for multivendor internetworking. The TCP protocol specification does not try to nail down exactly how the internals of an implementation will work. It is left to vendors to find the mechanisms that suit their own environment best.

Even RFC 1122, the *Host Requirements* document, leaves ample room for variations in implementations. Each function is tagged with a suggested level of conformance:

- MUST
- SHOULD
- MAY
- SHOULD NOT
- MUST NOT

TABLE 10.6   Partner's Closing State Transitions

Closer's partner's state	Event	Description
ESTABLISHED	TCP receives FIN/ACK.	
CLOSE-WAIT		A FIN has arrived.
	TCP sends ACK.	
		TCP waits for its application to issue a close. The application may optionally send more data.
	Local application issues close.	
	TCP sends FIN/ACK.	
LAST-ACK		TCP is waiting for the final ACK.
	TCP receives ACK.	
CLOSED		All information about the connection is deleted.

Unfortunately, occasionally there will be a product that does not implement a MUST. Its users suffer from degraded performance as a result.

Some good implementation practices are not mandated in the standards. For example, to improve security, it is a good idea to restrict the use of well-known ports to privileged system processes, if the local operating system supports this notion. To improve performance, implementations should copy and move sent or retrieved data as little as possible.

A standard application programming interface was *not* defined, as a matter of policy. The intention was to leave the field free for experimentation with different toolkits. However, this also could have led to different programming interfaces on every platform and to no portability of application software across platforms.

In fact, vendors chose to base their toolkits on the Berkeley Socket programming interface. The value of a standard programming interface was proven when the WINSock (Windows Socket) programming interface was introduced. This led to a proliferation of new desktop applications that could run on top of any WINSock conformant TCP/IP stack.

10.19   RECOMMENDED READING

The original TCP standard is defined in RFC 793. Updates, corrections, and conformance requirements are specified in RFC 1122. Karn and Partridge published their article, *Improving Round-Trip Estimates in Reliable Transport Protocols,* in the *Proceedings of the ACM SIGCOMM 1987.* The Jacobson article, *Congestion Avoidance and Control,* appeared in the *Proceedings of the ACM SIGCOMM 1988 Workshop.* Jacobson has published several RFCs refining performance algorithms.

11

Configuration with BOOTP and DHCP

11.1 INTRODUCTION

One of the most remarkable changes in the use of computers in recent years has been the spread of TCP/IP network connectivity to desktops across an entire enterprise. The infrastructure needed to support the growing network—routers, bridges, switches, and hubs—has been expanding at a similar rate.

Support staffs have struggled to keep up with the demand for connectivity and the frequent moves, changes, and network reconfigurations that characterize today's volatile environment. These circumstances created a need for a mechanism that could automate node configuration and distribute operating system and network software. The most effective way to do this is to store configuration parameters and software images at one or more network *boot* servers. At start-up time, systems interact with a boot server, get start-up parameters, and optionally retrieve an appropriate software download.

In this chapter, we will describe two protocols. The initial Bootstrap Protocol, *BOOTP,* assigned IP addresses from a table of physical addresses and matching IP addresses. An administrator had to create the table at the BOOTP server manually. BOOTP's updated version, the *Dynamic Host Configuration Protocol* (DHCP), can automate IP address assignment fully and has several other useful enhancements.

11.2 BOOTSTRAP PROTOCOL REQUIREMENTS

Some computers require only a handful of configuration variables in order to get started. For others, it may be convenient to provide long, detailed lists of parameter values. Occasionally, desktop stations, Unix hosts, or other operating systems need complete operating system software downloads. Other systems, such as routers, bridges, switches, or even hubs may need boot configuration information and software downloads.

An initialization protocol must be robust and flexible. Depending on network size, topology, and availability requirements, it might be most convenient to centralize boot information at a single server, disperse it across several servers, or replicate it.

11.3 BOOTP CAPABILITIES

BOOTP was the first standard for automatic booting in a TCP/IP environment. After the protocol had undergone several rounds of extensions, BOOTP was able to provide systems with all of their basic configuration parameters—as well as quite a few specialized parameters. BOOTP also enabled a system to find out where it could obtain an appropriate software download.

Configuring a desktop client to use BOOTP or DHCP is very simple. Figure 11.1 shows how BOOTP or DHCP can be selected from a *Chameleon* setup menu. The pop-up box enables the user to enter the address of a BOOTP server, if it is known. If no address is entered, boot requests will be broadcast.

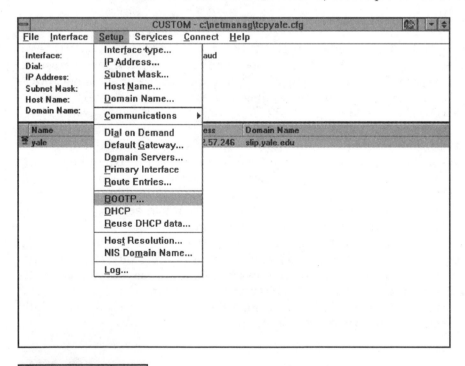

Figure 11.1 Configuring BOOTP at a desktop client.

11.4   THE NEED FOR DHCP

Field use of BOOTP made administrators aware that some more features were needed. Administrators wanted *automated* IP address configuration so that they would not have to type in (and maintain) long lists of hardware addresses and matching IP addresses. They also wanted *foolproof* IP address configuration so that if a user unplugged a system, moved it onto a desk in a different building, and plugged it into the network, the system would receive valid configuration data and could be run immediately.

Dynamic Host Configuration Protocol extended BOOTP to have these features, as well as many other capabilities. DHCP also cleaned up a number of ambiguities in BOOTP that sometimes led to less than optimal interworking.

11.5   INITIAL VERSION OF BOOTP

BOOTP was originally designed with diskless workstations in mind. The modest goal was to enable automatic booting even if the system started out with nothing more than some basic IP, UDP, and trivial file transfer code in Read-Only Memory. Hence, the original scenario (shown in Figure 11.2) was:

- The client broadcasts a request for information in a UDP message.

- The server returns the client's IP address, and optionally, the location of a file to be downloaded.

- The client uses the *Trivial File Transfer Protocol*[1] (TFTP) to download software into its memory and executes the software.

Administrators quickly realized that it would be beneficial to use BOOTP to get more configuration data and to configure systems that had disks and did not require the software download step.

---

[1] See Chapter 14.

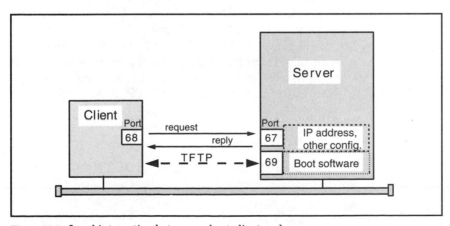

Figure 11.2   Local interaction between a boot client and server.

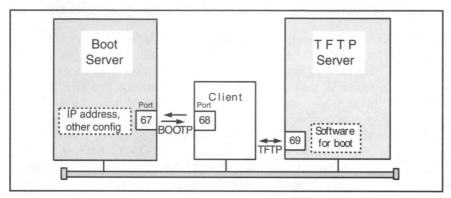

Figure 11.3   Using separate servers for parameters and software.

For systems that *did* need TFTP software downloads, it often was convenient to use one server for BOOTP parameters and one or more separate servers for downloading software (as shown in Figure 11.3). For example, operating system software might best be obtained from a server with the same type of operating system as the client.

11.6   EVOLUTION OF BOOTP

BOOTP became very flexible:

- A client may have absolutely no information to start with, or the client may already have been partially configured.
- The client may be willing to accept information from any boot server, or it may wish to choose one specific server.
- The client may wish no software download, a default download, or a specific download file.

It did not take long for additional parameters—such as the subnet mask, addresses of default routers, addresses of Domain Name Servers, as well as other information—to be included in BOOTP messages.

The parameter list grew and grew.[2] Eventually, so many parameters were defined that not even a fraction of them could fit into a reasonably sized UDP message. To solve this problem, overflow values were entered into a configuration file that could be downloaded to the client via TFTP. A parameter that identified this file was added to the list.

11.7   BOOTP PROTOCOL

Now let's take a look at the Bootstrap Protocol. BOOTP is a simple request/reply UDP application.

---

[2]The list of BOOTP parameters defined at the time of this writing may be found in Table 11.1, at the end of this chapter.

- The client transmits a *bootrequest* message from client port 68 to server port 67.

- The server responds with a *bootreply* message to client port 68.

Since UDP is unreliable, the client will retransmit if no reply is received within a timeout period.

11.7.1 BOOTP Message Format

The same overall message format is used for *bootrequest* and *bootreply* messages. Some fields will be zeroed out in the request. Figure 11.4 shows the message format.

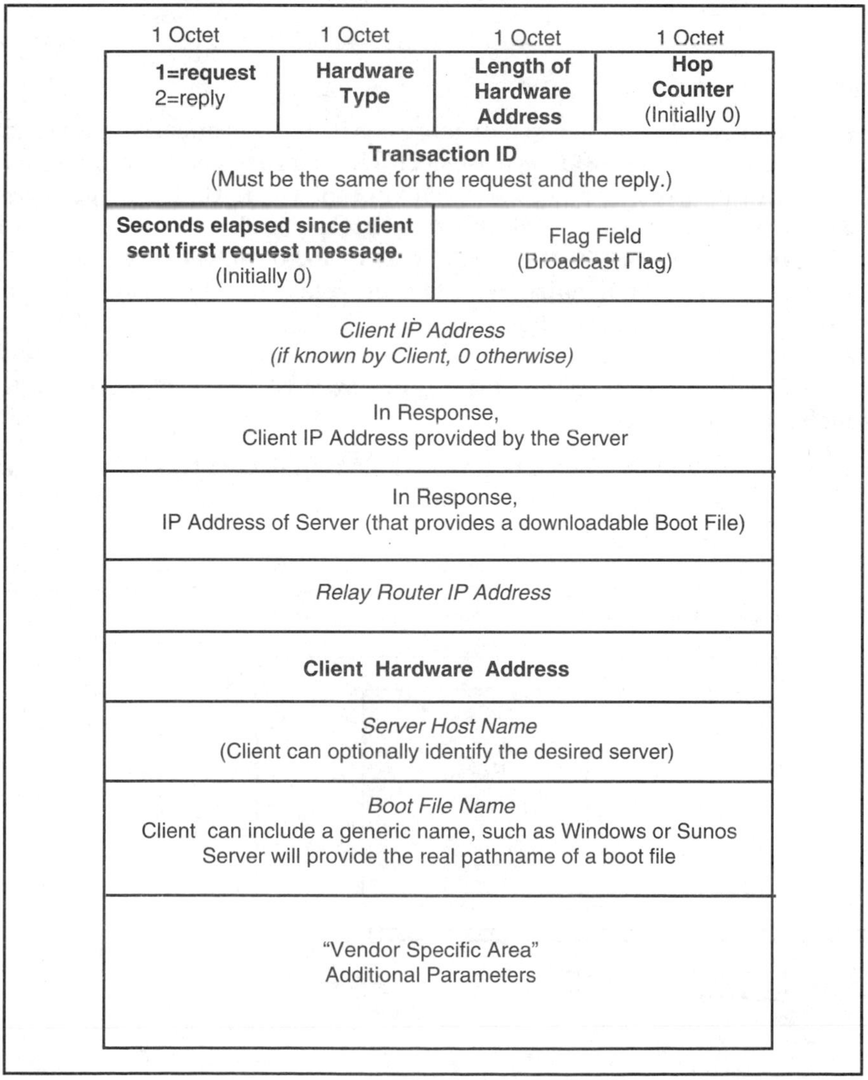

1 Octet	1 Octet	1 Octet	1 Octet
1=request 2=reply	**Hardware Type**	**Length of Hardware Address**	**Hop Counter** (Initially 0)
Transaction ID (Must be the same for the request and the reply.)			
Seconds elapsed since client sent first request message. (Initially 0)		Flag Field (Broadcast Flag)	
Client IP Address *(if known by Client, 0 otherwise)*			
In Response, Client IP Address provided by the Server			
In Response, IP Address of Server (that provides a downloadable Boot File)			
Relay Router IP Address			
Client Hardware Address			
Server Host Name (Client can optionally identify the desired server)			
Boot File Name Client can include a generic name, such as Windows or Sunos Server will provide the real pathname of a boot file			
"Vendor Specific Area" Additional Parameters			

Figure 11.4 *Bootrequest* and *bootreply* message format.

Fields that must be filled in for a *bootrequest* are shown in boldface. Fields that the client may *optionally* specify in a request are shown in italics. For detailed encoding of messages and their parameters, see the documents listed in Section 11.13.

11.7.2   Delivering a Client Request to a Server

A client that has no prior information will send a request using an IP source address of 0.0.0.0 and a destination address of 255.255.255.255.

A server (or servers) on the client's Local Area Network (LAN) will hear the request. If the client has filled in a *Server Host Name,* only that server will respond, as shown in Figure 11.5. If no name is included, multiple servers might respond.

11.7.3   Using Relay Agents

It may be more convenient to use one or more centralized boot servers instead of placing one on each LAN. But if a BOOTP client broadcasts a request on its LAN, how can it reach a remote server? The answer is that a helper system will relay the client's request (as shown in Figure 11.6).

A *relay agent* is a helper system that relays local BOOTP requests to remote servers. Routers normally are used as relay agents (although the specification allows for the use of hosts).

Typically, a router will be configured with the IP address(es) of one or more boot servers to whom requests should be forwarded.[3] When a relay agent receives a client request, it:

- Checks the *Relay Router* field in the BOOTP request. If this is zero, the relay agent inserts the IP address of the interface on which it received the

---

[3]This is the best implementation, although the specification also allows a router to broadcast the request onto selected links in order to search for a boot server when its IP address is unknown. The *hops* field is included to prevent endless looping in this case.

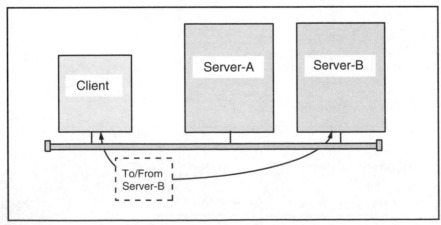

Figure 11.5 Selecting a specific server.

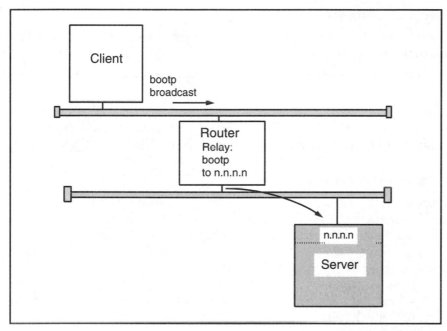

Figure 11.6   Relaying boot requests to an external server.

request. The BOOTP server will use this address to send the BOOTP server's reply back toward the client via the relay agent.

- The agent then sends the client request to one or more preconfigured server addresses.

11.7.4   Assigning an IP Address

An administrator configures a BOOTP server to assign IP addresses to systems by manually creating a table that maps client hardware type and hardware address combinations to IP addresses. Codes for hardware types are published in the *Assigned Numbers* document. For example, 1 = Ethernet. A tabulation could have a form such as:

Hardware type	Hardware address	IP address
1	02 60 8C 12 14 AA	128.121.2.5
1	08 00 20 D3 20 14	128.121.2.19

Most implementations include another column identifying host names in order to improve the readability of the table.

A simple scenario for a client that does not know its IP address is:

- The client broadcasts a request (directed to server port 67).

- A server receives the request.

- The server uses the client's hardware type and hardware address as a key and looks up the IP address in the table.

- If the client is local, the server broadcasts the response (directed to client port 68).

- If the client is remote, the response is sent to port 67 at the IP address that has been filled into the *Relay Router IP Address* field by the relay agent. The relay agent then broadcasts the response onto the client's LAN.

11.7.5   Booting Clients That Have IP Addresses

But suppose that a boot client has been preconfigured with an IP address or has stored an IP address that it was given in an earlier boot process. In this case the client will place that address into the *Client IP Address* field in the request.

According to the original BOOTP standard, the BOOTP server would allow the client to keep its old address and would use that address in the IP header that delivers the response to the client. Note that if, since its last boot, the computer has been moved to a new building and has been plugged into a completely different subnet or network, the client will not receive the response, which has been sent to its old subnet.

As we shall see, DHCP provides a very solid solution to this problem.

11.7.6   Configuring Software Downloads

Recall that, initially, the BOOTP model took the simple view that both the configuration data and the TFTP software download would be obtained from the same server. However, it is easy to separate the configuration service from the download service. The BOOTP configuration server simply can return the IP address of the TFTP server host, along with the pathname of the file to be downloaded.

But how are the TFTP server and download file selected by the BOOTP server? Software downloads might be dispersed across many TFTP servers.

The BOOTP server can be configured with a table that maps system nicknames to the IP address of the TFTP server system that has the download file, and the pathname for that file. For example:

Nickname	TFTP server IP address	File pathname
Sunos	128.121.50.2	/bin/vmunix

A BOOTP client sends an appropriate nickname to the server in the *Boot File Name* field. (DHCP provides a separate *Class Identifier* option field for this purpose instead.) The server looks up the nickname in the table, places the complete file pathname into the *Boot File Name* field, and writes the TFTP server's IP address into the *Server IP Address* field in the message.

A client sends a zeroed *Boot File Name* field if the client does not need a boot file or can accept preconfigured default values.

11.7.7 "Vendor Specific Area"

The *Vendor Specific Area* originally was included in the message to carry miscellaneous information that system vendors might want to enclose. However, early in the use of BOOTP, it became clear that a lot of useful configuration data—such as the subnet mask and address of a default router—had been left out of the formal part of the message. The space in the *Vendor Specific Area* was used for the additional configuration parameters as well as optional vendor information. Many parameter subfields have been defined for this area.

11.7.8 Sending the Response to an Addressless Client

If the client has not filled in a preexisting IP address, an address will be assigned by the server. The simplest way to deliver the response back to a client is to:

- Use an IP header with the newly assigned IP address as the destination address
- Wrap the datagram in a frame addressed to the client's physical address

But some older clients are implemented so that they cannot receive an IP datagram with an explicit IP address unless they previously have been configured with that address. This is called the "chicken and egg" problem.

However, such a client will be able to receive a datagram with destination port 68 and IP broadcast address 255.255.255.255. Newer BOOTP clients signal their preference to receive a response via the broadcast IP address by setting the *broadcast flag* (located in the flag field) equal to 1 in their requests.

11.7.9 Seconds Elapsed

When a client sends its initial request for boot data, the *Seconds Elapsed* field is set to 0. If no response is received, the client times out, updates the seconds elapsed field, and tries again. The client uses randomized timeouts that increase until the average backoff reaches 60 seconds.

This field has no specific use. It could be examined by servers or network monitors that wish to find out how long a client has been attempting to boot. A server could use the information to prioritize requests, but currently, real implementations ignore this field.

11.8 DHCP FEATURES

The Dynamic Host Configuration Protocol extends BOOTP's capabilities significantly. The most important enhancements are:

- Easier administration
- Automated configuration
- Support for moves and changes
- Ability of the client to ask for specific parameter values
- New DHCP message types that support robust client/server interactions

Another very important feature is that a BOOTP client can access a DHCP server. The standard is backward-compatible.

11.8.1 Administration and Automated Configuration

DHCP is capable of greatly reducing the administrative effort required to configure systems. If desired, an administrator can simply identify a block of IP addresses that the DHCP server can allocate to LAN clients. It is an easy job for an administrator to enter other critical parameters such as the subnet mask, Domain Name Server addresses, and address of the default router for the LAN. If needed, the administrator also can enter additional system-specific parameters.

In this case, a DHCP client that requests boot information from the server is automatically allocated an IP address from the block and is sent an appropriate set of parameters.

DHCP inherits BOOTP's ability to provide detailed and specific configuration data and to identify a software image to be downloaded. A shortcoming of BOOTP was that a client had no control over which parameters it got. DHCP enables clients to request specific parameters.

11.8.2 Moves and Changes

What happens when a user unplugs a computer, moves it to another building, and plugs it into a new subnet or network? During the boot process, a computer that uses DHCP will automatically change its IP address and update its subnet mask, default router, and Domain Name Systems (DNSs) if needed.

With DHCP, manual configuration can become a thing of the past.

11.9 DHCP MECHANISMS

11.9.1 Assigning IP Addresses

There are, in fact, three types of address allocation supported by DHCP:

- *Manual allocation.*   An IP address that has been manually entered at the server is permanently assigned to the client.
- *Automatic allocation.*   An IP address is selected from the server's available pool and is permanently assigned to the client.
- *Dynamic allocation.*   An IP address is assigned to a client for a limited amount of time—or until it is given up by the client.

For example, users who carry laptop computers and sporadically attach them to an office LAN may not need permanent, or even long-term, addresses.

11.9.2 Leases

The way that the allocation process works is that a client asks for the use of an address for a period of time (which might be "forever"). The server grants the client a *lease,* which identifies the period of time that the client may hold the address. The client can periodically renew the lease or let it expire. Expired IP addresses can be reused.

In order to renew a lease, a client has to identify the lease. When initially setting up a lease, the client can provide a *DHCP Client Identifier* value that will be used to name the client's lease in all future negotiations. Otherwise, the lease is indexed by the client's hardware type, hardware address, and assigned IP address.

11.9.3 Binding

A DHCP server stores the mappings between clients and their configuration parameters. A *binding* consists of an IP address and the set of other parameters that are assigned to a particular client.

11.10   COMPATIBILITY AND DIFFERENCES

In order to maintain compatibility with BOOTP, the DHCP message format is identical to the BOOTP format. As a result:

- BOOTP clients can use DHCP servers.
- DHCP clients can make use of BOOTP's relay service.

The simplest change is that the *Vendor Specific Area* has been renamed—appropriately—the *Options* field. Some additional options subfields have been added, including:

- The *DHCP Class Identifier* parameter. The client sends this parameter to the server, which uses it as a key for selecting correct parameters for the client.
- The option of identifying a client's lease and bind[3] information by means of a *Client Identifier,* rather than the client's hardware type and hardware address.
- A parameter that announces the maximum size message that the client can receive from the server.

A significant change in the protocol was the ability to negotiate configuration parameters. Several new message types are defined in order to do this. For

---

[4]A binding is the association of a set of parameters with a client.

example, suppose that a response does not include all of the parameters that the client wants. DHCP enables the client to ask for more.

11.10.1 Message Type Option

The message type is defined in the *DHCP Message Type* option field. DHCP clients include this option; BOOTP clients do not. Types include:

DHCPDISCOVER	The client sends a message to discover servers.
DHCPOFFER	Servers respond to the client and offer an IP address and parameters.
DHCPREQUEST	The client selects one server and sends a request message. Optionally, the client may ask this server for additional parameters.
DHCPACK	The server responds and provides more parameters if they have been requested.
DHCPNAK	A server refuses a request—for example, a client may have requested an IP address that already is in use. The client must start the whole process over again.
DHCPDECLINE	The client refuses configuration parameters because one or more are invalid.
DHCPRELEASE	The client no longer needs its IP address and releases it.

11.10.2 Typical Initial Client/Server Message Scenario

A sample successful client/server initial interaction is described in fuller detail below:

1. The client sends a discover (*DHCPDISCOVER*) broadcast aimed at finding one or more servers.

2. Several servers may respond to the client. The client waits until it has received one or more (*DHCPOFFER*) responses. Each response includes an IP address, subnet mask, lease expiration date, the identity of the server (in the *DHCP Server Identifier* option field), and some configuration parameters for the client.

3. Based on the contents of the responses, the client selects the server that it wants to use. The client broadcasts a request (*DHCPREQUEST*) with that server's identifier in the *DHCP Server Identifier* option field. The client's message can include a *DHCP Parameter Request List* option, which identifies additional configuration data desired by the client.

4. The selected server saves the *binding* for the client in permanent storage, indexed by an appropriate key. The server sends parameters to the client in a *DHCPACK* message. The client should use an ARP request to make sure that no other station is using its assigned IP address.

11.10.3 Renewing

Clients can keep their leases alive by means of a quick request response interaction:

- A previously configured client can send a *DHCPREQUEST* that includes its assigned IP address.

- A server or servers that have stored the client's configuration respond with *DHCPACK* if all is well.

- If the client's information is no longer valid (e.g., the user's workstation has been connected to a different LAN), servers respond with *DHCPNAK,* and the client must restart a full configuration procedure.

- The client also must start over if the information sent back from the server in the *DHCPACK* was not valid.

11.11 BOOT PARAMETERS

The parameters in Table 11.1 may be included in either BOOTP or DHCP response messages. The parameters in Table 11.2 only may be used with DHCP.

It is possible that more parameters will be added to the list. Check the most recent version of the *Assigned Numbers* RFC for the current status.

Many of the values are lists of IP addresses. The IP addresses should appear in a list in order of preference.

11.12 OTHER AUTOMATIC CONFIGURATION METHODS

There have been a number of attempts to automate individual parts of the configuration process. A LAN-attached system can use *Reverse ARP* (RARP) to discover its IP address. The ICMP *Address Mask* request and reply can provide the subnet mask. But there is no particular benefit to using several separate protocols and messages to get information that can be obtained via a single BOOTP or DHCP response.

The ICMP router discovery mechanism does provide an advantage, since it provides continually updated information on available routers.

11.13 RECOMMENDED READING

The references that follow were valid at the time of this writing:

- BOOTP is defined in RFC 951.

- RFC 1533 defines DHCP options and BOOTP vendor extensions.

- RFC 1534 describes Interoperation between DHCP and BOOTP.

- RFC 1542 provides clarifications and extensions for the BOOTP.

- DHCP is defined in RFC 1541.

TABLE 11.1   Parameters For BOOTP And DHCP

IP configuration parameters	
Subnet mask	
Time Offset Difference	Difference in seconds between local time and Coordinated Universal Time (UTC).
Client Host Name	With or without local Domain Name.
Domain Name	To be used when resolving host names.
Enable/Disable IP Forwarding	Indicates whether the system will route datagrams.
Enable/Disable Non-Local Source Routing	Indicates whether source routed datagrams will be forwarded.
Policy Filter	A list of IP addresses and masks used to filter incoming source routes.
Maximum Datagram Reassembly Size	Largest incoming datagram that the client should be prepared to reassemble.
Default IP Time-to-Live	Initial setting for TTL field.
Lists of IP addresses	
Routers	
Time Servers	
IEN 116 Name Servers	(Obsolete)
Domain Name Servers	
Log Servers	
Cookie Servers	Quote of the day "Fortune Cookie" servers
LPR Servers	Line printer servers
Imagen Impress Print Servers	
Resource Location Servers	RFC 887 servers
Miscellaneous parameters	
Boot File Size	Number of 512-octet blocks in the boot file.
Dump File	Pathname of a file to hold a core image dump if the client crashes.
Swap Server	IP address of swap disk server.
Root Path	Pathname that contains the client's root disk.
Extensions Path	Pathname of a file containing configuration parameters that can be retrieved via TFTP.

TABLE 11.1   Parameters For BOOTP And DHCP (Continued)

Maximum Transmission Unit (MTU) parameters	
Path MTU Aging Timeout	Timeout for use of MTU values determined via Path MTU discovery.
Path MTU Plateau Table	Series of MTU sizes to try when executing Path MTU Discovery and routers do not include usable MTU sizes in ICMP messages.
IP per interface parameters	
Interface MTU	MTU: Biggest datagram that can be sent on the interface.
All Subnets Are Local	Indicates whether all subnets of the network support the same MTU as the local subnet.
Broadcast Address for the Interface	
Perform Mask Discovery	Indicates whether the client should use ICMP subnet mask discovery.
Mask Supplier	Indicates whether the client should respond to ICMP subnet mask requests.
Perform Router Discovery	Indicates whether the client should use the Router Discovery procedure.
Router Solicitation Address	Provides the address to which the client should transmit router solicitation requests.
Static Routes	List of static routes (destination/router pairs) for the client's routing table.
Link Layer parameters per interface	
Trailer Encapsulation	Whether to negotiate the (obsolete) use of trailers when using ARP.
ARP Cache Timeout	Timeout to flush the ARP table.
Ethernet Encapsulation	Ethernet Version 2 (DIX) or IEEE 802.3.
TCP parameters	
TCP Default TTL	Time-To-Live to be used when sending TCP segments.
TCP Keep-Alive Interval	Timeout for sending Keep-Alive messages on an otherwise inactive session. (0 means don't send Keep-Alives unless they are requested by an application.)
Send TCP Keep-Alive Garbage Octet	Include a "garbage" octet in Keep-Alive messages.

TABLE 11.1   Parameters For BOOTP And DHCP (Continued)

Application and Service parameters	
NIS Domain	Name of Network Information Service Domain (when running NIS database services).
Network Information Server (NIS) Addresses	
Network Time Protocol Server Addresses	
Vendor Specific Information	The vendor is identified in the class-identifier option.
List of NetBIOS over TCP/IP Name Servers	
NetBIOS over TCP/IP Datagram Distribution Server	
NetBIOS over TCP/IP Node Type	
NetBIOS over TCP/IP Scope	
X Window System Font Server	List of IP addresses.
X Window System Display Managers	List of IP addresses.

TABLE 11.2   Parameters for DHCP

DHCP-only extensions	
Requested IP Address	The client asks for a specific IP address.
IP Address Lease Time	In a request, the client can ask for a particular lease time. In the reply, the server sets the actual lease time granted.
Option Overload	Signals that the *DHCP Server Host Name* or *Boot File Name* fields are carrying DHCP options in addition to their standard contents.
DHCP Message Type	For example, *DISCOVER, OFFER,* or *REQUEST*.
Server Identifier	Used to differentiate between servers so that the client can identify whose lease was accepted.
Parameter Request List	List of option codes, enabling a client to request values for specific parameters.
Message	Used as an error message in a server response (e.g., reporting that no IP addresses are available). The client can use it to indicate why it is declining an offered set of parameter values.
Maximum DHCP Message Size	Largest DHCP message that the client is willing to accept.
Renewal (T1) Time Value	The time interval from address assignment until the client tries to contact its original server and renew its lease.
Rebinding (T2>T1) Time Value	If the client cannot get a response from its original server, the time interval from address assignment until the client tries to renew its lease by contacting *any* server.
Class Identifier	A locally assigned identifier used by the client to identify its type and configuration. Some parameters can be returned based on class.
Client Identifier	A unique identifier for the client, included in the *DHCPDISCOVER* message. The ID could be a DNS name or some other assigned ID. It is used to associate a client with its lease.

Domain Name System

12.1 INTRODUCTION

Often an end user will know a host's name but not its address. But IP needs to know a host's address in order to communicate with the host. Either the end user—or the application that the end user has invoked needs a way to look up the address for a given host name.

Small, isolated networks can meet this need by maintaining a central name-to-address translation table. Individual systems on the network can stay up to date by copying this table to their own disks periodically.

A central table was used across the Internet in the early days of the TCP/IP. The Department of Defense Network Information Center (DOD NIC) maintained the master version of the Internet name-to-address translation table, and other systems retrieved a copy on a regular basis. As time went by, this method became burdensome and inefficient.

12.2 PURPOSE OF THE DOMAIN NAME SYSTEM

The Domain Name System (DNS) was set up to provide a better method of keeping track of Internet names and addresses. DNS databases provide auto-mated name-to-address translation services. The system works well, and many organizations that are not connected to the Internet use DNS database software to track their own internal computer names.

DNS is a *distributed* database. Internet names and addresses are kept at servers all over the world.[1] The organization that owns a Domain Name (such as *yale.edu*) is responsible for running and maintaining the name servers that translate its own names to addresses. Local personnel enter node additions, deletions, and changes quickly and accurately at the domain's *primary*

---

[1]Later, we shall see that name servers also contain important mail routing information.

server. Because name-to-address translation is so important, the information is replicated at one or more *secondary servers.*

12.3 BIND SOFTWARE

Many computer vendors provide free software that lets their systems function as name servers. Usually this software has been adapted from the *Berkeley Internet Domain* (BIND) package. Periodically, new versions of BIND are made freely available on the Internet.

An organization can use this free software to run a private name-to-address translation service for its own internal use. If the organization wishes to connect to the Internet, it must set up at least two public name servers that will become part of the Internet Domain Name System.

12.4 RESOLVERS

A client program capable of looking up information in the Domain Name System is a standard part of TCP/IP products and is called a *resolver.* In normal use, a resolver works quietly in the background, and users don't even notice it. For example, below, a user requests a *telnet* connection to *minnie.jvnc.net.* The user's *telnet* application calls on a local *resolver* program that looks up the IP address of that site:

```
> telnet minnie.jvnc.net
Trying 128.121.50.141 ...
Connected to minnie.jvnc.net.
```

When TCP/IP is installed at a host that will use Domain Name database lookups, the host's configuration information must include the IP addresses of two or more Domain Name Servers. Resolver programs need to know the addresses of Domain Name Servers that they can query.

The demonstration that follows was run at *tigger,* a mail, news, and *telnet* server operated by *Global Enterprise Systems.* Like most Unix systems, *tigger* has a configuration file called */etc/resolv.conf,* which identifies the local domain name and the IP addresses of two Domain Name Servers for the domain.

```
> more /etc/resolv.conf
domain jvnc.net
128.121.50.2
128.121.50.7
```

Desktop TCP/IP systems need Domain Name Server information too. As shown in Figure 12.1, the *Chameleon* TCP/IP software package for Microsoft Windows provides a pop-up configuration menu in which to enter the same information.

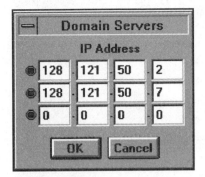

Figure 12.1 Configuring DNS.

12.5 LOOKING UP HOST ADDRESSES

As we have seen earlier, many systems provide an interactive resolver program that enables a user to communicate directly to a Domain Name Server, sending queries and getting back responses. The dialogue that follows uses the Unix *nslookup* resolver program. In the dialogue:

1. Immediately after the user types *nslookup,* the local default server identifies itself, displaying its name and address. In this case, the server's name is *r2d2.jvnc.net* and the server's address is 128.121.50.2.

2. The user types in the name of a host whose address is desired.

3. The request is sent to the server.

4. After each query, the server (*r2d2*) identifies itself and then provides the answer.

5. If the user has asked for local information, the server extracts the answer from its own database.

6. If the user has asked for information about an external host, the server first checks its *cache* of recent queries to see if the information is available, and if not, interacts with a remote authoritative server to get the answer.

7. When an answer comes back from the remote authoritative server, the answer is saved in the local server's disk cache for future reference and then is sent to the requesting user.

Each step in the dialogue is explained by comments on the right. Note that a response retrieved from the server's cache is marked *nonauthoritative.*

```
> nslookup
```

```
Default Server:
r2d2.jvnc.net                    The local server's name and address is displayed.
Address: 128.121.50.2
```

```
> mickey.jvnc.net.               User makes a query whose answer is in the local database.
```

```
Server: r2d2.jvnc.net          Server ID and address again.
Address: 128.121.50.2
Name: mickey.jvnc.net          The name in the query.
Address: 128.121.50.143        The answer.

> www.novell.com.              User makes a query about an external host.

Server: r2d2.jvnc.net          Server ID and address again.
Address: 128.121.50.2

Name: www.novell.com           The name in the query.
Address: 137.65.2.5            The answer has been saved on r2d2's disk and sent to the user.

> www.novell.com.              User repeats query about an external host.

Server: r2d2.jvnc.net          Server ID and address again.
Address: 128.121.50.2

Non-authoritative answer:      This came from the local cache.
Name: www.novell.com           The name in the query.
Address: 137.65.2.5            The answer.
```

Why does the server keep identifying itself? Recall that an organization will operate two or more servers because one of them might be very busy, or even out of service (e.g., for maintenance). If the resolver cannot get an answer from the first system on the list, it tries the next. An administrator using *nslookup* can see immediately which server is answering a query.

Note that we have included a final period at the end of each queried name. Later in this chapter, you will find out why this is a good idea.

12.6 AUTHORITATIVE AND CACHED RESPONSES

All data is entered and updated at a *primary* name server. The data at a primary server is on its own hard disk. *Secondary* servers download their information from the primary.

When a system sends a query to a Domain Name Server, the requester does not care whether it is talking to a primary or a secondary server. All of an organization's name servers—whether primary or secondary—are *authoritative* for its domain.

Recall that to cut down on traffic, your local server will *cache* (save) the answers on its own hard disks for a while. Any user who repeats a query that has been made recently gets the locally cached answer.

How long will the information be cached? The maximum time is configured at the authoritative servers and is sent to the requester with the query result.

12.7 ADDRESS-TO-NAME TRANSLATIONS

The Domain Name System is versatile, and also enables you to perform address-to-name translations. The way that you do this when using *nslookup* looks a bit strange:

- Set the type of the query to be *ptr.*
- Write the address *backward,* followed by *.in-addr.arpa.*

For example:

```
> set type = ptr
> 143.50.121.128.in-addr.arpa.
Server:  r2d2.jvnc.net
Address:  128.121.50.2

143.50.121.128.in-addr.arpa    host name = mickey.jvnc.net
>
```

This oddity makes sense when you know how global reverse lookups were architected. The organization that owns a network address is responsible for recording all of its address-to-name translations in the DNS database. This is done in a table that is *separate* from the name-to-address mappings.

The special *in-addr.arpa* domain subtree shown in Figure 12.2 was created to point to all of these network tables. When addresses are placed into the tree, it makes sense to put the first number at the top and work down. That way, all addresses of the form 128.*x.x.x* are under node 128.

If we read the labels in the tree using the same bottom-to-top convention that we used for names, the address appears backward—namely, *143.50.121.128.in-addr.arpa.*

Of course, the user interface for *nslookup* could have hidden this technicality, but this is Unix! Figure 12.3 shows a friendlier desktop *NSLookup* program, provided by Ashmount Research Ltd. Queries are entered in a small box near the bottom of the window, and responses are shown in the display

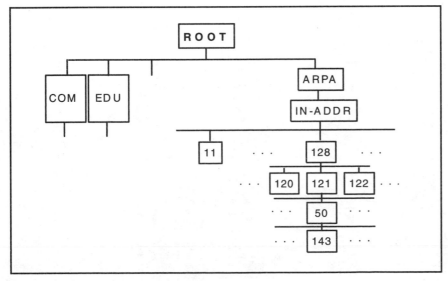

Figure 12.2   The in-addr domain subtree.

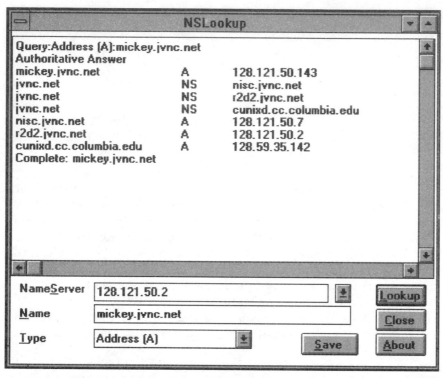

Figure 12.3 DNS queries.

area at the top. Note that both of the responses include the names and addresses of name servers that contain authoritative information relating to the queries.

12.8   LOCAL AND GLOBAL DOMAIN NAME SERVERS

If you have a stand-alone TCP/IP network, you can use the free DNS software to create a primary name translation database and replicate it at convenient points across your network. All user queries will be answered by your own name servers.

But if you connect your network to the Internet, your name servers will need to retrieve global information. How is this done? The key to making this work is that when an organization (e.g., *microsoft.com*) wishes to connect to the Internet, it registers with an appropriate *registration authority* (in this case, the InterNIC) and identifies the names and addresses of at least two Domain Name Servers that it operates. The InterNIC registration authority adds this information to its *root* list of Domain Name Servers.

This root list is replicated at several *root servers* which play a key role in processing remote queries. For example, suppose that a name-to-address translation query for *www.microsoft.com* is sent to *tigger*'s local Domain Name Server:

- The server checks to see if *www.microsoft.com* belongs to the local domain.

- It does not, so the server checks to see if the name is in its cache.

- If the name is not in the cache, the server sends a query to a root server.

- The root server returns the name and address of Domain Name Servers that contain information for *microsoft.com*.

To see the current list of root servers, we run *nslookup* and set the type of query to be *ns*. If we enter . (which stands for the root), the names and addresses of several root servers will be returned.

```
> nslookup

> set type = ns
> .
Server:  r2d2.jvnc.net
Address:  128.121.50.2

Non-authoritative answer:
(root)   nameserver = C.ROOT-SERVERS.NET
(root)   nameserver = D.ROOT-SERVERS.NET
(root)   nameserver = E.ROOT-SERVERS.NET
(root)   nameserver = I.ROOT-SERVERS.NET
(root)   nameserver = F.ROOT-SERVERS.NET
(root)   nameserver = G.ROOT-SERVERS.NET
(root)   nameserver = A.ROOT-SERVERS.NET
(root)   nameserver = H.ROOT-SERVERS.NET
(root)   nameserver = B.ROOT-SERVERS.NET

Authoritative answers can be found from:
C.ROOT-SERVERS.NET    internet address = 192.33.4.12
D.ROOT-SERVERS.NET    internet address = 128.8.10.90
```

```
E.ROOT-SERVERS.NET      internet address = 192.203.230.10
I.ROOT-SERVERS.NET      internet address = 192.36.148.17
F.ROOT-SERVERS.NET      internet address = 192.5.5.241
F.ROOT-SERVERS.NET      internet address = 39.13.229.241
G.ROOT-SERVERS.NET      internet address = 192.112.36.4
A.ROOT-SERVERS.NET      internet address = 198.41.0.4
H.ROOT-SERVERS.NET      internet address = 128.63.2.53
B.ROOT-SERVERS.NET      internet address = 128.9.0.107
>
```

The root servers provide direct referrals to servers for the second-level domains (such as *microsoft.com* or *yale.edu*) under *COM, EDU, GOV,* and the other original top levels. For example, here is part of the information about *3com.com* taken directly from the root list file:

```
3COM.COM.         172800²    NS     NS.3COM.COM.
NS.3COM.COM.      172800     A      129.213.128.2
3COM.COM.         172800     NS     TMC.EDU.
TMC.EDU.          172800     A      128.249.1.1
```

Note that the second *3com* name server is not located on *3com*'s own network. Organizations often have name servers that are located at their Service Provider's network or on a university campus.

We can get all of an organization's name server information via *nslookup* when *type = ns* has been specified:

```
> set type = ns
> 3com.com.

3com.com        nameserver = NS.3COM.COM
3com.com        nameserver = TMC.EDU
3com.com        nameserver = XANTH.CS.ODU.EDU
3com.com        nameserver = AEROSPACE.AERO.ORG
3com.com        nameserver = ANTARES.AERO.ORG
Authoritative answers can be found from:
NS.3COM.COM          inet address = 129.213.128.2
TMC.EDU              inet address = 128.249.1.1
XANTH.CS.ODU.EDU     inet address = 128.82.4.1
AEROSPACE.AERO.ORG   inet address = 130.221.192.10
>
```

12.9   DELEGATING

Rather than having the InterNIC try to maintain up-to-date lists of servers for organizations in Australia, Canada, or Switzerland, each country maintains its own registration service and publishes a list of domain servers at its own root servers.

When you look up name servers for a country code, the InterNIC root database will return a list of names and addresses of root servers for the country. The *nslookup* dialogue below shows the list of Canadian root servers:

```
> ca.

ca                                   nameserver = RELAY.CDNNET.CA
ca                                   nameserver = RS0.INTERNIC.NET
```

---

²The number is a timeout (measured in seconds) on the validity of the entry.

```
ca                              nameserver = CLOUSO.CRIM.CA
ca                              nameserver = SNORT.UTCC.UTORONTO.CA
ca                              nameserver = NS2.UUNET.CA
RELAY.CDNNET.CA                 inet address = 192.73.5.1
RS0.INTERNIC.NET                inet address = 198.41.0.5
CLOUSO.CRIM.CA                  inet address = 192.26.210.1
SNORT.UTCC.UTORONTO.CA          inet address = 128.100.102.201
NS2.UUNET.CA                    inet address = 142.77.1.5
>
```

In fact, the Domain Name System is very flexible and allows a long chain of referrals. A country could be divided into naming regions, and the national root could point to root servers for each region.

Similarly, any organization can set up a root tree of its own that points to Domain Name Servers that are authoritative for *parts* of its naming domain.

In actual practice, relatively little subdivision is done, so names usually are found in a few steps. Figure 12.4 illustrates the steps in resolving the name *viper.cs.titech.ac.jp:*

1. First the InterNIC root is consulted. It identifies servers for Japan.

2. One of the root servers for Japan is queried. It identifies servers for the Titech University domain.

3. The Titech server provides the address of the host.

Note that the local server took responsibility for finding the answer for the client. This is because clients normally ask for *recursive* name resolution, meaning, "keep going until you get the result."

In contrast, the local server itself worked *nonrecursively (iteratively)*. Each server that it queried returned a pointer for the next step. The local server then sent its next query directly to that database.

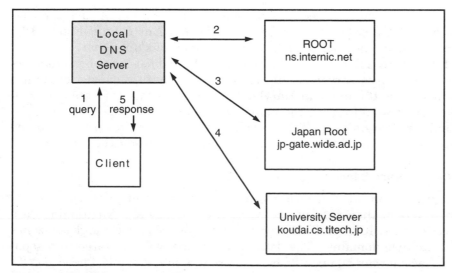

Figure 12.4 Resolving a name in Japan.

12.10   CONNECTING NAME SERVERS TO THE INTERNET

Connecting your own Domain Name Servers to the worldwide Internet database takes several steps:

1. Register for one or more blocks of IP addresses (and optionally an Autonomous System number).
2. Assign names and addresses to your hosts.
3. Obtain the list of root servers that tie the worldwide service together.
4. Set up a primary Domain Name Server and at least one secondary copy.
5. Test your servers.
6. Move to operational status.
7. Register your organization's domain name and servers with your region's registration services.

See Appendix C for more information about registration.

12.11   DESIGNING A NAME SERVER DATABASE

If you have a small organization, you can structure all of your information as a single database. This may not be convenient if you are a large, geographically distributed organization. For example, suppose that the company with Domain Name *fishfood.com* has headquarters in Maine and branches in Maryland and Georgia. It may make more sense to delegate management of the organization's naming tree to administrators at these locations and even run separate name servers at these locations.

12.11.1   Zones

An organization's naming tree consists of one or more *zones. A zone is a contiguous part of the naming tree that is administered as a unit.* Figure 12.5 shows a zone structure that could be used for domain *fishfood.com.*

The Internet root database would point to the *fishfood.com* headquarters name servers. These servers would respond directly to requests for names within their zone. If a name in Maryland (or Georgia) was requested, a headquarters server would return the names and addresses of the servers at the Maryland (or Georgia) site. The Domain Name Server that originated the request would then send the request to the correct server for the zone.

12.11.2   Location of Domain Name Servers

Many organizations find it easier to run a single set of primary and secondary servers in their internal network, even if they are partitioning data into several zones. It is totally acceptable to use one server for multiple zones (or even multiple domains). The data for each zone will be stored in a separate file. Each file can be updated by a different administrator, if needed.

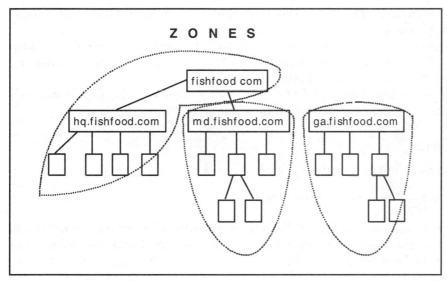

Figure 12.5 Defining zones.

12.11.3 Zone Transfer

A secondary server is set up to provide access to a copy of the information about one or more zones. It obtains its information from the primary server for a zone via a *zone transfer*.

A secondary server can be configured to retrieve information about several zones from different primary servers. Thus, one server can act as secondary for several different primaries. A server even can act as primary for some zones and secondary for others. The Domain Name System was designed to be flexible.

12.12 DNS DATA

What information does a Domain Name Server need in order to do its job? It needs at least:

- A list of the worldwide *root* servers to find out where to send external queries. A file containing this list can be copied from the registration InterNIC.
- A list of names and their corresponding addresses.
- A list of addresses and their corresponding names.

12.13 DNS ENTRIES

DNS data is stored as a series of text entries. An information entry record contains:

```
[name] [TTL] [class] Record-Type Record-Data [; comment]
```

The Time-To-Live (TTL) indicates how long a record may be cached after it has been retrieved.

If omitted, the *name* or *class* values default to the last previously included values. The *only* class currently used is "IN," for Internet, so this usually appears only once, in the first record.

The order of the *class* and *TTL* fields can be reversed. The *TTL* is numeric and hence cannot be confused with the class ("IN").

12.13.1 Resource Records

The part of a data entry consisting of:

```
[TTL] [class] Record-Type Record-Data
```

is called a *resource record,* or *RR.* There are several resource record types, and each is identified by a character or short acronym. Resource record types are listed in Table 12.1.

12.14   A SAMPLE NAME-TO-ADDRESS FILE

Figure 12.6 displays a file that translates names to addresses for our mythical *fishfood.com* domain. The file contains several comments, which are introduced by the ; character.

TABLE 12.1   Resource Record Types

Record type	Description
SOA	Start Of Authority—identifies the domain or zone and sets a number of parameters.
NS	Maps a domain name to the name of a computer that is authoritative for the domain.
A	Maps the name of a system to its address. If a system (e.g., a router) has several addresses, there will be a separate record for each.
CNAME	Maps an alias name to the true, *canonical name.*
MX	Mail Exchanger. Identifies the systems that relay mail into the organization.
TXT	Provides a way to add text comments to the database. For example, a *txt* record could map *fishfood.com* to the company's name, address, and telephone number.
WKS	Well-Known Services. Can list the application services available at the host. Used sparingly, if at all.
HINFO	Host Information, such as computer type and model. Rarely used.
PTR	Maps an IP address to a system name. Used in address-to-name files.

```
;        fishfood.com file
FISHFOOD.COM.  IN      SOA     NS.FISHFOOD.COM.(
                               postmaster.FISHFOOD.COM.
                                94101101 ; serial number
                                   86400 ; refresh after 24 hours
                                    7200 ; retry after 2 hours
                                 2592000 ; expire after 30 days
                                  345600 ; default TTL of 4 days
                               )

FISHFOOD.COM.  IN      NS      NS.FISHFOOD.COM.
FISHFOOD.COM.  IN      NS      NS2.FISHFOOD.COM

LOCALHOST      IN      A       127.0.0.1
NS             IN      A       172.66.1.1
NS2            IN      A       172.66.1.100
;
MAIL-RELAY     IN      A       172.66.1.2
               IN      TXT     www,ftp on mail-relay
               IN      TXT     gopher on mail-relay
               IN      HINFO   SUN UNIX  ;should not include
;
WWW            IN      CNAME   MAIL-RELAY
FTP            IN      CNAME   MAIL-RELAY
GOPHER         IN      CNAME   MAIL-RELAY
;
FISHFOOD.COM.  IN      MX      1       MAIL-RELAY
*              IN      MX      1       MAIL-RELAY
NS             IN      MX      1       MAIL-RELAY
;
;end of fishfood.com file
;
```

Figure 12.6 A sample name-to-address translation file.

12.14.1 The SOA Record

The first record of the file is the important Start of Authority (SOA) record:

```
FISHFOOD.COM.   IN     SOA     NS.FISHFOOD.COM.(
                               postmaster.FISHFOOD.COM.
                                94101101 ; serial number
                                   86400 ; refresh after 24 hours
                                    7200 ; retry after 2 hours
                                 2592000 ; expire after 30 days
                                  345600 ; default TTL of 4 days
                               )
```

The parentheses in the SOA record allow it to span several lines. Several timeout values, measured in seconds, are included in the record. This SOA record indicates that:

■ Server *ns.fishfood.com* is primary for domain *fishfood.com*.

- Problems should be reported to *postmaster@fishfood.com* (you have to change the first dot to an @).

Secondary servers will copy this whole file and will obtain important information from the next four items in the SOA record. The record shown above indicates that each secondary should:

- Connect to the primary every 24 hours.

- Check whether its current serial number is lower than the primary's. If it is, the primary has been updated, and the secondary needs to perform a *zone transfer*—that is, copy the entire database for this zone to its system.

- If the secondary cannot connect at the scheduled time, it should try again 2 hours later.

- If the secondary has not been able to contact the primary for 30 days, it should expire all of its data and stop answering queries.

The values shown in the example are the ones recommended[3] for top-level servers.

12.14.2 Time-To-Live

RFC 1035, which defines the DNS protocol, states that the *TTL* in the SOA record is the *minimum* timeout value permitted for all records. But in actual use, what an administrator wants to do is use the *TTL* in the SOA record as a default, putting in shorter values for specific hosts whose information is due to change soon. Implementations follow this common-sense definition instead of the one in the standard.

In the example, the default *TTL* is 4 days. Values tend to range from a day to a week, depending on how stable entries at a site tend to be.

12.14.3 Name Completion

A name that does not end in . is completed by adding the Domain Name for this zone, *fishfood.com*. Thus, in this file, *ns* corresponds to *ns.fishfood.com*.

12.14.4 Name Server (NS) Records

NS records identify the name servers for the domain. If there are subzones, entries for the "child" subzone name servers are needed so that the higher-level server can provide pointers to the lower-level servers. Address records also are needed so that these child servers can be accessed. These address records are called *glue* records.

Note that the higher-level server is *not* authoritative for these child servers; a different administrator may well be in charge of a child server. The

---

[3]See RFC 1537.

parent server administrator has to be careful to keep in touch with child administrators and update the list of child server names and addresses when changes occur.

12.14.5 Address Records

Address records simply map a name to an address. Thus, the address of *ns.fishfood.com* is 172.66.1.1.

12.14.6 CNAME Records

Recall that you can assign alias nicknames to server hosts so that users can guess their names. In the example, a World Wide Web service, file transfer service, and gopher service all run at the same machine that supports mail relay service. The canonical name (*CNAME*) records define the host's nicknames and enable users to type *www.fishfood.com, ftp.fishfood.com,* and *gopher.fishfood.com.*

12.14.7 Mail Exchanger Records

Mail Exchanger (MX) servers relay electronic mail to and from networks.[4] There are three *MX* records in the file that identify MX servers for *fishfood.com:*

```
fishfood.com.    IN MX 1 MAIL-RELAY
*                IN MX 1 MAIL-RELAY
ns               IN MX 1 MAIL-RELAY
```

These actually indicate that:

- Mail addressed to *somebody@fishfood.com* should be relayed to *mail-relay.fishfood.com.*
- The wild card (*) allows you to relay mail that has been addressed to specific hosts that are *not* listed in the DNS directory. Mail addressed to *somebody@anyhost.fishfood.com* would be passed to *mail-relay.fishfood.com.*
- Mail addressed to *somebody@ns.fishfood.com* should be passed to *mail-relay.fishfood.com.*

It looks as if the wild card should take care of *ns.fishfood.com,* so why do we need a separate statement? The rules say that the wild card can only apply to systems for which there are *no* other records in the DNS database.

The numbers that appear after MX are called *preference numbers* and are discussed in Chapter 16, which deals with electronic mail.

---

[4]See Chapter 16.

12.14.8 TXT and HINFO Records

TXT records have no real function, but they enable the administrator to embed some comments in the database.

HINFO records can be used to identify a system's hardware and operating system type. Since users are able to read this data via a program such as *nslookup,* many administrators feel that HINFO records should *not* be included in the database; they help crackers to find systems for which some kind of vulnerability is known.

12.15 ADDRESS-TO-NAME TRANSLATIONS

Why would anyone want to do a reverse lookup and translate an address to a name? Some system programs invoke reverse lookups to improve the appearance of administrative displays. For example, the *netstat* display below shows all or part of host names instead of IP addresses:

```
> netstat -a
Active Internet connections (including servers)
Proto   Recv-Q  Send-Q  Local Address   Foreign Address        (state)
tcp     0       121     tigger.nntp     c3po.4809              ESTABLISHED
tcp     0       0       tigger.smtp     news.std.com.1472      TIME_WAIT
tcp     0       925     tigger.1176     sun3.nsfnet-rela.smtp  ESTABLISHED
tcp     0       0       tigger.pop-3    ringotty8.16284        TIME_WAIT
```

Reverse lookups also are used by file transfer and World Wide Web servers, which create logs that record the names of systems that are using the service. *Some services refuse clients whose IP addresses do not correspond to some record in the Domain Name database.*

Recall that addresses have been placed into a special domain called *IN-ADDR.ARPA* and that since the tree must grow down from the most general part of the address to the least general, the order of the numbers in each address is reversed. For this reason, the subtree for network 172.66 is called 66.172.in-addr.arpa. Figure 12.7 shows reverse lookup records.

```
66.172.in-addr.arpa.   IN    SOA    NS.FISHFOOD.COM. (
                                     postmaster.FISHFOOD.COM.
                                     94101101  ; serial
                                        86400  ; refresh after 24 hours
                                         7200  ; retry after 2 hours
                                      2592000  ; expire after 30 days
                                       345600  ; default TTL of 4 days
                                     )
66.172.in-addr.arpa.   IN    NS     NS.FISHFOOD.COM.
1.1                    IN    PTR    NS.FISHFOOD.COM.
66.172.in-addr.arpa.   IN    NS     NS2.FISHFOOD.COM.
100.1                  IN    PTR    NS2.FISHFOOD.COM.
2.1                    IN    PTR    MAIL-RELAY.FISHFOOD.COM.
```

Figure 12.7 Reverse lookup table.

Entries are "backward" too. For example, the entry for 100.1 corresponds to address 172.66.1.100.

12.16   THE FORMAT OF DNS MESSAGES

The query and response messages exchanged between clients and DNS servers have a simple format. A server adds answer information to the original query and sends it back. The overall message format is shown in Figure 12.8.

12.16.1   Header Section

The header section contains the fields listed in Table 12.2.

12.16.2   Query Section

A query contains the fields listed in Table 12.3. Normally, a message contains a single query, but it is permissible to concatenate several requests in the query section.

12.16.3   Response Sections

The answer, authority information, and additional information sections all are structured in the same way. They consist of a sequence of resource records. Resource records contain the fields listed in Table 12.4.

The authority information section identifies authoritative name servers for the domain. The additional information section provides information such as the IP addresses of the authoritative name servers.

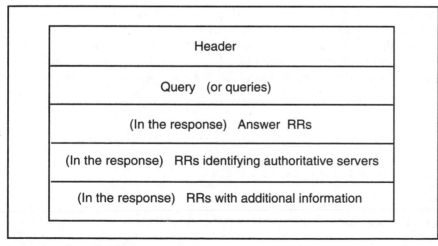

Figure 12.8   Overall DNS message format.

TABLE 12.2   Fields in a DNS Message Header

Field	Description
ID	An identifier used to match a response with its request.
Parameters	Query or response. Standard or inverse lookup. In responses, whether authoritative. In responses, whether truncated. Recursive or not. In responses, whether recursion was available. In responses, error code.
Number of queries	Provided in a query and a response.
Number of answers	Provided in a response.
Number of authority records	Provided in a response. Information in authority records includes the names of servers that hold the authoritative data.
Number of additional records	Provided in a response. Information includes the addresses of the authoritative servers.

12.17   TRANSPORTS

DNS queries and responses normally are transmitted via UDP, but TCP also is permitted. TCP is used for zone transfers.

12.18   VIEWING EXAMPLES

Some implementations of the *nslookup* program enable you to view message exchanges in detail. Below, we run *nslookup* at a host at Yale University and turn on detailed debugging via the *set d2* command.

The query asks for name-to-address translation for *www.microsoft.com,* and the response provides two addresses. It appears that two different computers are acting as Microsoft World Wide Web servers and are sharing the

TABLE 12.3   Fields in a DNS Query

Field	Description
Name	A domain name or IP address in the *IN-ADDR.ARPA subtree*
Type	Type of query, for example, A or NS
Class	IN for Internet, represented as 1

TABLE 12.4   Fields in Resource Records

Field	Description
Name	Name of the node for this record.
Type	Record type, such as SOA or A, indicated via a numeric code.
Class	IN, represented as 1.
TTL	Time-to-Live, a 32-bit signed integer indicating how long the record may be cached.
RDLENGTH	Length of the resource data field.
RDATA	The information; for example, for an address record, it is the IP address. For an SOA record, it includes extensive information.

client traffic between them. If a client does not get a response from the first address (because the system is too busy), it will try the second one.

```
> nslookup
Server: DEPT-GW.cs.YALE.EDU
Address: 128.36.0.36

> set d2
> www.microsoft.com.
Server: DEPT-GW.cs.YALE.EDU
Address: 128.36.0.36

res_mkquery(0, www.microsoft.com, 1, 1)
------
SendRequest(), len 35
HEADER:
    opcode = QUERY, id = 5, rcode = NOERROR
    header flags: query, want recursion
    questions = 1, answers = 0, auth. records = 0, additional = 0

QUESTIONS:
    www.microsoft.com, type = A, class = IN

------
------
Got answer (67 bytes):
HEADER:
    opcode = QUERY, id = 5, rcode = NOERROR
    header flags: response, auth. answer, want recursion,
    recursion avail.
    questions = 1, answers = 2, auth. records = 0, additional = 0

QUESTIONS:
    www.microsoft.com, type = A, class = IN
ANSWERS:
    -> www.microsoft.com
        type = A, class = IN, ttl = 86400, dlen = 4
        inet address = 198.105.232.5
    -> www.microsoft.com
        type = A, class = IN, ttl = 86400, dlen = 4
        inet address = 198.105.232.6
```

The response provided by our local server did not include any authority records or additional records. However, the local server did receive authority and additional information from the servers that *it* queried and has cached this information.

If we enter the query again, the answer is retrieved from the local server's cache. Since the information is not authoritative, the local server includes the names and addresses of the authoritative servers for *microsoft.com* in its response.

```
> www.microsoft.com.
Server: DEPT-GW.cs.YALE.EDU
Address: 128.36.0.36
res_mkquery(0, www.microsoft.com, 1, 1)

------

SendRequest(), len 35
HEADER:
    opcode = QUERY, id = 8, rcode = NOERROR
    header flags: query, want recursion
    questions = 1, answers = 0, auth. records = 0, additional = 0

QUESTIONS:
    www.microsoft.com, type = A, class = IN

------
------
Got answer (194 bytes):
HEADER:
    opcode = QUERY, id = 8, rcode = NOERROR
    header flags: response, want recursion, recursion avail.
    questions = 1, answers = 2, auth. records = 3, additional = 3

QUESTIONS:
    www.microsoft.com, type = A, class = IN
    ANSWERS:
    -> www.microsoft.com
        type = A, class = IN, ttl = 86392, dlen = 4
        inet address = 198.105.232.5
    -> www.microsoft.com
        type = A, class = IN, ttl = 86392, dlen = 4
        inet address = 198.105.232.6
AUTHORITY RECORDS:
    -> MICROSOFT.COM
        type = NS, class = IN, ttl = 172792, dlen = 7
        nameserver = ATBD.MICROSOFT.COM
    -> MICROSOFT.COM
        type = NS, class = IN, ttl = 172792, dlen = 16
        nameserver = DNS1.NWNET.NET
    -> MICROSOFT.COM
        type = NS, class = IN, ttl = 172792, dlen = 7
        nameserver = DNS2.NWNET.NET
ADDITIONAL RECORDS:
    -> ATBD.MICROSOFT.COM
        type = A, class = IN, ttl = 187111, dlen = 4
        inet address = 131.107.1.7
    -> DNS1.NWNET.NET
        type = A, class = IN, ttl = 505653, dlen = 4
        inet address = 192.220.250.1
    -> DNS2.NWNET.NET
        type = A, class = IN, ttl = 505653, dlen = 4
        inet address = 192.220.251.1
```

Note that for both of the queries above, *"www.microsoft.com."* was entered with the final period included. If you omit the final period, the query initially will be sent with the local domain name appended.

This demonstration was run at a Yale computer, and the name cs.yale.edu was tacked on. The example that follows shows how this occurs. That query will fail, but subsequently a query automatically would be sent out without extra labels at the end.

```
> www.microsoft.com
Server: DEPT-GW.cs.YALE.EDU
Address: 128.36.0.36

res_mkquery(0, www.microsoft.com.CS.YALE.EDU, 1, 1)
------
SendRequest(), len 47
HEADER:
    opcode = QUERY, id = 6, rcode = NOERROR
    header flags: query, want recursion
    questions = 1, answers = 0, auth. records = 0, additional = 0

QUESTIONS:
www.microsoft.com.CS.YALE.EDU, type = A, class = IN
. . .
```

12.19   ADDITIONAL RECORD TYPES

One way to extend the usefulness of the Domain Name System is to define new record types. Over the years, a number of types have been proposed. Those that are useful are absorbed into the system; others never get beyond the experimentation stage.

There is a record type that is used at some locations. The Open Systems Interconnect (OSI) layer 3 Connectionless Network Protocol (CLNP) is implemented in part of the Internet.

OSI uses Network Service Access Point (NSAP) addresses to route data to hosts. Since there is a need for name and address mappings for OSI hosts, name-to-address records with type NSAP have been defined for use in DNS databases. Address-to-name mappings are provided, as usual, via records of type PTR.

Later, we shall see that a new record type has been defined in order to translate names to IP version 6 addresses.

12.20   DNS PROBLEMS

The Domain Name System is a critical service. Incorrect database entries can make it impossible to reach application hosts. Since the database is distributed, and data is entered manually by many different administrators, it isn't hard to make mistakes. Typical problems include:

- Omitting a final period in a complete name.

- Missing NS records. Sometimes a new name server does not get listed everywhere that it needs to be listed (e.g., in a parent domain's database).

- The opposite problem—*lame delegations.* A lame delegation is an NS record for a name server that no longer exists. This can cause a *lot* of trouble.

- Failure to update *glue* records (which provide the addresses of name servers for child zones) when the name servers in the child zone change.

- Incorrect MX records that point to systems that do not act as Mail Exchangers for the domain.

- Forgetting that wild-card MX records do *not* apply to a system that already has an entry in the database. Separate MX records are needed for these systems.

- Alias names that point to aliases.

- Aliases that point to unknown host names.

- Address records without corresponding PTR records.

- PTR records without corresponding address records.

Fortunately, there are several free tools available that assist in DNS debugging. These are described in an RFC appropriately titled *Tools for DNS Debugging.*

12.21 RECOMMENDED READING

There are many RFC documents that deal with the Domain Name System. We will mention the most important ones here.

RFC 1034 defines Domain Name concepts and facilities. RFC 1035 describes the implementation and protocol specification for the Domain Name System. It should be consulted if further details on the message formats are required.

RFC 1713 describes several tools available for DNS debugging. RFCs 1912, 1536, and 1537 describe common DNS configuration errors, implementation errors, and fixes.

The book *DNS and BIND* by Albitz and Liu (O'Reilly & Associates) provides an excellent exposition of the Domain Name System.

13

Telnet

13.1 INTRODUCTION

What use is a network with a rich offering of applications if users cannot login to different computers and use the applications? TCP gives us computer-to-computer connectivity, but there are other obstacles to overcome. For a long time it seemed as if every computer vendor was determined to market a totally proprietary environment. An application at a vendor's host could be accessed only by special terminals manufactured by that vendor.

The *telnet* (terminal networking) protocol overcomes vendor dissimilarities and lets a user connected to any host in a network login to any other host. *Telnet* terminal emulation was the first TCP/IP application. The *telnet* protocol also was designed to be used as a basis for general application-to-application communications. As organizations have moved away from legacy terminal-based applications, *telnet* increasingly has been used as a toolkit for building client/server applications. In fact, *telnet* underlies client/server interactions for file transfer, electronic mail, and the World Wide Web (WWW).

In this chapter, we will explore *telnet*'s ability to help a user login to a remote application. We also will find out what *telnet* offers the client/server application builder.

13.2 USING TELNET FOR LOGINS

Telnet provides emulation of various types of terminals, so you can access Unix computers, VAX/VMS systems, or IBM mainframes. Some implementations of *telnet* support special authentication procedures.[1]

---

[1] *Kerberos,* from the Massachusetts Institute of Technology, is one of these. With Kerberos, passwords are never passed across the network, and a special encrypted authentication procedure is used. You may need to issue a special command, such as *kinit,* to get Kerberos authentication started. There are simpler procedures based on challenge handshakes.

If you run *telnet* from a multiuser system, you probably will operate a simple text-based user interface. Using a text-based *telnet* client is very easy. You just type:

```
> telnet hostname
```

Often, IBM 3270 emulation is packaged separately, and you access IBM hosts by typing:

```
> tn3270 hostname
```

Most people manage to use *telnet* successfully without knowing any more than this. Below, we show a login to *tigger* from a computer at Yale.

```
> telnet tigger.jvnc.net
Trying 128.121.50.145 ...
Connected to tigger.jvnc.net.
Escape character is '^]'.

SunOS UNIX (tigger.jvnc.net)

SunOS UNIX (pascal)

login: xxxxx
Password:
Last login: Wed Aug 23 19:24:02
TERM = vt100, PRINTER = lp
```

Desktop *telnet* products offer extra functionality, such as choosing the type of terminal to emulate from a list, saving all or part of a session in a log file,

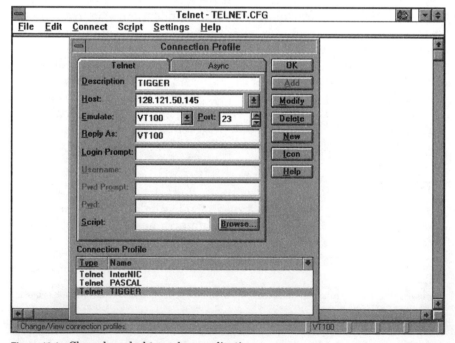

Figure 13.1 Chameleon desktop *telnet* application.

configuring your keyboard, or storing all of the information needed to access a frequently visited site. Some of these features are shown in Figure 13.1.

13.3 TELNETTING TO A SELECTED PORT

Port 23 is the standard well-known port for *telnet* terminal access. When a client connects to port 23, the normal response is a prompt for a login ID and password.

But since *telnet* was designed to be used as a tool for general application-to-application communications, it includes a magic carpet that can take a client to any port. For example, in the dialogue below, we connect to a popular weather reporting service at the University of Michigan that runs at port 3000 and requires no login ID or password:

```
> telnet madlab.sprl.umich.edu 3000
Trying 141.213.23.12 ...
Connected to madlab.engin.umich.edu.
Escape character is '^]'.

*                    University of Michigan                    *
*                    WEATHER UNDERGROUND                       *
==============================================================
*                                                             *
*         College of Engineering, University of Michigan      *
*      Department of Atmospheric, Oceanic, and Space Sciences *
*             Ann Arbor, Michigan 48109-2143                  *
*             comments: ldm@cirrus.sprl.umich.edu             *
*                                                             *
*   With Help from: The National Science Foundation supported *
*                    Unidata Project                          *
*        University Corporation for Atmospheric Research       *
*              Boulder, Colorado 80307-3000                   *
*                                                             *
*  Commercial, for-profit users should contact our data provider, *
*  Alden Electronics, 508-366-8851 to acquire their own data feed. *
*             comments: ldm@cirrus.sprl.umich.edu             *
*                                                             *
--------------------------------------------------------------
* NOTE:--> New users, please select option "H" on the main menu: *
* H) Help and information for new users                       *
--------------------------------------------------------------
Press Return for menu, or enter 3 letter forecast city code:
```

The ability to *telnet* to any port has proved to be a great convenience. It also has turned out to be a potential source of security problems, when crackers crash into a site via a poorly implemented program running openly at some port.

13.4 TELNET TERMINAL EMULATION MODEL

As shown in Figure 13.2, a user at a real terminal interacts with the local *telnet client* program. The *telnet* client program has to accept keystrokes from the user's keyboard, interpret them, and display output on the user's screen in a manner that is consistent with the emulation in use.

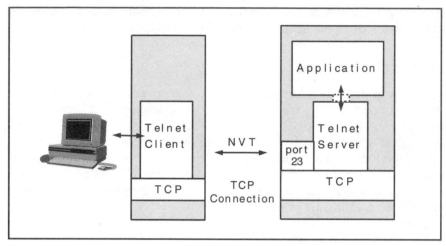

Figure 13.2 *Telnet* client and server.

The *telnet* client opens a TCP connection to the *telnet* server which is accessed at well-known port 23. The *telnet* server interacts with applications and assists in emulating a native terminal.

13.4.1 Network Virtual Terminal

In order to get the session going, both sides initially exchange information using a very simple protocol called *Network Virtual Terminal,* or NVT.

The NVT protocol was modeled on an old-fashioned half-duplex keyboard and printer operating in line-at-a-time mode. NVT has well-defined characteristics:

- NVT data is made up of 7-bit USASCII characters padded to 8 bits via an initial 0 bit.

- Data is sent a line at a time.

- Each line ends with an ASCII Carriage Return (CR) Linefeed (LF) combination.

- Bytes with an initial (high-order) 1 bit are used for command codes.

- The protocol is "half-duplex." After sending a line, the client waits to receive data from the server. The server sends its data and then sends a *Go Ahead* command, indicating that the client can now send another line.

13.5 COMMON TERMINAL TYPES

Usually, the client and server stay in NVT mode for a very short time—just long enough to negotiate some type of terminal to be emulated, such as an ASCII VT100 or IBM 3270.

In the years since *telnet* was introduced, many terminal emulations have been added to the protocol.

13.5.1 ASCII Terminals

ASCII terminals are used with Unix and Digital Equipment Corporation VAX computers. ASCII terminals are characterized by:

- *Remote echoing* of each character. That is, each character is sent to the remote host and then is sent back before it appears on the user's screen. (This is very hard on the network.)
- *Full duplex transmission.* Characters flow in both directions simultaneously. The server will *not* need to send *Go Ahead* control codes.
- Support for *interactive* full-screen applications—with a lot of network overhead.
- A larger ASCII character set than NVT.

Basic ASCII terminal characteristics are defined in standards ANSI X3.64, ISO 6429, and ISO 2022. There have been many ASCII terminal products, each offering a few extra features (for example, ANSI, VT52, VT100, VT220, TVI950, TVI955, and WYSE50). The VT100 terminal type is frequently emulated for remote logins to Unix computers.

13.5.2 Configuring Keyboards

A PC or Macintosh keyboard is not identical to a VT100 or 3270 keyboard. A *telnet* product usually provides a way to configure individual keys—or control key combinations—to perform functions normally available on the keyboards of emulated systems. For example, one problem that arises with VT*XXX* terminals is that the command character used to erase a mistyped character is not standardized. Some terminals use *Backspace,* while others use *Del.*

For Unix systems, you usually customize your terminal keyboard by referencing entries in a configuration file called */etc/termcaps.* The *Chameleon* Windows *telnet* configuration screen shown in Figure 13.3 is a lot easier to use. It enables you to drag keys from the top keyboard and drop them onto any convenient PC key. For example, if you wanted a PC's *Backspace* key to send the *Del* code, you would just drag the VT100 *Del* key onto the PC keyboard *Backspace* key.

13.5.3 IBM 3270 and 5250 Terminals

An IBM mainframe often needs to support hundreds or thousands of interactive terminals. For many years, proprietary 3270 terminal systems have been used to access IBM mainframes. These terminals were optimized for data processing applications.

3270 terminals operate in *Block Mode,* which means that a user works with a screen of data at a time. When the user presses ENTER or some other func-

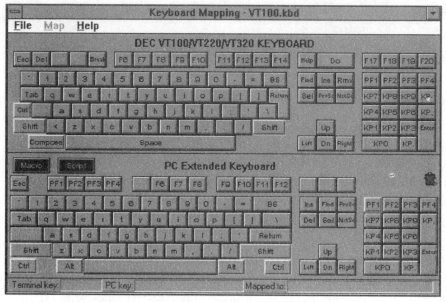

Figure 13.3   Drag-and-drop keyboard mapping.

tion key, the information on the screen is sent to the host. The keyboard locks, and the host processes the data. The host then sends back one or more screens of data. When the host is finished, it unlocks the keyboard. 3270s are characterized by:

- EBCDIC 8-bit codes
- Half-duplex communication
- Block mode

IBM 5250 terminals, used to access AS/400 computers, also have these characteristics.

13.6   OPTIONS

Terminal emulation characteristics are established by exchanging commands that negotiate *telnet options*. Either side may ask its partner to *DO* a particular option, such as "echo individual characters." The partner can accept or refuse. Either side can volunteer that it *WILL* perform some option. Again, the partner can accept or refuse.

There are four request/response exchanges that will be seen during option negotiation:

DO (*option code*)	Ask partner to perform option.
WILL (*option code*)	Partner agrees. Option now holds.
DO (*option code*)	Ask partner to perform option.

WON'T (option code)	Partner refuses. Status unchanged.
WILL (option code)	Indicates desire to begin option.
DO (option code)	Partner gives permission. Option now holds.
WILL (option code)	Indicates desire to begin option.
DON'T (option code)	Partner refuses. Status unchanged.
WON'T (option code)	Confirms that status will be unchanged.

At connection start-up, a swarm of option requests bounce back and forth between the partners. Occasionally options also are exchanged in mid-session. Some options signal the start of *subnegotiations* in which additional information is exchanged.

What happens if both sides refuse every option request? The session will stay in NVT mode.

13.6.1 Terminal Type Option

A very important option—*Terminal Type*—uses subnegotiation. A client sending *WILL TERMINAL TYPE* wants to tell the server about the types of terminals that it can emulate. If the server is willing to see this information, it responds with *DO TERMINAL TYPE*.

Subsequently, in a subnegotiation, the server asks the client to identify one of the terminal types that it can emulate, and the client responds. The server can repeat the query until either the client provides a type that is acceptable to the server or the client's list of supported types is exhausted. Formal terminal type identifiers such as DEC-VT100, HP-2648, or IBM-3278-2 are defined in the *Assigned Numbers* RFC document.

13.6.2 Negotiating VT100 Options

In the sample dialogue below, we start up *telnet* and enter *toggle options* to cause *telnet* to show us its negotiations. Then *open* is used to start a login. The partners negotiate an ASCII VT100 emulation by selecting the following characteristics:

- The server will not send *Go Aheads* because the session will be full-duplex.

- A *Terminal Type* subnegotiation will be used to identify the specific ASCII terminal to be emulated.

- The server will *echo* the client's characters.

Neither side needs to wait for an answer to an option request before sending another request. A negotiator doesn't even have to respond to options in the same order that they were received. As a result, a series of negotiations sometimes needs to be untangled before it can be understood.

```
> telnet
telnet> toggle options
Will show option processing.
```

```
telnet> open cantor.cs.yale.edu
Trying 128.36.12.26 ...
Connected to cantor.cs.yale.edu.
Escape character is '^]'.

SENT do SUPPRESS GO AHEAD
SENT will TERMINAL TYPE (reply)
RCVD do TERMINAL TYPE (don't reply)
RCVD will SUPPRESS GO AHEAD (don't reply)
RCVD will ECHO (reply)
SENT do ECHO (reply)

login:
```

13.6.3   Negotiating 3270 Options

A similar exchange is used to set up a 3270 emulation. The dialogue below shows option negotiation for a 3270 login to an IBM VM host. In this example, the remote host displays the Terminal Type subnegotiation on the screen. The partners negotiate an IBM 3278 Model 2 emulation by selecting the following characteristics:

- A Terminal Type *subnegotiation* identifies the specific 3270 terminal to be emulated, a 3278 model 2.

- The client and server both request *END OF RECORD* options in order to set up the Block Mode environment used by 3270s.

- Both sides agree to use 8-bit BINARY data to represent 3270 data stream characters.

```
> tn3270

tn3270> toggle options
Will show option processing.

tn3270> open uoft.utoledo.edu
Trying...
Connected to uoft.utoledo.edu.
RCVD do TERMINAL TYPE (reply)
SENT will TERMINAL TYPE (don't reply)
Received suboption Terminal type - request to send.
Sent suboption Terminal type is IBM-3278-2.
RCVD do END OF RECORD (reply)
SENT will END OF RECORD (don't reply)
RCVD will END OF RECORD (reply)
SENT do END OF RECORD (reply)
RCVD do BINARY (reply)
SENT will BINARY (don't reply)
RCVD will BINARY (reply)
SENT do BINARY (reply)

RUNNING
```

13.7   CONTROLLING A TEXT-BASED TELNET CLIENT

From time to time, you may want to interact with a text-based *telnet* client to set or display its parameters. You can find out about your own implementation by entering *telnet* and typing "?" or "help" to find out about your local commands.

```
> telnet

telnet> ?
Commands may be abbreviated. Commands are:

close     close current connection
display   display operating parameters
mode      try to enter line-by-line or character-at-a-time mode
open      connect to a site
quit      exit telnet
send      transmit special characters ('send ?' for more)
set       set operating parameters ('set ?' for more)
status    print status information
toggle    toggle operating parameters ('toggle ?' for more)
z         suspend telnet
?         print help information
```

Once you are within the *telnet* environment, the *open* command is used to connect to a remote host.

```
telnet> open plum.math.yale.edu
Trying 130.132.23.16 ...
Connected to plum.math.yale.edu.
Escape character is ''].

login: xxxxxxxx
Password: xxxxxxxx
Last login: Sat Dec 28 06:30:44 from golem.cs.yale.ed
Sun UNIX 4.2 Release 3.4 (Plum-EGP) #3: Tue Aug 2 10:25:24 EDT 1988
**********************************************************************
*                                                                    *
*       Welcome to the Yale Mathematics Department's Fabulous        *
*                         ** Plum **                                 *
**********************************************************************
You have mail.
```

13.7.1 An Important Control Sequence

How can a user change an active session's characteristics or abort the session? One keyboard control sequence always is reserved to mean *escape to telnet command mode*. The default sequence usually is CONTROL and] (sometimes denoted by ^]). This escape sequence can be redefined by the user. Note the reminder that was printed in the previous dialogue, three lines after the connection to *plum.math.yale.edu* was opened:

```
Escape character is '^]'.
```

At this point we will continue the dialogue. After the escape sequence is entered, we get a *telnet* prompt and can look at the status of the current session:

```
^]
telnet> status
Connected to plum.math.yale.edu.
Operating in character-at-a-time mode.
Escape character is '^]'.
```

After executing one command, we are returned to terminal emulation mode.

To enter another control command, we have to enter another escape
sequence. We will ask *telnet* to display its current attributes:

```
^]
telnet> display
will flush output when sending interrupt characters.
won't send interrupt characters in urgent mode.
won't map carriage return on output.
won't recognize certain control characters.
won't process ^S/^Q locally.
won't turn on socket level debugging.
won't print hexadecimal representation of network traffic.
won't show option processing.

[^E] echo.
[^]] escape.
[^?] erase.
[^O] flushoutput.
[^C] interrupt.
[^U] kill.
[^\] quit.
[^D] eof.
```

13.8   NVT FEATURES

In the sections that follow, we will investigate the structure of *telnet* in more
detail and get an idea of the features that *telnet* can offer a developer who is
building a client/server application.

After option negotiation is complete, a particular terminal emulation might
provide a rich repertoire of characters and graphic symbols for the interaction
between a user and an application.

However, when *telnet* is used as a building block for client/server applications, often most or all of the interaction occurs in simple NVT mode.
Therefore, let's take a look at the characteristics of a standard NVT session.

13.8.1   NVT Character Set

The octets sent on an NVT session represent *USASCII* characters and *telnet*
commands. There are 128 USASCII characters. 95 are printable letters, numbers, symbols, and punctuation marks; 33 are ASCII control characters (such
as *horizontal tab*). USASCII was designed as a 7-bit code set. The USASCII
characters are transmitted as octets whose high-order bit is set to 0.

13.8.2   NVT "Printer"

During a basic NVT session, the *telnet* server sends characters and controls to
the client's *NVT printer*—that is, the user's display. The NVT display is
restricted to showing the 95 printable USASCII characters (corresponding to
ASCII codes 32 through 126).

There is a small repertoire of ASCII control characters that the server can
use to manipulate the client's display. These are shown in Table 13.1. In the
table, the ASCII codes have been translated to decimal numbers.

TABLE 13.1   NVT Printer Controls

Description	ASCII code
Null (for filler time)	0
Bell, to sound an audible signal	7
Backspace, to move one space left	8
Horizontal tab	9
Line feed	10
Vertical tab	11
Form feed (move to top of next "page")	12
Carriage return	13

13.8.3   NVT Telnet Client/Server Interaction

Recall that an NVT interaction is half-duplex, which means that at any time, either the *telnet* client or the *telnet* server is in charge:

- After the *telnet* client has sent a line terminated by CR and LF, control switches to the server.

- The server sends lines of output to the client. At the end of each line of output, the server uses CR and LF to move to the next line of the client's display.

- The *telnet* client accepts output from the server and can enter input again after receiving a *Go Ahead* control code sequence from the server.

Note that lines sent across the *telnet* session end in CR LF, no matter what the client and server hosts may use as their local end-of-line characters. Each host translates its end of lines to and from the *telnet* end-of-line characters.

13.9   TELNET COMMANDS

Before networks became common, terminals were directly attached to computers. The keys that a user typed were immediately interpreted by the local computer's operating system.

There were some special control keys that caused the operating system to wake up and take notice. For example, an ASCII terminal user might hold down the CONTROL key and C at the same time (written ^C) to ask the operating system to kill the currently running application.

During a *telnet* session, control codes need to be translated to *telnet* commands and passed to the operating system at the *remote* end of a network connection. The *telnet* client program must therefore handle all of the raw keystrokes typed by the user, translate special control keys into *telnet* commands, and pass them to the *telnet* server.

Telnet commands are represented by an *Interpret As Command* (IAC) byte followed by one or more code bytes:

Interpret As Command is X'FF, or, in decimal, 255.

The *telnet* client sends command sequences to the server to support useful functions such as:

Break (BRK)	Send a break or attention signal to the remote application process.
Interrupt Process (IP)	Signal the remote operating system to stop the currently running remote application program (e.g., to stop a program that is in a loop).
Abort Output (AO)	Ask the server application not to send the rest of the output for the current operation.
Are You There? (AYT)	This means show evidence that the server is still running.
Erase Character (EC)	A user who makes a typing error while entering a line of data generally corrects it by using a *Backspace* or *Del* key. When operating in ASCII character-at-a-time mode, the characters have already been sent to the remote application, so EC must be sent across the connection.
Erase Line (EL)	Asks the remote application to erase the current line.

Commands can be sent even after negotiation, when partners are no longer in basic NVT mode. But suppose that negotiation has enabled the partners to send binary data. How can a command sequence be recognized? The way that this is done is that whenever X'FF appears as *data,* it is doubled by the sender. The receiver corrects the doubling. When a receiver sees a single incoming X'FF (or an odd number of them), the receiver knows that a command is arriving.

It is easy to see how *telnet* commands could be useful to a client/server product developer. For example, the result of clicking a World Wide Web browser's STOP button might be to send an *Abort Output* command, to halt the downloading of a big image or a very long document.

By the way, it is easiest to understand *telnet*'s capabilities by thinking of an end user working at a client end and an application at the server end. But it is important to note that when using *telnet* as a development toolkit, commands can be sent in either direction.

13.9.1   Synch Signal

For some functions, such as *Interrupt Process,* just sticking a command into the data stream does not really do the trick. When a *real* terminal sends an interrupt, the host operating system sees it right away. The host would stop the currently running application promptly.

But *telnet* runs over a TCP session which delivers data *in order.* And normally, the remote *telnet* server processes all data that it receives *in order.* It might be quite a while before the *telnet* server sees an interrupt command that is buried in an incoming data stream.

The client wants to get the server's attention and tell the *telnet* server, "Throw away all of the characters that you have buffered *up to here* except for the commands." The client gets the server to do this by sending a special TCP segment called a *Synch signal*.

- This segment is flagged as *Urgent Data*.
- The server will throw away everything except commands until it reaches a special command code called a *Data Mark* (DM).
- DM marks the spot where the server *stops* discarding data.

When the Synch signal segment arrives, the server extracts the NVT commands from the data stream and tosses away everything else—up to the Data Mark. The server executes the NVT commands. Normal processing resumes with data past the Data Mark.

13.9.2 Encoding Common Commands

Table 13.2 lists acronyms for some common commands along with the decimal values of their codes. Each would be preceded by 255 (X'FF) when sent across the *telnet* connection.

TABLE 13.2   Telnet Command Codes

Acronym	Command	Code
EOF	End of File	236
SUSP	Suspend Current Process	237
ABORT	Abort Process	238
EOR	End of Record	239
NOP	No Operation	241
DM	Data Mark	242
BRK	Break	243
IP	Interrupt Process	244
AO	Abort Output	245
AYT	Are You There	246
EC	Erase Character	247
EL	Erase Line	248
GA	Go Ahead	249

13.9.3 Encoding Option Requests

Option requests are encoded with three bytes consisting of *IAC*, a request octet, and an option code. For example, the decimal representation of the sequence for *WILL TERMINAL TYPE* is:

IAC	WILL	TERMINAL TYPE
255	251	24

This is one of the options that opens the door to subnegotiation. Subsequently, there could be a subnegotiation exchange:

SERVER:

IAC	SB	TERMINAL TYPE	SEND	IAC	SE
255	250	24	1	255	240

CLIENT:

IAC	SB	TERMINAL TYPE	IS	DEC-VT220	IAC	SE
255	250	24	0	DEC-VT220	255	240

Table 13.3 displays the decimal values for the negotiation and subnegotiation codes. It also includes the code numbers assigned to some frequently used options. Subnegotiation parameters and additional option numbers are defined in a variety of RFCs dealing with *telnet* options and are listed in the *Assigned Numbers* RFC.

13.9.4 More About Options

More than 30 RFCs have been written detailing options that define specialized features. Some interesting options include:

- The ability to poll the partner for the current option settings. A status request and response are carried in a subnegotiation exchange.

- Negotiating Window Size. The partners agree that the client can use a subnegotiation to inform the server of the height and width of the window that will be used for the *telnet* session. This feature is useful when a *telnet* session is run at a windowing station.

TABLE 13.3 Negotiation and Option Codes

Request	Code
Negotiation codes	
WILL	251
WON'T	252
DO	253
DON'T	254
SB Start Subnegotiation	250
SE End Subnegotiation	240
Command Option	Code
Sample option codes	
Transmit Binary	0
Echo	1
Suppress Go Ahead	3
Status	5
Timing Mark	6
Output Line Width	8
Output Page Size	9
Extended ASCII	17
Data Entry Terminal	20
Terminal Type	24
End of Record	25
Window Size	31
Terminal Speed	32
Remote Flow Control	33
Linemode	34
Authentication	37
Encryption	38
Extended Options List	255

An implementation is not required to support all, or even most of, the defined options. Two of the options used for 3270 emulation have special features:

- *Transmit Binary.*   Begin transmitting 8-bit binary data. Recall that IBM 3270 sessions are conducted in binary.

- *End of Record.*   The partner receiving DO END-OF-RECORD will use a standard control code of IAC 239 to denote end of record in its data stream.

Recall that even after entering binary mode, *telnet* commands can be sent to the partner by doubling *IAC* escape characters.

13.10   TELNET USES

From the point of view of users who want to access applications via ASCII or IBM terminal emulations, the most important *telnet* feature is the ability to negotiate options and carry out emulations. But to application developers, a basic NVT *telnet* offers a sackfull of client/server functions that would be difficult—and tedious—to reproduce. As we have seen, a basic NVT has the ability to:

- Find out if the peer application is alive
- Signal a break
- Request that the current remote process be interrupted
- Use a Synch signal to tell the peer to discard all data other than *telnet* commands
- Tell the partner to discard pending buffered output instead of sending it to you

13.11   SECURITY ISSUES

Today, broadcast-style LANs are still in common use. Many organizations even use a broadcast-style Fiber Distributed Data Interface (FDDI) LAN as a backbone network.

It is very easy for a PC or Macintosh user to obtain software that turns a desktop system into a "snooper" that can eavesdrop on LAN traffic. Many Unix stations have the capability built in—it just has to be enabled by the station owner.

Traditionally, a user has "proved" its identity by sending a secret password to a host. But in a broadcast LAN environment, sending a userid and password in the clear across a network is almost as bad as yelling them in a crowded room. Somebody simply has to take the trouble to listen.

Just encrypting the password does not help. An eavesdropper can pick up the encrypted version of the password and send *that* in order to access your account. What is needed is a secure authentication mechanism.

13.11.1   Telnet Authentication

There is a *telnet authentication* option that enables *telnet* partners to agree that secure identification will be used and to negotiate the specific authentication method. The scenario is:

- The server sends *DO AUTHENTICATION.*
- The Client sends *WILL AUTHENTICATION.*

From this point onward, all of the information will be exchanged in subnegotiation messages.

- The Server sends a message containing a list of authentication pairs. Each pair includes a *type of authentication* to be used and a modifier providing extra information, such as whether only the client will send authentication or whether both the client and the server need to prove their identity.
- The client sends a simple userid or account ID.
- The client chooses one of the authentication pairs from the list. The client sends a message identifying the authentication type and including authentication data. Depending on the protocol, more than one message may be needed.
- The server accepts the authentication.
- If authentication will be mutual, the client asks for the server's authentication data.
- The server replies with a message including its authentication data.

Authentication types are registered with the Internet Assigned Numbers Authority (IANA) and are assigned code numbers. Current codes and types include:

Code	Type
1	KERBEROS_V4
2	KERBEROS_V5
3	SPX
6	RSA
10	LOKI
11	SSA

Challenge handshakes and secure ID cards are becoming popular in actual implementations.

13.12   PERFORMANCE ISSUES

Telnet has some serious performance disadvantages. When emulating an ASCII terminal such as a VT100, *telnet* is very inefficient. Segments sent by the client often contain a single character or just a few characters. Each character must be echoed. A lot of overhead is consumed in order to transfer very little data.

Each interactive application has a different user interface, with its own commands, control codes, and conventions. This means that users must be trained to use the application, and sometimes it takes a fairly long time to become proficient.

Today, many new applications are being built so that the user can access information via a standard client, such as a World Wide Web browser. The application developer needs to write an interface between a new application and a Web server. But then users can work with a uniform and familiar user interface.

13.13   X WINDOWS

Recently, many applications have been designed for a standard X-terminal interface, rather than a proprietary terminal. The X Window System, designed and developed at the Massachusetts Institute of Technology, enables a user to run several concurrent applications in windows at the user's display. X Windows does not care where an application is located. Each application may in fact be running in a different computer on the network.

The X Window protocol provides a uniform way for applications to handle input and output to a display window. It is designed to be independent of hardware, operating system, and network type. Current implementations run over TCP/IP.

The protocol can run in a workstation or on a multiuser computer that controls bit-mapped displays. There are many dedicated X Window display products. X Window support is very widespread, and there are highly functional application development tools available. X Window development tools frequently are bundled with TCP/IP products.

There are some disadvantages to the use of X Windows. In order to draw the screen, a lot of data flows to the station—it is a heavy user of bandwidth.

There also are security problems associated with X Windows. It is not easy to prevent a process from masquerading as a legitimate data source.

13.14   RECOMMENDED READING

RFC 854 defines the *telnet* protocol. Options defining different terminal types have been published in: RFC 1205 for 5250 emulation, RFC 1096 for X display location, RFC 1053 for the X.3 PAD option, RFC 1043 for data entry terminals, and RFC 1041 for 3270 regimes. The terminal-type option is explained in RFC 1091, and the window size option can be found in RFC 1073. RFC 1184 describes the *telnet* linemode option. RFCs 855 through 861 describe other frequently used emulation options.

RFC 1416 describes the *telnet* authentication option. RFC 1510 presents the Kerberos Network Authentication Service.

14

File Transfer Protocol

14.1 INTRODUCTION

In a networked environment, it is natural to wish to copy files between computer systems. Why isn't it always easy to do this? Computer vendors have devised hundreds of file systems. These differ in dozens of minor ways and in quite a few drastic major ways too! This is not just a multivendor problem. Sometimes it is difficult to copy files between two different types of computers manufactured by the same vendor.

Among the problems that may be encountered when dealing with a multi-system environment are:

- Different conventions for naming files
- Different rules for traversing file directory systems
- File access restrictions
- Different ways to represent text and data within files

The designers of the TCP/IP protocol suite did not try to create a very complicated general solution to every file transfer problem. Instead, they produced a fairly basic but elegant *File Transfer Protocol* (FTP) that is serviceable and easy to use.

The File Transfer Protocol is designed to be operated by interactive end users or by application programs. We will confine the discussion here to the familiar interactive FTP end-user service that is universally available with TCP/IP implementations.

The user interface developed for the Berkeley Unix file transfer client has been ported to many types of multiuser computers. In this chapter we will see end-user dialogues based on that text-based interface, as well as some graphical desktop user interfaces.

The core file transfer functions enable users to copy files between systems, view directory listings, and perform housekeeping chores—such as renaming or deleting files. These functions are part of the standard TCP/IP protocol suite.

The *Trivial File Transfer Protocol* (TFTP), discussed at the end of this chapter, is a bare-bones file transfer protocol used for special situations such as downloading software to routers, bridges, or diskless workstations.

14.2   PUBLIC AND PRIVATE FTP

Computer systems usually require a user to provide a login ID and a password before the user can view or manipulate files. However, there are times when it is useful to create a public file area. FTP accommodates both public information sharing and private file access by offering two kinds of services:

- Access to public files by means of "anonymous" logins
- Access to private files, which is restricted to users with system login identifiers and passwords

14.2.1   An Introductory Dialogue

The dialogue that follows demonstrates how a file may be copied from the public repository of RFC documents at the AT&T InterNIC Data Services site.

Today, many people have desktop Graphical User Interfaces (GUIs) for file transfer, and one of these will be shown later. However, the text interface gives a very good idea of how the protocol actually operates, so initially we will connect to the InterNIC using a text-based client.

The InterNIC file archive is public, so we will enter the login identifier *ftp*. Traditionally, public systems were set up to accept login identifier *anonymous*. Now, most accept *ftp,* which is easier to type. Public file transfer servers expect you to enter your electronic mail identifier at the password prompt.

The prompt *ftp>* will be displayed whenever the local FTP application is waiting for user input. Lines that start with numbers contain messages from the remote file server.

`> ftp ftp.internic.net`	The ftp command starts the FTP client user interface program. The user wants to connect to remote host ftp.internic.net.
`Connected to ftp.ds.internic.net.`	The local FTP client reports that it has connected successfully.
`220- InterNIC Directory and Database Services`	This message comes directly from the remote system.
`220- . . .`	We omit the greeting message.
`220 ds.internic.net FTP server ready.`	
`Name (ftp.internic.net:sfeit):` `ftp` `331 Guest login ok, send ident as password.`	The local FTP client prompts for a userid. The InterNIC will accept ftp.
`Password:`	The local FTP client prompts for a password. We enter the polite response, which is our email identifier.

```
230 Guest login ok, access
restrictions apply.
ftp>                                    The ftp> prompt means "What do you want to do
                                        next?"

ftp> cd rfc                             The user changes to remote directory rfc, which con-
                                        tains RFC documents.

250 CWD command successful.             The cd command was sent to the server as the formal
                                        CWD (change working directory) command. The server
                                        directory now has been changed to rfc and we are
                                        ready to get an RFC document.

ftp> get rfc1842.txt myrfc              We ask for a copy of rfc1842.txt. A second connection
                                        will be created to copy the file.

200 PORT command successful.            The local FTP client has obtained a second port and
                                        has sent a PORT command to the server, telling the
                                        server to connect to this port.

150 Opening ASCII mode data             The data connection for the file transfer is opened.
connection for rfc1842.txt
(24143 bytes).

226 Transfer complete.                  The transfer is complete.

local: newfile remote: rfc1842.txt      A new local file has been created.

24818 bytes received in 0.53
seconds (46 Kbytes/s)

ftp> quit                               Enough work for now.

221 Goodbye.
```

Our first command asked the server to change to directory *rfc*. Then we copied the remote document *rfc1842.txt* to a local file called *myrfc*. If we had not entered a filename, the local file would have been assigned the same name as the remote file.

FTP allows us to write remote filenames just as users at the remote host would write them. When we copy a file to the local computer, we can assign it a local filename. If we do not assign a name, if necessary, FTP will translate the remote filename to a format that is acceptable to the local host. Sometimes this will cause characters to be translated from lower- to upper-case and names to be truncated.

The File Transfer Protocol has a distinctive style of operation. Whenever a file needs to be copied, a second connection is opened up and used to transfer the data. After the *get* command in the dialogue, the local FTP client acquired a second port and told the server to open a connection to that port. We did not see this outgoing command, but we did see the response:

```
200 PORT command successful.
150 Opening ASCII mode data connection for rfc1842.txt (24143 bytes).
```

In Figure 14.1, we access another public archive using the *Chameleon* Windows-based file transfer client, which provides a Graphical User Interface.

Files can be copied by dragging them from one window to another or by clicking a copy arrow. You have the option of entering a local filename in the box on the left, under the "Files" label.

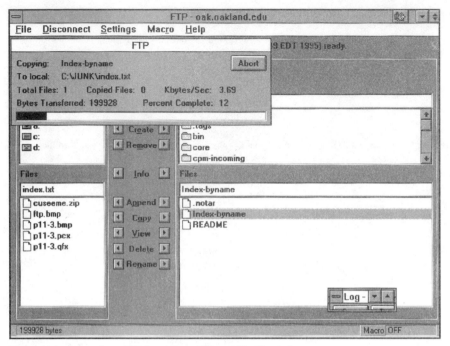

Figure 14.1 Accessing a file transfer archive with *Chameleon*.

The same site is accessed via the *Netscape* file transfer client in Figure 14.2. A file is copied by clicking on its name. A text file is displayed on the screen and can be saved by choosing *save* on the *file* menu. If a binary file is requested, a pop-up menu asks where it should be saved.

14.3 FTP MODEL

As we can see from the earlier dialogue, a user interacts with a *local client* FTP process. The local client software conducts a formal conversation with the *remote server* FTP process across a *control connection*. When an end user enters a file transfer or file management command, the command is translated into one of the special acronyms used on the control connection.

Under the covers, the control connection is just a simple NVT *telnet* session. The client sends commands to the server across the control connection, and the server sends responses back to the client across the control connection.

If the user requests a file transfer, a separate data connection is opened, and the file is copied across that connection. A data connection also is used to transmit directory listings. Figure 14.3 illustrates this model. The server normally uses port 20 for its end of a data connection.

During the dialogue in the previous section, the end user entered requests to change to a different directory and copy a file. These requests were trans-

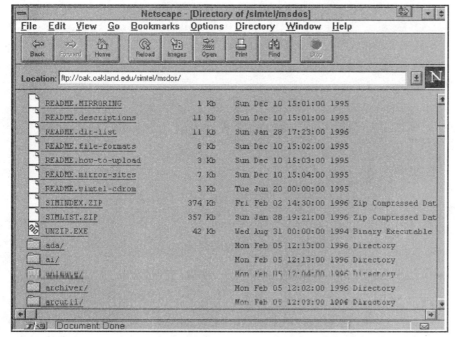

Figure 14.2 Accessing a file transfer archive using *Netscape*.

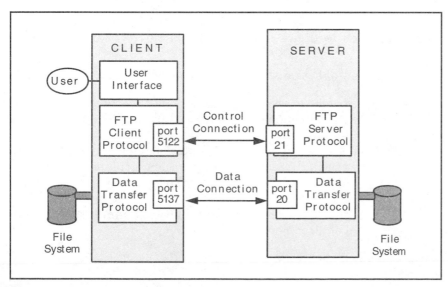

Figure 14.3 FTP control and data connections.

lated to formal FTP commands and passed across the control connection to the remote FTP server. The file transfer took place over the separate data connection that was created for the purpose.

14.4 FTP COMMANDS

What types of commands can be sent across the control connection? There are authentication commands that enable a user to identify the userid, password, and account to be used for a set of FTP activities.

There are file transfer commands that enable a user to:

- Copy a single file between hosts
- Copy multiple files between hosts
- Append a local file to a remote file
- Copy a file and append a number to its name so that the name is unique[1]

There are file management commands that enable the user to:

- List the files in a directory
- Identify the current directory and change directories
- Create and remove directories
- Rename or delete files

There are control commands that enable the user to:

- Identify whether ASCII text, EBCDIC text, or binary data is to be transferred
- Establish whether the file is structured as a series of bytes or as a sequence of records
- Describe how the file will be transferred—for example, as a stream of octets

The commands that are sent across the control connection have a standard form. For example, the *RETR* command is used to copy a file from a server to the client site.

FTP puts no restrictions on the kind of user interface that any vendor may provide, and as we have seen, desktop systems provide ingenious and easy-to-use clients. Thus, typing "get," dragging an icon, or clicking a filename all could be translated to an *RETR* command.

The user interface generally includes additional commands that let the user customize the local environment, such as:

- Ask FTP to ring a bell at the end of a transfer

---

[1]For example, a daily log file could be retrieved and automatically called log.1, log.2, and so forth.

- For a text-based user interface, ask FTP to print a hash symbol, #, for each block of data transferred

- Set up automatic translation of the case of letters in a filename or set up a table for automatically translating characters in the names of transferred files

The complete set of functions supported by a particular host can be viewed using the FTP client's help facility. Consult your system's manuals for full information.

14.4.1 Using Commands in a Text-Based Dialogue

Although many users prefer the Graphical User Interfaces available at their desktop systems, a text-based user interface reveals a lot about the inner workings of File Transfer Protocol.

The text-based dialogue that follows starts with a help display. There are a number of commands that have synonyms, such as *ls* or *dir* to ask for directory information, *put* or *send* to copy a file to a remote host, *get* or *recv* to copy a file from the remote host, and *bye* or *quit* to leave FTP.

Multiple files can be copied using *mget* or *mput* using "global wild card" naming. For example, *mget* a* retrieves a copy of every file whose name starts with the letter a. This will work if *globbing* is on; you turn global wild card naming on and off by typing *glob*.

In the dialogue, we turn debugging on in order to get some insight into how the protocol works:

- Lines starting with --> show the messages that the local host sends across the control connection.

- Lines starting with a number show the messages sent from the remote server to report the outcome of a command.

```
plum-feit > ftp
ftp> help
Commands may be abbreviated. Commands are:

!          cr          macdef      proxy       send
$          delete      mdelete     sendport    status
account    debug       mdir        put         struct
append     dir         mget        pwd         sunique
ascii      disconnect  mkdir       quit        tenex
bell       form        mls         quote       trace
binary     get         mode        recv        type
bye        glob        mput        remotehelp  user
case       hash        nmap        rename      verbose
cd         help        ntrans      reset       ?
cdup       lcd         open        rmdir
close      ls          prompt      runique

ftp> debug
Debugging on (debug = 1).

ftp> open tiger.jvnc.net
Connected to tiger.jvnc.net.
```

```
220 tigger.jvnc.net FTP server (Version wu-2.4(1) Fri Apr 15 13:54:36 EDT
1994)
ready.
```

A real userid and password are entered so that private files can be accessed.

```
Name (tigger.jvnc.net:sfeit): feit
--> USER feit
331 Password required for feit.
Password:
--> PASS abcd1234
230 User feit logged in.
```

The *status* command shows the current settings in effect for the FTP session. Many of these will be explained later in this chapter. For now, note that the data *Type* is ASCII. FTP often assumes by default that we want to transfer text files.

```
ftp> status
Connected to tigger.jvnc.net.
No proxy connection.
Mode: stream; Type: ascii; Form: non-print; Structure: file
Verbose: on; Bell: off; Prompting: on; Globbing: on
Store unique: off; Receive unique: off
Case: off; CR stripping: on
Ntrans: off
Nmap: off
Hash mark printing: off; Use of PORT cmds: on
```

Next, we will request a directory listing. Directory listings can be long, so FTP sends directory listings on a data connection:

```
ftp> dir
```

FTP needs a port for the data transfer. The client sends a *PORT* command that identifies its IP address (4 bytes) and the new port (2 bytes) to be used for data transfer. The bytes are translated to decimal and separated by commas. IP address 128.36.4.22 is written as 128,36,4,22. Port 2613 is translated to 10,53.

```
--> PORT 128,36,4,22,10,53
200 PORT command successful.
```

The server will open a connection to this socket address. *LIST* is the formal message that asks for a detailed directory listing:

```
--> LIST
```

The server now opens the connection to the client's announced port:

```
150 Opening ASCII mode data connection for /bin/ls.
total 531
-rw-r--r-- 1 feit     tigers       0 Oct 24 1994  .addressbook
-rw-r--r-- 1 feit     tigers    2808 Sep 23 1994  .article
```

```
-rw-r--r-- 1 feit     tigers    397 Mar 14 1993  .cshrc
. . .
-rw-r--r-- 1 feit     tigers   3113 Jul 31 13:29 subnets
-rw-r--r-- 1 feit     tigers  59901 Jun  5 17:48 typescript
226 Transfer complete.
2239 bytes received in 0.31 seconds (7 Kbytes/s)
```

As soon as the directory listing has been transmitted, the data connection is closed. Next, we'll get a file.

```
ftp> get subnets
```

The client identifies a new socket address for the file transfer. Note that this time, client port 2614 (10,54) is used.

```
---> PORT 128,36,4,22,10,54
200 PORT command successful.
---> RETR subnets
150 Opening ASCII mode data connection for subnets (3113 bytes).
226 Transfer complete.
```

As soon as the file transfer is complete, the data connection is closed.

```
local: subnets remote: subnets
3187 bytes received in 0.27 seconds (11 Kbytes/s)
ftp> quit
--> QUIT
221 Goodbye.
plum-feit>
```

Note that the scenario for the data connection was:

- The local client got a new port and used the control connection to tell the server FTP what the port number was.
- The FTP server connected to the client's new data port.
- The data was transferred.
- The connection was closed.

An alternative scenario is possible. If the client sends a *PASV* command, the server will send back a port number and then will listen for a data connection from the client. In the past, the use of *PORT* predominated. A client needs to use PASV when files are transferred through a simple filtering security firewall that does not permit incoming connections to be set up. We'll describe this more fully later in the chapter.

Sometime in your lifetime, you will request the transfer of a massively huge file and then realize that it was not the file that you wanted to transfer. A good implementation will enable a user to interrupt a transfer. For a text-based user interface, this usually is done by signaling interrupt via a combination such as CONTROL-C. A Graphical User Interface should provide an *Abort* button.

14.5   DATA TYPE, FILE STRUCTURE, AND TRANSFER MODE

The two ends of a file transfer dialogue need a common understanding of the format of the data that will be transferred. Is it text or binary? Is there any structuring of the data into records or blocks?

Three attributes are used to define the transfer format: *data type, file structure,* and *transmission mode.* The values that can be assigned to these attributes are described in the sections that follow. The most common case is:

- Either ASCII text or binary data is transferred.
- The file is unstructured and is viewed as a string of bytes.
- The mode of transfer is to send a file as a stream of bytes.

However, there are notable exceptions. Some hosts structure text files as a series of records. IBM hosts use EBCDIC encoding for their text files. And IBM hosts often prefer to exchange files with one another as a series of formatted blocks, rather than as a stream of bytes.

In the sections that follow, we will describe the choices available for data type, file structure, and transfer mode.

14.5.1   Data Type

A file may contain ASCII text, EBCDIC text, or binary image data.[2] A text file may contain ordinary text or text formatted for a printer. A print text file contains vertical format codes that are either:

- *Telnet* NVT vertical format controls (i.e., <CR>, <LF>, <NL>, <VT>, <FF>)
- ASA (FORTRAN) vertical format controls

ASCII non-print text is the default data type. The data type is changed by sending the formal *TYPE* command across the control connection.

14.5.2   Transferring ASCII TEXT

Although ASCII text is a "standard," computers manage to use ASCII differently. The most frequently encountered problem is that computers use different codes to represent the end of a line. Unix systems use <LF>, PCs use <CR><LF>, and Macintoshes use <CR>.

To prevent problems, the sending FTP converts a local ASCII text file to NVT format, and the receiver converts the NVT ASCII to its own local format. For example, if a text file is copied from a Unix system to a PC, all of the Unix end-of-line <LF> codes will be stored as <CR><LF> at the PC.

---

[2]There also is a type called *local* or *logical byte* to accommodate computers that have unusual byte sizes such as 11-bit bytes.

14.5.3 Transferring EBCDIC Text

Hosts that support EBCDIC will provide a suitable user interface command that will cause the *TYPE E* command to be sent across the control connection. EBCDIC text characters are sent in their normal 8-bit form. Lines are terminated with the EBCDIC new line <NL> character.

14.5.4 Transferring Binary Data

It is easy to switch from default ASCII to binary Image data. With a text user interface, you just type the *binary* command. With a Graphical User Interface, you just click a button labeled "binary." The client will change the type by sending the formal *TYPE I* command across the control connection.

What happens if you forget to switch from ASCII to binary when copying a binary file? Good implementations of FTP will warn you that you are about to do something peculiar and give you a chance to change the setting. Unfortunately, many implementations will just go ahead and "helpfully" change all binary bytes that look like end-of-line codes, stuffing in or removing bytes. A few really bad implementations start the transfer and crash in the middle.

14.5.5 File Structure

The two supported structures[3] are:

- *File-structure,* which means no structure at all. The file is viewed as a sequence of bytes.

- *Record-structure,* which applies to a file made up of a sequence of records.

File-structure is the most common and is the default for the protocol. The structure is changed to *Record* by sending the formal *STRU R* command across the control connection.

14.5.6 Transmission Mode

The transmission mode combined with the file structure determines how the data will be formatted for transfer. The three transmission modes are *stream, block,* and *compressed.*

- For stream mode and file-structure, the file is transmitted as a stream of bytes. FTP relies on TCP to provide data integrity, and no headers or delimiters are inserted into the data. The only way to signal that the end of the file has been reached is by a normal close of the data connection.

---

[3]A third type, *page-structure,* was introduced to accommodate DEC TOPS-20 files and now is obsolete.

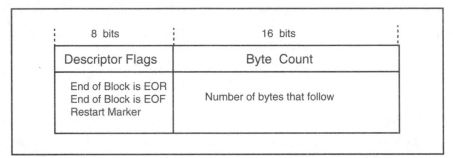

Figure 14.4   Header format used for FTP block mode transfers.

- For stream mode and record-structure, each record is delimited by a 2-byte End Of Record (EOR) control code. Another 2-byte code is used to represent End Of File (EOF).[4]

- In block mode, a file is transmitted as a series of data blocks. Each block starts with a 3-byte header. The header has the format shown in Figure 14.4.

- Compressed mode is rarely supported because it provides only a very crude method of collapsing strings of repeated bytes. Normally, the user will call on one of the excellent compression programs that are available to compress a file prior to starting a transfer. The file would then be transferred as binary data.

A block may contain an entire record, or alternatively, a record can span several blocks. Note that the descriptor contains:

- An End Of Record flag that is used to identify record boundaries

- An End Of File flag that indicates whether the block is the last one in the file transfer

- A Restart Marker flag that indicates whether this block contains a text string that can be used to identify a restart point if the transfer fails at a later point

Stream mode is the most commonly used and is the default. The mode is changed to block by sending the formal *MODE B* command across the control connection.

An advantage of using record-structure or block mode is that the end of the file will be marked clearly so that a data connection could be held and reused for multiple transfers.

---

[4]EOR is X'FF 01 and EOF is X'FF 02. For the last record in a file, EOR and EOF can be represented as X'FF 03. If the file data includes an all 1s byte, it must be sent as X'FF FF.

In the dialogue shown earlier, the response to the status command included the statement:

```
Mode: stream; Type: ascii; Form: non-print; Structure: file
```

That is, the default setting for data transfer mode was stream, the data type was ASCII non-print, and the organization of the file was file-structure, which really means unstructured.

14.6 FTP PROTOCOL

There are several elements that make up the file transfer protocol, including:

- The command words and related parameters sent on the control connection
- The numeric codes returned in response to the commands
- The format of the data to be transferred

The set of FTP commands that can be sent over the control connection is summarized in the sections that follow. The set of commands has been growing steadily for years and has become quite large. However, hosts need not implement all of the commands listed below.

Sometimes a local user interface will not directly support a command that the user wants to send—and that actually is supported at the remote host. A good implementation will offer the *quote* command, which lets you enter the exact formal command that you want to send. Your input will be transmitted across the control connection, exactly as you entered it. Therefore, it can be helpful to know the formal commands and parameters.

14.6.1 Access Control Commands

The commands and parameters that define a user's access to a remote host's filestore are defined in Table 14.1.

14.6.2 File Management Commands

The commands in Table 14.2 permit a user to execute typical directory positioning and file management functions at a remote host. The working directory is the one in which you are currently located.

14.6.3 Commands that Set Data Formats

The commands in Table 14.3 are used to establish the combination of data format, file structure, and transmission mode that will be used when copying files.

14.6.4 File Transfer Commands

The commands in Table 14.4 set up data connections, copy files, and support restart recovery.

TABLE 14.1   Commands Authorizing a User to Access a Filestore

Command	Definition	Parameter(s)
USER	Identify the user	Userid
PASS	Provide a password	Password
ACCT	Provide an account to be charged	Account ID
REIN	Reinitialize to start state	None
QUIT	Logout	None
ABOR	Abort the previous command and its associated data transfer	None

14.6.5   Miscellaneous Commands

The final set of commands in Table 14.5 provide helpful information to an end user.

14.6.6   Site Command

Many Unix file transfer servers use file transfer server software developed at Washington University in St. Louis, called WU-FTP. This implementation accepts a *SITE* command that can be used to execute a variety of special-pur-

TABLE 14.2   Directory Selection and File Management Commands

Command	Definition	Parameter(s)
CWD	Change to another server directory	Directory name
CDUP	Change to the parent directory	None
DELE	Delete a file	Filename
LIST	List information about files	Directory name, list of files, or else none to get information about the working directory
MKD	Make a directory	Directory name
NLST	List the files in a directory	Directory name or none for the working directory
PWD	Print the name of working directory	None
RMD	Remove a directory	Directory name
RNFR	Identify a file to be renamed	Filename
RNTO	Rename the file	Filename
SMNT	Mount a different file system	Identifier

TABLE 14.3   Commands that Define the Type, Structure, and Mode

Command	Definition	Parameter(s)
TYPE	Identify the data type and optionally, the print format (if any) for the transfer	A (ASCII), E (EBCDIC), I (binary Image) N (non-print), T (*telnet*), C (ASA)
STRU	Organization of the file	F (file) or R (record)
MODE	Transmission format	S (stream), B (block), C (compressed)

pose programs at the file transfer server. For example, users can first gain access with userid *ftp* and then can use the *SITE* command to provide a group login ID and password. The group login ID authorizes the user to access more files than an ordinary anonymous user can.

14.6.7   Error Recovery and Restart

Many organizations have a need to transfer very large files. Suppose that a system that is executing the transfer of a very large file fails. The FTP restart service was designed to solve this problem. Implementation of the restart service is optional. Unfortunately, at the time of this writing, few if any TCP/IP products include this service. However, we will take an optimistic view of the future and describe it here.

TABLE 14.4   Commands that Support File Transfer

Command	Definition	Parameter(s)
ALLO	Allocate (reserve) enough storage for data that follows	Integer number of bytes
APPE	Append a local file to a remote file	Filenames
PASV	Asks server to identify an IP address and port for a data connection to be initiated by the client	None. Server will return an IP address and port number
PORT	Identify a network address and port to be used for a data connection to be initiated by the server	IP Address and port number
REST	Identify a restart marker (to be followed by the transfer command to be restarted)	Marker value
RETR	Retrieve or get a file	Filename(s)
STOR	Store or put a file	Filename(s)
STOU	Store unique: create a version of a file with a unique name	Filename

TABLE 14.5   Miscellaneous User Information Commands

Command	Definition	Parameter(s)
HELP	Return information about the server implementation	None
NOOP	Asks server to return an "OK" reply	None
SITE	Used for server-specific subcommands that are not part of the standard but may be needed at the server's site	None
SYST	Asks the server to identify its operating system	None
STAT	Requests parameter information and connection status	None

If block mode transfer is used and the restart service is supported, the sending FTP can transmit blocks containing restart markers at convenient points of the data transfer. Each marker is a printable text string. For example, successive markers could be 1, 2, 3, etc. Whenever the receiver gets a marker, the receiver writes the file data onto nonvolatile storage and keeps track of the marker's position in the data.

If the client is the receiving system, the end user is notified of each marker as soon as the data has been stored. If the remote server is the receiving system, a message is sent back to the user on the control connection indicating that data up to the marker has been safe-stored.

After a system failure, the user can invoke a restart command with a marker value as its argument. This is followed by the command that was being executed (get or put) when the system failed.

14.6.8   Reply Codes

Each command in a dialogue is answered with a reply code and message. For example:

```
ftp> get subnets
--> PORT 128,36,0,22,10,54
200 PORT command successful.
--> RETR subnets
150 Opening ASCII mode data connection for subnets (3113 bytes).
226 Transfer complete.
```

Reply codes consist of three digits. Each has a specific purpose:

- Codes in the 200s indicate successful completion of a command.

- Codes in the 100s indicate that an action is being started.

- Codes in the 300s indicate that an intermediate point has been reached successfully.

- Codes in the 400s signal transient errors.
- Codes in the 500s are really bad news and announce a permanent error.

The second and third digits of a reply code classify the reply more precisely.

14.7 SECURITY ISSUES

14.7.1 Checking the Client's Host Name

Sometimes, users are mysteriously refused access to an anonymous file archive. If it only happens occasionally, the cause is that the server just is busy. If it happens all of the time, you may have a Domain Name problem.

Some file transfer servers will not accept a client from any system that is not listed in the Domain Name System. The FTP server will perform a reverse lookup on each incoming IP address. If it is not in the DNS database, access is refused. The only solution is to contact your DNS administrator and arrange to have your computer listed. Some servers do a *double lookup*. They translate a client address to a name and then translate that name to an address and compare the result with the original address. This means that a wild card DNS entry is not sufficient.

14.7.2 PASV Versus PORT

Organizations protect their networks by configuring a firewall system that applies filtering criteria to datagrams and discards any traffic that has not been specifically enabled. Often, a simple filtering firewall allows users on a LAN to initiate connections to external locations but rejects all incoming connection attempts.

The original FTP specification defined the *PORT* command as the default way to set up a data connection. As a result, many implementations were built for which this was the only way to set up a data connection. But use of *PORT* requires an external file server to open a connection *to* the client, which normally will be forbidden by a firewall.

Fortunately, many newer implementations support the *PASV* command, which tells the server to obtain a new port for a data connection, send the server's IP address and port number back in the response, and listen on the port. The client then will open the data connection to the server.

14.7.3 Proxy Relay Firewalls

Some organizations want a further level of protection. Every request is actually relayed to an intermediate firewall host. The firewall host, which is set up with close attention to security, is the only system that actually communicates directly with the outside world. A client is prepared to use a firewall by entering:

- The name or IP address of the firewall
- A userid and password to gain access to the firewall

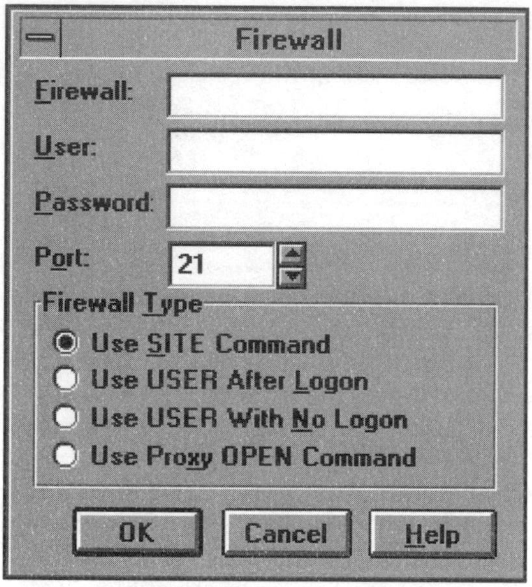

Figure 14.5 Configuring a client to use a firewall.

- The port at which file transfer users will access the firewall (it does not have to be 21)

- Additional information, depending on how the firewall has been implemented

Figure 14.5 illustrates a client's firewall configuration screen. Once the configuration data is entered, the end user operates the file transfer application in the normal way. The relaying process happens behind the scenes, depending on the type of firewall that is in use. Some firewalls require local users to connect to the firewall and enter a userid and password before any transactions can be carried out.

14.8 PERFORMANCE ISSUES

The efficiency of file transfer operations depends on a number of factors:

- Host file system and disk efficiency
- Processing required to reformat data
- Underlying TCP service

Note that a brief throughput report usually is printed at the end of a file transfer:

```
226 Transfer complete
local: rfc1261 remote: rfc1261
4488 bytes sent in 0.037 seconds (1.2e + 02 Kbytes/s)
```

A bulk transfer can be initiated in order to obtain a rough measure of TCP and FTP performance.

14.9 TRIVIAL FILE TRANSFER PROTOCOL

There are file copying applications that require a very simple level of functionality. For example, an initial download of software and configuration files to a router, hub, or diskless workstation that is booting is best carried out using a very simple protocol.

The Trivial File Transfer Protocol has proved to be very useful for basic file copying between computers. TFTP transfers data via UDP datagrams.[5] Very little communications software—only IP and UDP—need to be running in a system that participates in a TFTP download. TFTP has proved to be very useful for initializing network devices such as routers, bridges, and hubs.

The Trivial File Transfer Protocol:

- Sends 512-octet blocks of data (except for the last)
- Prefixes each data block with a simple 4-octet header
- Numbers the blocks starting with 1
- Supports ASCII or binary octet transfers
- Can be used to read or write a remote file
- Has no provision for user authentication

One partner in a TFTP interaction sends numbered, uniformly sized blocks of data, and the other partner acknowledges (ACKs) the data as it arrives. The sender must wait for an ACK of a block before sending the next block. If no ACK arrives within a timeout period, the current block is resent. Similarly, if the receiver does not get any data during a timeout period, an ACK is retransmitted.

14.9.1 TFTP Protocol

A TFTP session starts with a *Read Request* or a *Write Request*. The TFTP client starts off by obtaining a port and then sends the Read Request or Write Request message to port 69 at the server. The server should then identify a different server port which it will use for the remainder of the file transfer. The server directs its messages to the client's port. Data transfer proceeds with an exchange of data blocks and ACKs.

All blocks except for the last must contain 512 octets of data, and this is how End of File is signaled. If the file's length is a multiple of 512, the final block

---

[5]TFTP could be run on top of a packet delivery service for another protocol family.

consists of a header and no data. Data blocks are numbered, starting with 1. Each ACK contains the block number of the data that it is acknowledging.

14.9.2   TFTP Protocol Data Units

There are five types of protocol data units:

1. Read Request (RRQ)
2. Write Request (WRQ)
3. Data (DATA)
4. Acknowledgment (ACK)
5. Error (ERROR)

Error messages signal conditions such as "file not found" or "no space to write file on disk."

Each TFTP header starts with an operation code identifying its Protocol Data Unit (PDU) type. The formats for the PDUs are displayed in Figure 14.6.

Note that the lengths of Read Requests and Write Requests vary, depending on the length of the filename and mode fields, each of which contains an ASCII text string terminated by a 0 byte. The mode field contains "netascii" or "octet."

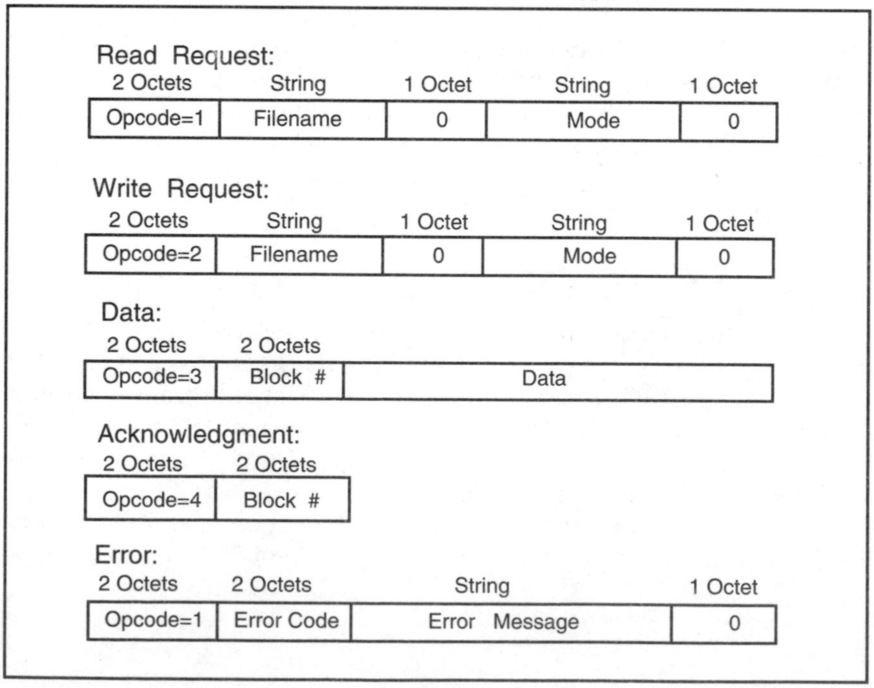

Read Request:

2 Octets	String	1 Octet	String	1 Octet
Opcode=1	Filename	0	Mode	0

Write Request:

2 Octets	String	1 Octet	String	1 Octet
Opcode=2	Filename	0	Mode	0

Data:

2 Octets	2 Octets	
Opcode=3	Block #	Data

Acknowledgment:

2 Octets	2 Octets
Opcode=4	Block #

Error:

2 Octets	2 Octets	String	1 Octet
Opcode=1	Error Code	Error Message	0

Figure 14.6   Formats for TFTP protocol data units.

14.9.3 TFTP Options

An extension to the TFTP protocol that would enable the negotiation of options via read or write requests has been proposed. The main motivation was to allow the client and server to agree on a block size that is larger than 512 bytes, enabling more efficient transfers.

14.9.4 TFTP Scenario

The protocol can be illustrated by means of a simple scenario. Figure 14.7 illustrates how a TFTP originator reads a remote file. After the responder sends a block to the reader, the responder will wait until an ACK for the block arrives before sending the next block.

14.10 RECOMMENDED READING

RFC 959 defines File Transfer Protocol, and RFC 1350 describes the Trivial File Transfer Protocol.

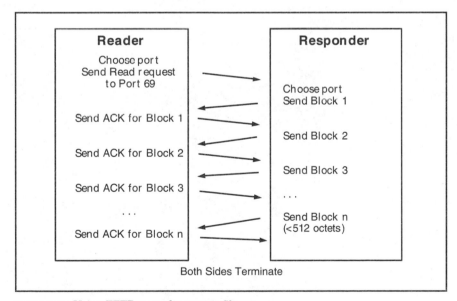

Figure 14.7 Using TFTP to read a remote file.

15

RPC and NFS

15.1 INTRODUCTION

The computer environment has changed over the past 10 years. Instead of dumb terminals tied to a central computer, we find intelligent desktop stations, one or more servers, and local area networks (LANs).

Users have been drawn to the convenience, availability, and control that comes with a personal system. But they also need to access common information and share printers. And someone still must be responsible for configuration, maintenance, and backup chores. As a result, today's systems managers coordinate software updates, supervise resources, schedule backups, and configure network parameters for large numbers of computers.

Over a period of years, many organizations turned to a network operating system for resource sharing and central management. More recently, client/server computing has elevated network interactions to the application layer.

15.1.1 Purpose of the Network File System

Sun Microsystems introduced its *Network File System* (NFS) to support resource sharing services for Unix workstation LANs. NFS makes remote file directories appear to be part of the local directory system—remote files are accessed by end users and programs exactly as if they were on a directly attached disk. NFS offers many benefits.

For example, a single copy of software or of important data can be kept at a server and shared by all users. Updates can be installed at the server rather than at multiple computers across the network. Figure 15.1 shows a LAN with one central server that provides NFS services.

15.1.2 Relationship between NFS, RPC, and XDR

NFS was built on top of a *Remote Procedure Call* (RPC) framework that was designed from the start to support general client/server application develop-

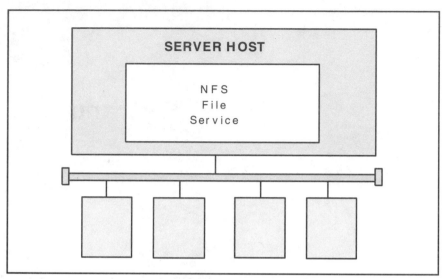

Figure 15.1   NFS server on a LAN.

ment. In this chapter, we will outline NFS services and describe the *Open Network Computing* (ONC) *Remote Procedure Call* architecture.

The *eXternal Data Representation* (XDR) standard is an important part of the RPC architecture. XDR includes a datatype definition language and a method of encoding datatypes in a standard format. This enables data to be exchanged between different types of computers, such as Unix hosts, PCs, Macintoshes, Digital Equipment Corporation VAX VMS systems, and IBM mainframes.

15.1.3   RPC As an Internet Standard

Sun Microsystems published RFCs describing Remote Procedure Calls in 1988 and NFS in 1989. However, Sun maintained control of both protocols until 1995, when new versions were published. At that point, Sun's Open Network Computing Remote Procedure Call and its supporting protocols were turned over to the Internet Engineering Task Force (IETF) and submitted to the Internet standards process. Sun cooperates with the X/Open consortium in releasing new versions of NFS.

15.1.4   RPC and NFS Implementations

RPC and NFS have been implemented by most Unix vendors and also have been ported to many proprietary operating systems. For example, IBM VM, IBM MVS, and DEC VAX VMS systems can act as NFS file servers.

Some vendors bundle NFS client or server software with their TCP/IP products, while others market NFS as an option available for an additional fee. Many vendor NFS products include an RPC programming library.

Most TCP/IP products for Windows offer an option that lets a Windows system be an NFS client. Some implementations allow a Windows system to act as a server as well. Recent enhancements of Novell's NetWare support NFS along with NetWare file and print services. Any client that can speak either of these protocols can access the server. In particular, DOS, Macintosh, and Unix clients are supported.

15.2 REMOTE PROCEDURE CALL MODEL

Open Network Computing client/server applications are built on top of Remote Procedure Calls. A Remote Procedure Call is modeled on an ordinary subroutine call. For example, in the C programming language, an ordinary subroutine procedure call commonly has the form:

```
return-code = procedure-name(input_parameters, output_parameters)
```

Before the procedure is invoked, input values are stored in the input parameters. If the procedure executes correctly, results are stored in the output parameters. On completion, the return-code indicates whether the procedure completed successfully.

A Remote Procedure Call is similar. The local system sends a call request to a remote server. The request identifies the procedure and includes values for the input parameters. The remote server executes the procedure. When its work is complete, the remote server replies, indicating whether the procedure succeeded. If the procedure completed successfully, the response contains the procedure's output parameters. Figure 15.2 illustrates the exchange of request and reply messages. The Remote Procedure Call protocol defines the mechanisms that make this happen.

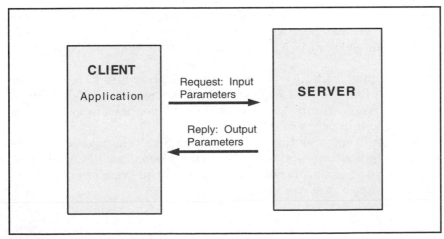

Figure 15.2 An RPC interaction.

TABLE 15.1   RPC Number Assignments

0–1fffffff	Defined by Sun (rpc@sun.com)
20000000–3fffffff	Numbers used only within one site
40000000–5fffffff	For applications that generate program numbers dynamically
60000000–7fffffff	Reserved
80000000–9fffffff	Reserved
a0000000–bfffffff	Reserved
c0000000–dfffffff	Reserved
e0000000–ffffffff	Reserved

15.3 RPC PROGRAMS AND PROCEDURES

The components of the RPC framework are easy to understand:

- An RPC service is implemented by one or more *programs* that run at a server. For example, there are separate programs that implement file access and file locking services.

- Each program can execute several *procedures*. The idea is that a procedure should perform one simple well-defined function. For example, there are separate NFS file access procedures that read, write, rename, and delete files.

- Each program is assigned a numeric identifier.

- Each procedure in a program is assigned a numeric identifier.

Up to the time of writing, Sun Microsystems administered the assignment of unique program numbers.[1] Number assignment ranges are shown in Table 15.1. Numeric IDs are assigned to a program's procedures by the program's designers. For example, *read* is NFS procedure 6 and *rename* is NFS procedure 11.

A client RPC request identifies the program and procedure to be run. For example, to read a file, the RPC request will ask for program 100003 (*NFS*) and procedure 6 (*read*). Figure 15.3 shows a client application accessing a remote procedure within program 100003.

Experience shows that over time, programs change. The procedures are refined and more procedures are added. For this reason, an RPC call must identify a program's version. It is not unusual for multiple versions of an RPC program to be running at a server host.

---

[1]The Internet Assigned Numbers Authority (IANA) might have this job in the future.

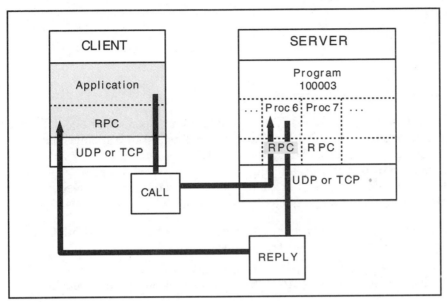

Figure 15.3 A client application accessing a remote procedure.

A Remote Procedure Call is sent from a client to a server in a formatted message. RPC does not care what transport protocol is used to carry its messages. In the TCP/IP world, RPC runs over both the UDP and TCP, but it could be implemented over other transports.

Although we usually think of a client interacting with a unique server, RPC requests may also be multicast or broadcast.

15.4 TYPICAL RPC PROGRAMS

NFS is the best-known RPC program. The related *mount* program that enables a client to glue a remote directory into a local directory system also is implemented as an RPC program. There are *lock manager* and *status* programs that provide a crude locking apparatus when users want to update shared files at an NFS server.

Spray is an example of a very simple RPC program. A *spray* client sends a batch of messages to a remote system and gets a report on the result. The command below sends 100 datagrams to host *plum:*

```
> spray -c 100 plum
sending 100 packets of lnth 86 to plum ...
in 10.1 seconds elapsed time,
29 packets (29.00%) dropped by plum
Sent: 9 packets/sec, 851 bytes/sec
Rcvd: 7 packets/sec, 604 bytes/sec
```

The *rusers* program finds out who is logged on at either a selected list of hosts or at all hosts on the local network. An *rusers* client broadcasts its RPC call on the LAN. Responses contain a hostname and the host's logged-in users.

```
> rusers
zonker.num.cs.yale.edu    leonard jones harris
mark.num.cs.yale.edu      davis   sherman
duke.num.cs.yale.edu      burry   victor
. . .
```

15.5   DEALING WITH DUPLICATE RPC REQUESTS

When an RPC service is based on TCP, requests and responses will be delivered reliably. TCP takes care of assuring that nothing gets lost in transmission.

If the service is based on UDP, the client and server must provide their own timeout, retransmission, and duplicate detection strategies. These will vary, depending on the needs of the application. The application developer can adopt either of the following client strategies:

- If no response is received within a timeout, just return an error message to the end user. Let the end user try the service again.

- If no response is received within a timeout, resend the request. Repeat until a reply is received or a maximum retransmission limit is reached.

If the client resends, the application developer must provide the server with a strategy to deal with duplicates. The server might:

- Keep no record of what was done in the past. If a request arrives, execute its procedure, even though it may be a duplicate. Note that there are procedures—such as reading a particular set of bytes from a file—for which this would be a harmless way to proceed. Of course, the client may end up receiving duplicate replies but can discard duplicates by keeping track of recently completed transactions.

- Keep a copy of the replies that you have sent during the past few minutes. If a request with a duplicate transaction identifier arrives from a client and the server already has performed the procedure and sent back a response, the server could send back a copy of the original response. If the server is currently performing the procedure, it would discard the duplicate request.

Keep in mind that each client/server application can incorporate whichever strategies fit it best.

15.6   RPC PORTMAPPER

Many client/server programs have been written, and more are being produced all of the time. The supply of well-known ports is limited—how will clients be able to identify the growing family of services?

15.6.1   Role of the Portmapper

The RPC architecture introduced a method for dynamically discovering the port at which a service can be accessed. At each server host, a special RPC program, called the *portmapper* (or, in newer versions, *rpcbind*), acts as a clearinghouse for information about the ports that *other* RPC programs use. In the remainder of this discussion, we will refer to the *portmapper,* but *rpcbind* performs the same functions.

The *portmapper* maintains a list of:

- The local active RPC programs
- The programs' version numbers
- Transport protocol or protocol
- Ports at which the programs are operating

The *portmapper* program is started when an RPC server computer initializes. As shown in Figure 15.4, when an RPC program starts up, it gets an unused port from the operating system and then tells the *portmapper* that it is ready to go to work; that is, it registers its port, program number, and version with the *portmapper.*

The *portmapper* (or *rpcbind*) listens at well-known port 111. When a client wants to access an RPC service, the client sends an RPC message to the *portmapper* at port 111. The message contains the service's program number, version, and transport protocol (UDP or TCP). The *portmapper* responds, giving the current port for the service.

The *portmapper* also enables some RPC services to be based on broadcast. In this case, a client broadcasts an RPC request on a link. For example, a broadcast *rusers* RPC call asks every machine on a LAN to report its logged-on users.

But note that the *rusers* program at each host could be operating out of a different port. What port number should a client put into the request message that it broadcasts?

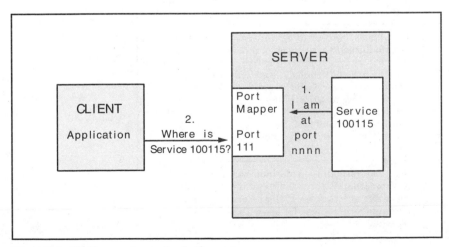

Figure 15.4 Finding service ports via the portmapper.

The answer is that the client wraps its request inside a special *indirect request* call to the *portmapper* and sends the request to port 111. The *portmapper* relays the request to the service and then relays the response back to the client. The port number for the service is added to the response so that future (individual) calls can be sent directly.

15.6.2 Portmapper Procedures

The procedures executed by a *portmapper* program are listed in Table 15.2.

15.6.3 Viewing Portmapper RPC Services

The Unix *rpcinfo* command displays useful information about RPC programs by sending RPC queries to a *portmapper*. Other operating systems that support RPC clients provide similar commands.

The *rpcinfo -p* display that follows asks the *portmapper* at host *bulldog.cs.yale.edu* for a list of the Remote Procedure Call programs that are running at that host. (That is, it sends a request to the *portmapper's PMAP-PROC_DUMP procedure*.)

TABLE 15.2 Portmapper Procedures

Procedure	Description
PMAPPROC_NULL	Returns a response that shows that the portmapper is active.
PMAPPROC_SET	Used by a service to register (i.e., add a local program, version, protocol, and port number to the list of active servers).
PMAPPROC_UNSET	Used by a service to unregister (i.e., remove a local program from the list of active servers).
PMAPPROC_GETPORT	Used by a client to find out a server's port number. Input includes a specific program number, version, and transport protocol (UDP or TCP).
PMAPPROC_DUMP	Returns a list of all local RPC programs, their versions, their communications protocols, and their ports. (Used by *rpcinfo -p*.)
PMAPPROC_CALLIT	Relays an incoming client's indirect request to a local RPC program. Returns the result if the procedure completes successfully. It also returns the program's port number. Intended for use with broadcast requests.

The output includes the program number, version, transport protocol, port, and identifier for each program at the server. We see that the *portmapper* itself is listed first:

```
> rpcinfo -p bulldog.cs.yale.edu
program vers   proto    port
100000  2      tcp      111        portmapper
100000  2      udp      111        portmapper
100029  1      udp      657        keyserv
100005  1      udp      746        mountd
100005  2      udp      746        mountd
100005  1      tcp      749        mountd
100003  2      udp      2049       nfs
100005  2      tcp      749        mountd
100026  1      udp      761        bootparam
100024  1      udp      764        status
100024  1      tcp      766        status
100021  1      tcp      767        nlockmgr
100021  1      udp      1033       nlockmgr
100021  3      tcp      771        nlockmgr
100021  3      udp      1034       nlockmgr
100020  1      udp      1035       llockmgr
100020  1      tcp      776        llockmgr
100021  2      tcp      779        nlockmgr
100021  2      udp      1036       nlockmgr
100011  1      udp      1070       rquotad
100001  2      udp      1111       rstatd
100001  3      udp      1111       rstatd
100001  4      udp      1111       rstatd
100002  1      udp      1124       rusersd
100002  2      udp      1124       rusersd
100012  1      udp      1127       sprayd
100008  1      udp      1132       walld
```

Note that we have used a Remote Procedure Call application to find out about the activities of other Remote Procedure Call applications.

The *rpcinfo -b* command broadcasts on the network, asking for all servers running a specific program and version. Below, we ask who is running version 1 of *spray*. *Spray* is program number 100012.

```
> rpcinfo -b 100012 1
128.36.12.1 casper.na.cs.yale.edu
128.36.12.28 tesla.math.yale.edu
128.36.12.6 bink.na.cs.yale.edu
. . .
```

Every RPC program includes a null procedure, number 0, which does nothing but return an "I am alive" response. The *rpcinfo -u* command below sends a message to the null procedure of the *spray* program at *bulldog.cs.yale.edu:*

```
> rpcinfo -u bulldog.cs.yale.edu 100012
program 100012 version 1 ready and waiting
```

15.7   RPCBIND

Recall that recent versions of RPC have replaced the *portmapper* program with a program called *rpcbind*. The original *portmapper* was tied to UDP and TCP. *Rpcbind* is independent of the transport that is used. It returns an

ASCII string containing address information. This transport-independent information is called the *universal address format*.

15.7.1 Role of Rpcbind

Rpcbind operates on exactly the same principles as the *portmapper*. When an RPC program initializes, it obtains one or more dynamically assigned transport addresses. It registers these addresses with the *rpcbind* program, which makes the addresses available to clients.

As before, a client query contains a program number and version number. But an *rpcbind* response contains the universal address, which might provide information for NetWare SPX/IPX, SNA, DECnet, or AppleTalk, rather than TCP or UDP. The type of transport address that is provided in the response depends on the transport that was used for the query.

Just like the *portmapper, rpcbind* is accessed at well-known port 111 for UDP or TCP. Appropriate predefined access locations must be used for other communications protocols.

Like the *portmapper, rpcbind* supports broadcast RPC services. Broadcasts are directed at the well-known transport access point for the *rpcbind* service—for example, port 111 for UDP or TCP. Each *rpcbind* program that hears the broadcast calls the desired local service program on behalf of the client, gets the response, and forwards it to the client. Version 4 of RPC enables clients to get the same kind of indirect service via *rpcbind* when a query is unicast, rather than broadcast.

15.7.2 Rpcbind Procedures

Rpcbind version 4 procedures are listed in Table 15.3.

15.8 RPC MESSAGES

An RPC client sends a call message to a server and gets back a corresponding reply message. What should these messages contain so that the client and server understand each other?

A transaction identifier is needed to match a reply with its call. The client's call must identify the program and procedure that it wants to run. The client may need some way to identify itself via credentials that prove its right to invoke the service. Finally, the client's call will carry input parameters. For example, an NFS *read* call identifies the file to be accessed and the number of bytes to be read.

In addition to reporting the results of successful calls, the server must let the client know when its requests are rejected and why. A call may be rejected for reasons such as mismatched versions or a client authentication failure. The server needs to report errors caused by a bad parameter or a failure such as "unable to find file."

TABLE 15.3   Rpcbind Procedures

Procedure	Description
RPCBPROC_SET	Used by a service to register a program with a local RPCBIND.
RPCBPROC_UNSET	Used by a service to unregister a local program.
RPCBPROC_GETADDR	Returns a program's universal address to a client.
RPCBPROC_GETVERSADDR	Includes a desired version number in a universal address request.
RPCBPROC_GETADDRLIST	Provides a list of addresses for a program. The client then may choose from several available transports.
RPCBPROC_DUMP	Lists all entries in RPCBIND's database (i.e., provides information for an *rpcinfo* display).
RPCBPROC_BCAST	Supports broadcast requests—RPCBIND passes the request to a local program.
RPCBPROC_INDIRECT	Supports indirect requests that are unicast—RPCBIND passes the request to a local program and sends the result or an error indication back.
RPCBPROC_GETTIME	Returns the local time at the server measured in seconds since midnight of the first day of January, 1970.
RPCBPROC_UADDR2TADDR	Converts universal addresses to transport specific addresses.
RPCBPROC_TADDR2UADDR	Converts transport-specific addresses to universal addresses.
RPCBPROC_GETSTAT	Provides statistics on the number and kind of requests that have been received.

Figure 15.5 illustrates a client interacting with a server program. The client sends a call. When the requested procedure completes, the server program returns a reply. As shown in Figure 15.5, a request includes:

- A Transaction Identifier
- The current RPC version number
- The program number
- The program version

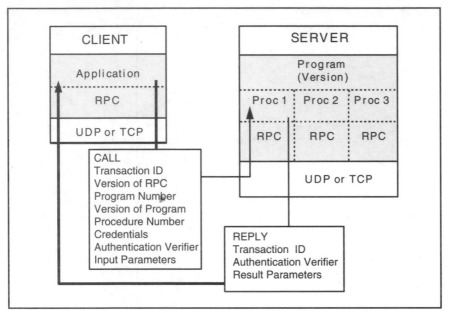

Figure 15.5 Remote Procedure Call messages.

- The procedure number
- Authentication credentials
- An authentication verifier
- Input parameters

If the procedure is executed successfully, the reply contains the results. If there was a problem, the reply will contain information that describes the error.

15.9 RPC AUTHENTICATION

Some services do not need any security protection. An RPC service that lets users find out the time of day at a server can safely be left open to the public. However, a client that wishes to access private data needs to provide some authentication information. In some cases, it also is important to have a server authenticate itself. You might not want to send a credit card number to an on-line order entry service without some assurance that the server was valid. Thus, there are times when some kind of authentication information must be included in both requests and responses.

In a call message, RPC authentication information is carried in two fields:

- The *credentials* field contains identification information.
- The *verifier* field contains additional information and validates the identity. For example, the verifier could contain an encrypted password and timestamp.

There is no single standard for authentication. It is left to each program's designer to decide what is needed for that program. However, there is an ongoing effort to provide standards in this area.

Currently, each method of authentication is called a *flavor*. The flavor used in a credentials or verifier field is identified by an integer flavor value at the start of the field. New authentication flavor values can be registered in the same way that new programs are registered. The credentials and verifier fields each start with a flavor number.

15.9.1 Null Authentication

Null authentication is just what it sounds like. No information is provided— the call message's credentials and verifier and the reply's verifier all contain a 0 flavor number, meaning that no more information is included.

15.9.2 System Authentication

System authentication provides information modeled on Unix user information. The System credentials include:

stamp	An arbitrary ID generated by the calling computer
machinename	Name of the caller's machine
uid	The caller's effective user ID number
gid	The caller's effective group ID number
gids	A list of groups that the caller belongs to

The caller's verifier is null.

The verifier returned by the server may be null or may have a flavor of "short," which means that a system-specific octet string has been returned. In some implementations, this octet string will be used as credentials by the caller in subsequent messages, instead of the user and group information.

Note that this method does not provide any security. The next two methods use encryption to protect authentication information. However, there is a trade-off between providing secure RPC services and achieving satisfactory performance. Encryption of even a single field can inflict a substantial overhead burden on a high-performance service such as NFS.

15.9.3 DES Authentication

The *Data Encryption Standard* (DES) is a symmetric encryption algorithm. DES is a Federal Information Processing Standard (FIPS) that was defined by the United States National Bureau of Standards.[2]

---

[2]Now called the National Institute of Standards and Technology (NIST).

RPC DES authentication is based on a mixture of asymmetric public and private keys and symmetric DES encryption:

- A username is associated with a public key.

- The server encrypts a DES session key with the public key and sends it to the user's client process.

- The DES session key is used to encrypt client and server authentication information.

15.9.4   Kerberos Authentication

Kerberos authentication is based on the use of a Kerberos security server at which user and server keys (based on passwords) are stored. Kerberos authenticates an RPC service by:

- Using the secret client and server keys that are registered at the Kerberos security server to distribute a DES session key to the client and server

- Using the DES session key to encrypt client and server authentication information

15.10   SAMPLE VERSION 2 RPC MESSAGES

Figure 15.6 shows a Network General *Sniffer* display of the UDP header and RPC fields for an NFS call message that requests file attributes. The data link and IP headers have been omitted from the display.

Note that a call is an RPC message of *type 0*. The reply will have *type 1*. The RPC protocol is periodically updated, and so the RPC protocol version is stated. In this call, the version is 2.

The caller uses *Unix credentials,* which identify its machine and include an effective Unix userid and groupid. One additional groupid is included. The *stamp* is an arbitrary identifier created by the caller. The *verifier* field has flavor 0 and so provides no further information. NFS often is implemented with sketchy authentication because fuller protection would slow down performance.

The parameters for program 100003 (NFS) and procedure 4 (look up filename) appear next in the message. The parameters are a *file handle* and a filename.

A *file handle*[3] is a special identifier associated with a directory or file at a server. We are interested in a file named *README* located in the directory identified by the file handle.

The fields in the call message are encoded using simple XDR format rules which will be discussed in the next section.

---

[3]A version 2 file handle is a 32-byte fixed-length string. A version 3 file handle is a variable-length string at most 64 bytes in length.

```
UDP:  --- UDP Header ---
UDP:
UDP:  Source port = 1023 (Sun RPC)
UDP:  Destination port = 2049
UDP:  Length = 124
UDP:  No checksum
UDP:
RPC:  --- SUN RPC header ---
RPC:
RPC:  Transaction id = 641815012
RPC:  Type = 0 (Call)
RPC:  RPC version = 2
RPC:  Program = 100003 (NFS), version - 2
RPC:  Procedure = 4 (Look up file name)
RPC:  Credentials: authorization flavor = 1 (Unix)
RPC:  len = 32, stamp = 642455371
RPD:  machine = atlantis
RPC:  uid = 0, gid = 1
RPC:  1 other group id(s):
RPC:  gid 1
RPC:  Verifier: authorization flavor - 0 (Null)
RPC:  [Verifier: 0 byte(s) of authorization data]
RPC:
RPC:  [Normal end of "SUN RPC header".]
RPC:
NFS:  --- SUN NFS ---
NFS:
NFS: [Params for Proc = 4 (Look up file name) follow]
NFS:  File handle = 0000070A00000001000A0000000091E3
NFS:              5E707D6A000A0000000044C018F294BE
NFS:  File name = README
NFS:
NFS:  [Normal end of "SUN NFS".]
NFS:
```

Figure 15.6   Format of an RPC message carrying an NFS request.

We can get a feeling for the way that XDR works by looking at some of the hexadecimal encodings that appear in the call message:

Type = 0 is encoded (in hex) as:

```
00 00 00 00
```

RPC version = 2 is encoded as:

```
00 00 00 02
```

Machine = atlantis is encoded as:

```
(length of string = 8)  a  t  l  a  n  t  i  s
      00 00 00 08      61 74 6C 61 6E 74 69 73
```

```
RPC:   --- SUN RPC header ---
RPC:
RPC:   Transaction id = 641815012
RPC:   Type = 1 (Reply)
RPC:   Status = 0 (Accepted)
RPC:   Verifier: authorization flavor = 0 (Null)
RPC:   [Verifier: 0 byte(s) of authorization data]
RPC:   Accept status = 0 (Success)
RPC:
RPC:   [Normal end of "SUN RPC header".]
RPC:
NFS:   --- SUN NFS ---
NFS:
NFS:   Proc = 4 (Look up file name)
NFS:   Status = 0 (OK)
NFS:   File handle = 0000070A00000001000A000000005AC9
NFS:                 3298621C000A0000000044C018F294BE
NFS:   File type = 1 (Regular file)
NFS:   Mode = 0100644
NFS:    Type = Regular file
NFS:    Owner's permissions = rw-
NFS:    Group's permissions = r-
NFS:    Others; permissions = r-
NFS:   Link count = 1, UID = 303, GID = 1
NFS:   File size = 130, Block size = 8192,
        No. of blocks = 2
NFS:   File system id = 1802, File id = 23241
NFS:   Access time       =  23-Oct-95 16:35:01 GMT
NFS:   Modification time =  20-Oct-95 12:10:43 GMT
NFS:   Inode change time =  20-Oct-95 12:10:43 GMT
NFS:
NFS:   [Normal end of "SUN NFS".]
NFS:
```

Figure 15.7 Format of an RPC message carrying an NFS reply.

The reply shown in Figure 15.7 has a matching transaction identifier. Null authentication information is included. The call has been accepted and has completed successfully.

The reply contains a lot of useful information about file *README:*

- Its *file handle* identifier is returned. Any further operations on this file will refer to the file using this file handle.

- Its *mode* describes the type of file and indicates who may access the file (owner, group, or world). The mode also declares whether users can read or write the file. If the file is application software, the mode shows whether users can execute the application.

- Additional file attributes are included, such as the file size, last time accessed, and last time updated. We would expect these attributes to be maintained in just about any file system.

15.11   XDR

When heterogeneous machines want to operate in a client/server environment, how can they understand one another's data? For example, an NFS client may want to ask a server to read 1000 bytes of data from some position in a file. How are the parameters of that request encoded? Typical parameters include file or directory names, file attributes such as file size, and integers specifying a number of bytes or position in a file.

All of the parameters in Sun RPC messages are *defined* and *encoded* using XDR, the *eXternal Data Representation* protocol. Specifically:

- The XDR data description language is used to define the datatypes that appear in calls and replies.

- The XDR encoding rules are applied to these definitions to format the data for transmission.

A large part of the RPC programming library consists of calls that translate datatypes to and from the XDR network format.

15.11.1   XDR Data Description Language

XDR definitions are similar to programming language datatypes, and they are quite easy to understand. There are a number of basic XDR datatypes such as unsigned and signed integers, enumerated integers, ASCII strings, booleans, and floating point numbers. The *opaque* datatype is used to carry general octet strings. Encrypted information can appear in an opaque field. More complicated array, structure, and union datatypes are built from the basic datatypes.

An enumerated integer type assigns a meaning to each number on a short list of integers. A simple example of an enumerated integer datatype is the message type (*msg_type*), which identifies whether a message is a call or a reply:

```
enum msg_type {
  CALL = 0,
  REPLY = 1
  };
```

Only one of the enumerated values, 0 or 1, may appear in this field. Entering any other integer into the field would be an error.

The structure that defines the body of an RPC call message is:

```
struct call_body {
    unsigned int rpcvers;     /* The version must be equal to two */
    unsigned int prog;        /* This is the program number       */
    unsigned int vers;        /* This is the program version      */
    unsigned int proc;        /* This is the specific procedure   */
    opaque_auth cred;         /* Credentials, e.g., userid        */
    opaque_auth verf;         /* Verifier for the credentials     */
                              /* This might be an encrypted field */
    /* procedure specific parameters start here   */
};
```

15.11.2 XDR Encoding

Call and reply messages for a given version of a program and procedure have a fixed format. You know what kind of data will be in a field by its position in the message. The length of every field must be a multiple of 4 bytes.

There are many parameters represented by unsigned integers, which occupy 4 bytes. For example, *Procedure = 5* is represented by:

```
00 00 00 05
```

ASCII strings are encoded as a 4-octet integer that contains the string length, followed by the ASCII characters and padded so that the field's length is a multiple of 4. For example, the string *README* appears as:

```
(string length = 6)  R  E  A  D  M  E  (pad)
     00 00 00 06      52 45 41 44 4D 45 00 00
```

The OSI *Abstract Syntax Notation 1* (ASN.1) data definition standard and *Basic Encoding Rules* (BER) encoding standard provide an alternative method for defining and encoding data. ASN.1 and BER are used by some TCP/IP applications—most notably, the Simple Network Management Protocol (SNMP).

The standard BER encoding precedes the data contents of a field with an identifier and length for the field. ASN.1 and BER will be discussed in Chapter 20. The advantage of XDR is that data is encoded with far fewer bytes. The disadvantage of XDR is that each field must be in a predetermined position.

15.12 RPC AND XDR PROGRAMMING INTERFACE

RPC client/server applications are developed using a library of subroutines that create, send, and receive RPC messages. Other library routines are used to convert between the local data representation of message parameters and their XDR format. A typical RPC subroutine is:

```
int callrpc (host, prognum, versnum, procnum, inproc, inparams, outproc, out-
params)
```

The *host* parameter identifies the server computer, *prognum* identifies the program, and *procnum* is the procedure to be executed. The input parameters to be sent in the call message are in structure *inparams*. The *inproc* routine will convert the input parameters to XDR format. When the reply arrives, the *outproc* routine will convert the XDR reply parameters to a local format and store them in structure *outparams*.

NetWise and Sun have developed a toolkit that greatly simplifies the development of RPC client/server applications and hides the underlying RPC calls from the developer.

15.13   INTRODUCTION TO NFS

The Network File System is a file server architecture that is portable across different hardware, operating systems, transport protocols, and network technologies. However, it was designed with the Unix file system in mind.

A client host prepares to use NFS by *mounting* a remote directory subtree into its own file system. A client accomplishes this by sending an RPC request to the *mount* program at the server.

An end user or application is not aware of NFS. When a call is made to perform a file operation (such as open, read, write, copy, rename, delete, etc.) and the file happens to be located at a remote computer, the operating system redirects the request to NFS. The request is transmitted in a Remote Procedure Call request message. The input and output parameters are encoded using XDR.

Figure 15.8 shows the components that support NFS calls. NFS usually is implemented over a UDP transport, but some recent products also run over TCP connections. UDP works well when a client and server reside on the same LAN. TCP should be used when communicating across a wide area network, where retransmission timeout calculations and congestion recovery are needed.

NFS typically is implemented by running many service processes at the server so that many clients can be handled concurrently.

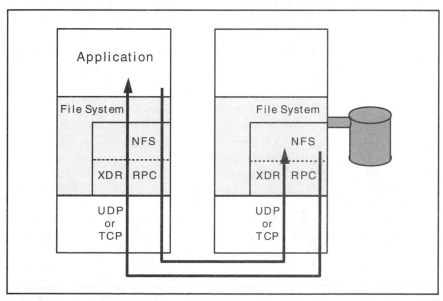

Figure 15.8   Components supporting NFS.

15.14   NFS FILE MODEL

NFS works best for clients and servers that have a Unix-like file structure. Unix files are stored in a hierarchical tree of directories.[4] A Unix file is accessed as a sequential stream of bytes.

Unix directories and files are identified by pathnames that are formed by listing the names along the path from the root, separating the names by a slash (/). For example, */etc/hosts* and */usr/john/abc* are pathnames.

The syntax used to write pathnames on non-Unix systems may be different. For example, *E:\WP\LETTER.DOC* is a DOS pathname. NFS assumes that every file can be identified by a pathname.

15.14.1   Source of the NFS Model

Pieces of a Unix directory system can reside on different hard disks. For example, files and subdirectories under */etc* may be on one physical disk, while */var* and all of its subdirectories are on another. The Unix *mount* command is used to glue a piece such as */var* into the overall directory tree. A typical Unix mount command to do this is:

```
mount /dev/xy0b /var
```

The files on the physical device *xy0b* are identified with the files in directory */var*.

In designing NFS, it was natural to simply extend the mount command so that remote subtrees also could be glued into a computer's directory tree. For example, suppose a network administrator wants the user files for host *tiger* to be physically located at computer *bighost,* where they will safely be backed up every night. The administrator creates directories for the user's files at *bighost,* say under */users*. From *tiger,* the administrator issues the command:

```
mount -t nfs bighost:/users /usr
```

Server directory */users* and all of its subdirectories are logically glued on top of *tiger*'s directory */usr*. To end users, *tiger*'s directory system appears as shown in the left half of Figure 15.9. However, the file */usr/john/abc* actually is stored in */users/john/abc* at server *bighost*.

Another way of putting this is that whenever a local user asks for a file that is in a tree under */usr,* the operating system knows that the file really is at *bighost* under */users*.

While Unix directory systems form a single tree, DOS directory systems have multiple trees (a forest?) rooted at devices A:, B:, C:, and so on. DOS computers glue a remote NFS directory onto a device such as E:.

---

[4]There are successful implementations for servers without a hierarchical directory tree. For example, an IBM VM server has a flat file structure.

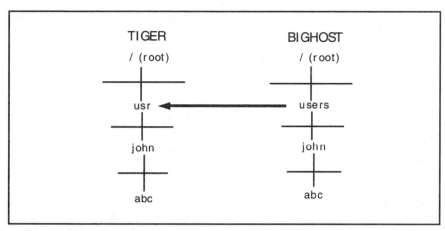

Figure 15.9   Mounting a remote directory.

Many other operating systems have hierarchical directories. Those that do not are mapped onto this model by incorporating restrictions on the depth of the tree that can be attached and on the directory and filename syntax.

15.15   MOUNT PROTOCOL

Mount, which is used to glue a remote directory into the local file system, is RPC program number 100005, and its port is advertised by the *portmapper.* *Mount* runs over both UDP and TCP.

Before a computer can mount a directory from a server, the server must be configured to *export* the directory. The way that this is typically done is that an administrator edits a file that lists the server directories to be exported, the hosts allowed to access them, and the access restrictions to be imposed. For example, a Unix */etc/exports* configuration might have the form:

```
/man     -ro
/bin     -ro,access = tiger:lion
/users   -rw,access = tiger
```

The first directory is accessed read-only (ro), and is accessible from any host. Hosts *tiger* and *lion* can read the second directory. Host *tiger* has read-write access to the third directory.

A server can export only its own directories. It cannot export a directory that it has mounted from another NFS server. A client system can mount directories from as many servers as it likes. Of course, a directory can be mounted only if the server's export restrictions make the directory available to the client.

The client needs to identify all of the remote servers and directories that it wants to mount. Typically, this is done by executing a sequence of *mount* commands at system start-up. Sometimes client *mount* command information is read out of a configuration file.

There are many optional parameters for a *mount* command. The most important select:

- Whether the directory should be mounted read-only or read-write.
- Whether to periodically retry failed mounts in the background and what limit to set on the number of retries.
- Whether a user can interrupt an NFS RPC call that is taking a very long time to complete.
- Whether to use a version of NFS that is based on secure Remote Procedure Calls.

A *mount* command causes an *Add Mount Entry* RPC message to be sent to the server. *In the response message, the mount protocol returns the file handle that the client will use to identify that directory in all future calls.* Recall that a file handle is a string that has meaning for the server system and identifies a corresponding directory or file. For example, when we mount */users* as local directory */usr,* the response to the mount request contains a file handle for directory */users.*

15.15.1 Mount Procedures

The procedures supported by a *mount* server program are listed in Table 15.4.

15.16 STATELESSNESS AND IDEMPOTENTS

The spirit of NFS is that the server should be as *stateless* as possible. What this means is that an NFS server should have a minimum of client information to remember so that recovery from a client or server crash is painless and simple.

TABLE 15.4 Mount Procedures

Procedure	Description
0	*Null:* Respond showing the program is active.
1	*Add Mount Entry:* An entry for the client host is added to the mount list, and a file handle for the mounted directory is returned.
2	*Return Mount Entries:* Reports to the client on the currently mounted pathnames.
3	*Remove Mount Entry:* Removes the mount of a specified directory.
4	*Remove All Mount Entries:* Removes all of the client's mounts.
5	*Return Export List:* Returns a list consisting of directories and the hosts allowed to access each.

A client knows that the NFS server has completed all of the work for a request when a reply is received. But NFS often runs on top of UDP, and UDP is unreliable. What should be done if no reply arrives? NFS repeats requests after a timeout period.

However, it is possible that the original request actually got through to the server but the reply was lost. For this reason, NFS servers usually are not perfectly stateless. They cache some recent replies so that duplicate requests can be handled correctly.

Which replies should be kept? Some operations—such as a *read* or *lookup*—are *idempotent*. This means that they can be executed more than once and still return the same result. Others, such as *remove file* or *create directory,* are not idempotent. If the original result was lost, a second execution will return a misleading error message. Caching the results of nonidempotent operations enables NFS to send the appropriate reply to a retransmitted request.

15.17   NFS PROTOCOL

The most current NFS version is 3, but many version 2 implementations can be expected to persist for quite a long time. The NFS server program number is 100003. By convention, an NFS server grabs port 2049 when it initializes.

15.17.1   More About File Handles

Recall that when a client mounts a directory, the mount protocol returns a file handle that identifies that directory in all future calls. The directory that is mounted may have subdirectories, and these may have subdirectories. A file's pathname can be nested several levels deep. For example, before a user at the client can update the file:

```
/usr/john/book/chapter3
```

it is necessary to obtain the file handle of the corresponding file, */users/john/book/chapter3*, from the server. The way that NFS does this is to look up one pathname at a time. For our example, the NFS client would:

- Send a lookup call to the server that includes the file handle for */users* and pathname component *john*. The reply contains a handle for */users/john*.

- Send a lookup call to the server that includes the file handle for */users/john* and pathname component *book*. The server returns a file handle for */users/john/book*.

- Send a lookup call to the server that includes the handle for */users/john/book* and filename *chapter3*. The reply contains the file handle that we want.

The result of this long-winded approach to getting to a file handle that we want is that NFS clients issue lots of lookup calls.

15.17.2 NFS Procedures

There are NFS procedures that let a client access, read, and write files. The client can find out about the organization and capacity of the remote file system and can ask to see the attributes of individual files. The client can delete or rename files. Some of the procedures deal with features provided specifically by the Unix file system, such as linking an alias name to a file. The NFS procedures for versions 2 and 3 are described briefly in Table 15.5.

15.17.3 Special Utilities

Ideally, NFS should be transparent to a user. Server files are opened, read, written, and closed as if they are local. Ordinary local commands are used to copy, rename, or delete server files.

When the client and server have the same operating system, this is straightforward. Sometimes a few special additional commands make NFS work better when the client and server file conventions are very different. An example may help to clarify this.

When a DOS client uses a Unix system as a file server, files that are created and named by the DOS client will conform to DOS conventions and will be a natural part of the client's file system.

If the DOS client wants to read a text file created by a Unix user, there are some problems. First of all, DOS names consist of up to eight characters optionally followed by a dot and up to three more characters. When a DOS user types a filename, all letters are converted to uppercase. For example, COMMAND.COM is a DOS filename. Unix names can be much longer and can contain a mixture of upper- and lowercase letters. For example, *aLongerName.More* is a valid Unix filename.

How can a DOS user access the Unix file? Vendors usually perform automatic name translation and also include a utility that enables users to view the native filename.

There is one other hurdle to get over. Lines in a DOS text file end with carriage return (CR) and line feed (LF) characters while lines in a Unix text file end with an LF. Some vendors perform automatic translation, while others provide utilities that enable a user to convert a text file to the local format.

15.17.4 File Locking

There are some files that must be accessed by several users. For example, a configuration file may be read by many application processes. A user who needs to update a shared file will want to obtain exclusive access to the file—that is, lock the file—during the update activity.

File locking in an NFS environment is handled by two RPC services: the *lock manager* and the *status* program. The *lock manager* handles client requests for file locks. A server's *status* monitor tries to keep track of which client hosts currently hold locks. If a server crashes, it will send notification to the status monitors at registered client hosts, asking them to resubmit their lock requests.

TABLE 15.5   NFS Versions 2 and 3 Procedures

Procedure	Version 2	Version 3
0	Null procedure for testing.	Null procedure for testing.
1	Get file attributes.	Get file attributes.
2	Set file attributes.	Set file attributes.
3	Obsolete operation.	Lookup a filename. That is, given the file handle for a directory and the name of a subdirectory or file, return the file handle for the subdirectory or file.
4	Lookup a filename.	Check access permission.
5	Read information associated with a symbolic link.	Read information associated with a symbolic link.
6	Read data from a file.	Read data from a file.
7	Not used.	Write data to a file. The call can indicate whether the write can be to cache or whether it must be committed to stable storage before a reply can be sent.
8	Write data to a file.	Create a file.
9	Create a file.	Create a directory.
10	Remove a file.	Create a symbolic link.
11	Rename a file.	Create a node (i.e., a special device).
12	Create a link to a file.	Remove (delete) a file.
13	Create a symbolic link.	Remove a directory.
14	Create a directory.	Rename a file or directory.
15	Remove a directory.	Create a link to an object.
16	Read filenames and fileids from a directory.	Read filenames and fileids from a directory.
17	Get file system information such as the block size and number of free blocks.	Read filenames, fileids, attributes, and handles from a directory.
18		Get dynamic file system information such as the total size and amount of free space.
19		Get static file system information such as the maximum sizes for read and write requests.
20		Retrieve POSIX information, such as attributes and the maximum length of a filename.
21		Commit—forces data that previously was written to memory cache to be written to stable storage.

15.17.5 NFS Implementation Issues

A program may repeatedly ask its operating system to read or write a few bytes of data. Accessing a hard disk frequently for small amounts of data is not efficient. Normally, operating systems read in entire blocks of data ahead of time and respond to read calls using data stored in memory. Similarly, writes are saved in memory and periodically written to disk.

Frequent accesses to a remote NFS server for small amounts of data are even more inefficient than local disk accesses. Client NFS implementations perform read-aheads of blocks of data.

An NFS server improves its performance by keeping directory and file attribute information in memory and by reading ahead to anticipate client calls. Version 3 supports writes to memory cache, which can be forced onto permanent media by calling a *commit* procedure.

15.17.6 Monitoring NFS

The Unix *nfsstat* command results in a report on NFS activities. Similar commands are available at other operating systems. In the example shown below, the local system is acting as both a server and as a client. There is very little server activity reported. However, the system's users are making a large number of client calls.

The display shows the number of uses of each type of call over the monitoring period. Note the large number of *lookups*. Recall that these calls are used to walk down a file's pathname one step at a time in order to obtain the file handle.

```
> nfsstat

Server rpc:
calls          badcalls     nullrecv     badlen          xdrcall
25162314       0            0            0               0

Server nfs:
calls      badcalls
25162314   491
null       getattr      setattr      root    lookup         readlink
478 0%     9689121 38%  380591 1%    0 0%    5596396 22%    5992775 23%
read
1009813 4%

wrcache    write        create       remove       rename       link         symlink
0 0%       1146142 4%   627381 2%    66180 0%     13089 0%     6042 0%      265 0%
mkdir      rmdir        readdir      fsstat
1718 0%    66 0%        626437 2%    5820 0%

Client rpc:
calls      badcalls  retrans   badxid    timeout   wait     newcred   timers
3931394    2069      0         42        2037      0        0         1697

Client nfs:
calls      badcalls     nclget       nclsleep
3929178    32           3929357      0
null       getattr      setattr      root         lookup         readlink
0 0%       2221718 56%  6689 0%      0 0%         1423702 36%    93498 2%
read       wrcache      write        create       remove         rename
54110 1%   0 0%         19501 0%     7362 0%      6493 0%        158 0%
```

link	symlink	mkdir	rmdir	readdir	fsstat
5 0%	0 0%	28 0%	12 0%	95804 2%	98 0%

15.18 RECOMMENDED READING

At the time of this writing, the *portmapper* and *RPCBIND* were defined in RFC 1833, Remote Procedure Call Protocol version 2 in RFC 1831, and XDR in RFC 1832.

Version 2 of NFS was described in RFC 1094, and version 3 was described in RFC 1813. A very complete specification of version 2 of NFS can be found in *X/Open CAE Specification: Protocols for X/Open Internetworking: XNFS*, published by the X/Open Company, Ltd.

16

Electronic Mail

16.1 INTRODUCTION

Of all of the TCP/IP applications, electronic mail engages the largest number of people. When an organization offers good access to mail, usage grows explosively. Mail attracts users who never dreamed that they would use a computer.

Electronic mail is a convenient way to reach people and is easy to use. The dialogue below shows a very simple interaction with a bare-bones Unix mail program. The program prompts for the Subject:, and the user signals the end of the message by typing a period as the only character on a line.

```
> mail fred
Subject: New Materials
The manuals have arrived.
Let's discuss them next week.
.
```

There are mail programs that are far more elegant, with full-screen user interfaces and point-and-click options. For example, Figure 16.1 shows the interface for the *Chameleon* Windows electronic mail application, and Figure 16.2 shows the Macintosh *Eudora* electronic mail program.

The formal name for an end-user mail program is a *User Agent* (UA). A User Agent is expected to perform several chores, such as:

- Display information about incoming mail messages that are waiting in a user's mailbox
- Save incoming or outgoing messages in folders or local files
- Provide a good editor for entering message text

The style of User Agent that an individual prefers has always been viewed as a matter of personal taste and not subject to standardization. The important

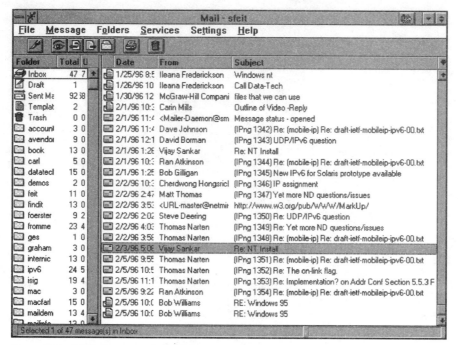

Figure 16.1 A Windows user interface.

thing is that the end result always is the same—mail items are sent and delivered.

Let's return to our original mail transaction. It all looks very easy, but there is a lot of muscle hiding behind the scenes. As it happens, "fred" is a *nickname* or *alias* that I have defined in my private address book. When my User Agent looks it up, it discovers that the real recipient identifier is fred@microsoft.com.

This identifier has a format that is typical for Internet mail. However, vendors of proprietary electronic mail software and public mail service providers have expressed a lot of individuality in designing their own recipient formats. There are *mail gateways* that are kept very busy converting between these formats.

How is mail delivered? In earlier times, mail was transferred across a direct TCP connection between a source host and recipient host. But today, mail is likely to be relayed via one or more intermediate hosts. We'll have a lot more to say about relaying later.

16.2 INTERNET MAIL PROTOCOLS

Mail is heavily used, and many Internet protocols have evolved to meet the requirements of electronic mail users. Figure 16.3 illustrates the Internet mail protocols.

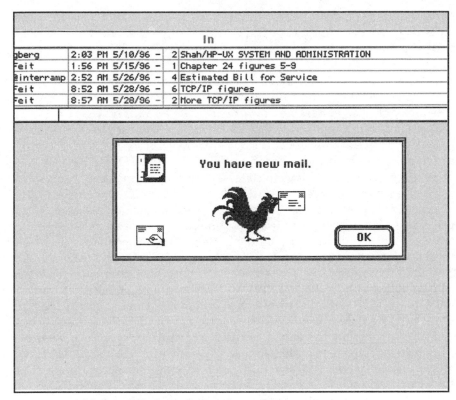

Figure 16.2   A Macintosh Eudora user interface.

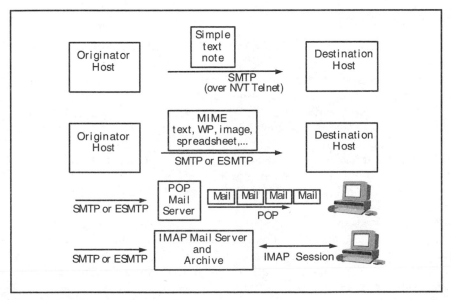

Figure 16.3   Internet mail protocols.

The *Simple Mail Transfer Protocol* (SMTP) is the classic Internet standard for moving mail between computers. SMTP was designed to carry simple text notes and was implemented on top of a simple Network Virtual Terminal (NVT) *telnet* session.

When mail arrives, a User Agent will need to understand message elements such as the sender's identifier, date sent, subject, and the information part of the message. The venerable *Standard for the Format of ARPA Internet Text Messages* provides the durable format for simple Internet text mail messages.

A series of more recent standards define *Extensions to SMTP* (ESMTP) that enable it to carry any type of information. New, multipart message bodies are described in *Multipurpose Internet Mail Extensions* (MIME) standards. Many different types of information can be delivered, such as documents created by word processors, Macintosh Binhex files, images, videos, encoded sounds, spreadsheets, executable code, or whatever. New MIME types are defined as needed and are registered with the Internet Assigned Numbers Authority.

Another set of standards was designed to fit the way that many people work today. The *Post Office Protocol* (POP) enables a desktop client to download mail from a mail server. An alternative choice, the *Internet Message Access Protocol* (IMAP), enables a user to read, copy, or delete messages that are stored at a server, but the server is the authoritative repository for messages. This is helpful for users who want to take advantage of administrative services (such as daily backup), save desktop disk space, or access their mail when they are traveling. Mail is delivered to a server via SMTP or ESMTP.

Some organizations relay mail by means of the OSI X.400 protocols, which will be discussed briefly later in this chapter.

16.3   MODEL FOR MAIL TRANSMISSION

Figure 16.4 shows the elements of a mail system. Mail is prepared with the help of a User Agent application. The User Agent typically queues mail to a separate application, called a *Message Transfer Agent* (MTA), which is responsible for setting up communications with remote hosts and transmitting the mail. *User Agent* and *Message Transfer Agent* are terms used in the X.400 message system standards, but the terms describe components that are valid for SMTP mail as well.

The mail may be sent directly between the source and destination MTAs or relayed via intermediate MTAs. When a mail item is relayed, the entire message is transmitted to an intermediate host, where it is stored until it can be forwarded at a convenient time. Mail systems that use relaying are called *store-and-forward* systems.

At the recipient host, mail is placed on an incoming queue and later is moved to a user's *mailbox* storage area. When a recipient user invokes a User Agent program, the User Agent usually displays a summary of the incoming mail that is waiting in the mailbox.

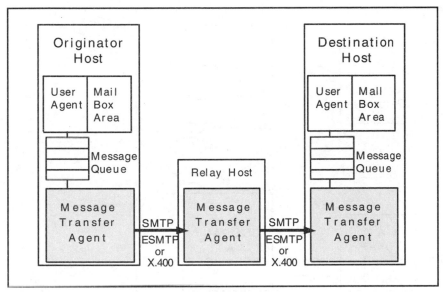

Figure 16.4   Components of an electronic mail system.

16.4   RELAYING MAIL

Why would a Message Transer Agent ever want to relay mail rather than connect directly to the recipient host? When a host uses a direct connection, it can be sure that mail has reached its destination. Relayed mail uses intermediate storage resources and requires multiple connections. To relay mail, we have to design a store-and-forward mail relaying road map, and if we do not do a good job, mail will wander around in an inefficient manner.

16.4.1   A Mail Relay Scenario

In order to see why store-and-forward is so prevalent, let's walk through the scenario that is illustrated in Figure 16.5. Fred, who works for ABC Industries, is sending a mail item to Mary, who works for JCN Computers. Fred's computer is a LAN workstation that is powered down much of the time. The workstation sends and receives mail via a relay server on the LAN.

Both ABC Industries and JCN Computers are very security-minded. They allow mail to be exchanged with the outside world only via designated *Mail Exchanger* relay hosts. Each company is attached to the outside world by a router that blocks all traffic except for connections to the mail port (25) at the company's Mail Exchanger.

A proprietary LAN electronic mail product is used on Fred's LAN. TCP/IP electronic mail protocols are used at Mary's site.

As shown in Figure 16.5, mail is transferred from Fred's desktop to a LAN

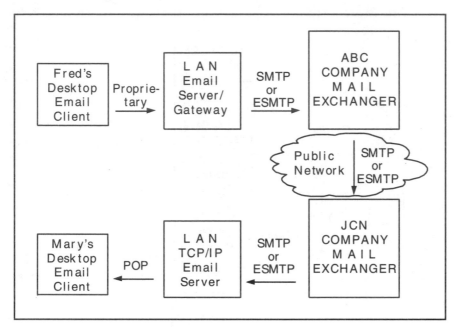

Figure 16.5 Relaying electronic mail.

server using a proprietary mail protocol. The LAN server has gateway software that translates between the proprietary mail format and Internet message format. Then the mail is forwarded to ABC's Mail Exchanger. From there, it is transmitted across an external network (e.g., the Internet) to JCN's Mail Exchanger. It is relayed again to Mary's LAN electronic mail server, where it will be stored until Mary connects and picks up her mail via the Post Office Protocol.

This scenario illustrates that relaying offers a number of benefits:

- PCs and workstations can depend on a LAN server system to forward outgoing mail and hold their incoming electronic mail for them.
- A company's employees can use electronic mail but still maintain security by funneling mail through a Mail Exchanger.
- Cost savings can be realized by batching mail from a relay at favorable times.
- A mail relay can perform mail format translations.

In the sections that follow, we will take a closer look at the mechanisms that have evolved within the TCP/IP protocol family to support an expanding electronic mail universe.

16.5   MAIL RECIPIENT IDENTIFIERS AND MAIL EXCHANGERS

Internet mail recipients are identified by names following the general pattern:

local-part@domain-name

We shall see that this format is quite flexible. For many years, the prevalent format for Internet-style names was:

userid@hostname

For example:

smithm@sales.chicago.jcn.com.

Today, far more convenient formats are used, such as:

firstname-lastname@mail-domain-name

For example:

Mary-Smith@jcn.com

In this identifier, *Mary-Smith* is not a userid, and *jcn.com* is not the name of a computer—it is a logical name assigned to a Mail Exchanger. So how does this mail get delivered? The mail relaying architecture depends on the Domain Name System. The way that it works is:

- One or more computers are selected to act as Mail Exchangers for an organization.
- A logical name—usually the organization's Domain Name—is selected for the Mail Exchanger, and a Mail Exchanger (MX) entry is added to the DNS database.
- A Message Transfer Agent program looks up the mail-domain-name part of the recipient identifier in the DNS, retrieves the real name and address of a Mail Exchanger, and relays the mail to the Mail Exchanger.

A demonstration will make this clearer. Below, we start the *nslookup* program and ask for the identity of Cisco's Mail Exchanger. We discover that actually there are two—*hubbub.cisco.com* and *beasley.cisco.com*. It is a good idea to run two or more mail servers to ensure availability of the service.

Note the *preference* numbers (5 and 10). The server with the *lower* number, *hubbub,* is more preferred. *Hubbub* would be contacted first. If it was not possible to connect to *hubbub, beasley* would be tried. By the way, the actual preference numbers used do not matter, only their relative size does.

```
> nslookup

Default Server: DEPT-GW.cs.YALE.EDU
Address: 128.36.0.36

> set type = mx

> cisco.com

cisco.com          preference = 5,  mail exchanger = hubbub.cisco.com
cisco.com          preference = 10, mail exchanger = beasley.cisco.com
hubbub.cisco.com        inet address = 198.92.30.32
beasley.cisco.com       inet address = 171.69.2.135
dennis.cisco.com        inet address = 171.69.2.132
ns1.barrnet.net  inet address = 131.119.245.5
noc.near.net     inet address = 198.112.8.2
noc.near.net     inet address = 192.52.71.21
```

The next query shows how some organizations build in an extra layer of security. Note that there are three Mail Exchangers, but two of them actually belong to the UUNET Service provider:

```
> clarinet.com
Server:  DEPT-GW.cs.YALE.EDU
Address:  128.36.0.36

clarinet.com  preference = 10,  mail exchanger = looking.clarinet.com
clarinet.com  preference = 100, mail exchanger = relay1.uu.net
clarinet.com  preference = 100, mail exchanger = relay2.uu.net
looking.clarinet.com    inet address = 192.54.253.1
relay1.uu.net    inet address = 192.48.96.5
relay2.uu.net    inet address = 192.48.96.7
>
```

Clarinet could set up its network so that incoming mail would first be funneled through one of the UUNET Mail Exchangers and then would be relayed to *looking.clarinet.com*.

Figure 16.6 shows how this is done. A filtering router has been set up to

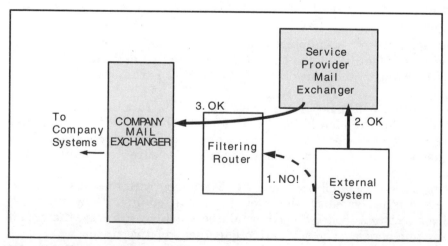

Figure 16.6 Forcing mail along a path.

refuse connections from all systems except for the UUNET service provider's Mail Exchangers. An external system will try to connect to the most preferred site, *looking.clarinet.com*. But in step 1, the filtering router prevents the connection from going through. Therefore, one of the less preferred sites is tried, and the mail is forwarded to *relay1* or *relay2*. The UUNET system now can forward the mail to Clarinet's Mail Exchanger.

When mail reaches a company's Mail Exchanger, the local-part of the logical name will be looked up in an alias file and converted to a userid and host name—or whatever type of mail identifier is used in the destination network. Thus, a Mail Exchanger also can act as a gateway to non-Internet-style mail services.

There is one more problem to be solved in order to route mail through a Mail Exchanger. Suppose that users at host *sales.clarinet.com* had mail identifiers of the form *username@sales.clarinet.com*. What would happen to mail with an address like *jonesj@sales.clarinet.com*? A few more entries in the Clarinet DNS database take care of this problem:

```
*.clarinet.com.  IN  MX   10  looking.clarinet.com.
*.clarinet.com.  IN  MX  100  relay1.uu.net
*.clarinet.com.  IN  MX  100  relay2.uu.net
```

These "wild card" entries direct mail that is addressed to the old-style *userid@hostname* to the Mail Exchangers.

Organizations are replacing their old *userid@hostname* identifiers with logical names, which do not reveal userids to outsiders. In addition to improving the security of a network, logical names also permit users to acquire new userids or move to different computers without changing their mail identifiers.

16.6   SIMPLE MAIL TRANSFER PROTOCOL

The Simple Mail Transfer Protocol (SMTP) defines a straightforward way to move mail between hosts. There are two roles in the SMTP protocol: sender and receiver. The sender acts as client and establishes a TCP connection with the receiver, which acts as server. The well-known port used for a receiver is 25. Even though the sender is a mail service program (a "Message Transfer Agent"), the sender acts as client and uses a temporary port from the pool.

During an SMTP session the sender and receiver exchange a sequence of commands and responses. First, the receiver announces its host name. Then the sender:

- Announces its host name
- Identifies the message originator
- Identifies one or more recipients
- Transmits the mail data
- Transmits a line containing ". <CR><LF>"—which indicates that the item is complete

Note that an item can be delivered to several recipients at a host via one transaction because multiple recipients can be listed. At the end of a transaction, the sender can:

- Start another transaction
- Quit and close the connection

The standard includes a *TURN* command that enables the sender to reverse roles so it becomes the receiver. However, this is rarely (if ever) implemented.

16.6.1   Mail Dialogue

In the dialogue that follows, a sender transfers a message to a receiver. The sending host also happens to be a Mail Exchanger gateway for the computers in its department. The appearance of the mail item when it is delivered is:

```
Received: from PASCAL.MATH.YALE.EDU (MATH-GW.CS.YALE.EDU) by tigger.jvnc.net
with SMTP id AA08294
  (5.65c/IDA-1.4.4 for feit); Sun, 27 Aug 1995 08:02:55 -0400
Received: by PASCAL.MATH.YALE.EDU; Sun, 27 Aug 1995 08:01:44 -0400
Date: Sun, 27 Aug 1995 08:01:44 -0400
From: Sidnie Feit <feit-sidnie@MATH.YALE.EDU>
Message-Id: <199508271201.AA02330@PASCAL.MATH.YALE.EDU>
To: feit@tigger.jvnc.net
Subject: It's OK to talk to yourself!
Date: 08/26/95 1:29:59 PM

Hi there.
See you soon.
```

The *Received* item at the top of the message was added by the receiving MTA at *tigger*. The rest of the message was transmitted to *tigger* by *pascal*.

To transmit, the sender opens a connection to port 25 at the receiver. Then the receiver starts the dialogue by announcing its domain name.

The command/reply model that we have seen in the File Transfer Protocol (FTP) also applies here, and the encoding of the reply messages is similar. Hence, all messages from the remote electronic mail server start with a reply number.

Note that electronic mail identifiers are enclosed in angle brackets for this protocol (e.g., *<sfeit@pascal.math.yale.edu>*). Host names are not case-sensitive and may appear in upper- and lowercase. However, a username may be case-sensitive, depending on the user naming conventions in use by the electronic mail system.

`220 tigger.jvnc.net 5.65c/IDA-1.4.4`
`Sendmail is ready at Sun, 27 Aug 1995`
`08:02:55 -0400`
	Receiver identifies itself and announces the time.

`HELO MATH-GW.CS.YALE.EDU`	Sender identifies itself.
`250 Hello MATH-GW.CS.YALE.EDU, pleased` `to meet you`	
`MAIL FROM: <sfeit@pascal.math.yale.edu>`	The mail originator identifier is sent.
`250 <sfeit@pascal.math.yale.edu>.. Sender ok`	

```
RCPT TO: <feit@tigger.jvnc.net>
```
A recipient is identified. There can be multiple RCPT TO statements.

```
250 <feit@tigger.jvnc.net>.. Receiver ok
```

```
DATA
```
The message starts here.

```
354 Enter mail, end with "." on a
line by itself
```

```
Received: by PASCAL.MATH.YALE.EDU;
Sun, 27 Aug 1995 08:01:44 -0400
```
The headers are sent first.

```
Date: Sun, 27 Aug 1995 08:01:44 -0400
```

```
From: Sidnie Feit
<feit-sidnie@math.yale.edu>
```

```
Message-Id: <199508271201.AA02330
@PASCAL.MATH.YALE.EDU>
```

```
To: feit@tigger.jvnc.net
```

```
Subject: It's OK to talk to yourself!
```

```
Date: 08/26/95 1:29:59 PM
```

A blank line follows the headers.

```
Hi there.
```
This is the body of the message.

```
See you soon.
```

The message ends with .<CR><LF>

```
.
```

```
250 Ok
```

```
quit
```
More messages could be sent before quitting.

```
221 tigger.jvnc.net closing
connection
```

```
Connection closed by foreign host.
```

Note that the end of the message is signaled by a period on a line by itself.

Suppose that a user actually wants to send such a line. An additional period is inserted by the sending SMTP and deleted by the receiving SMTP.

16.7   TIMESTAMPS AND MESSAGE ID

When you receive mail, you might want to know what time it was sent and when it arrived at your computer. SMTP adds this information to your message. SMTP also keeps track of all of the hosts that relayed the message and the time that each received the message.

When a message is passed to an SMTP Message Transfer Agent, the agent inserts a timestamp at the beginning of the message. Each time that an item is relayed, another timestamp is inserted. Each timestamp shows:

- The identity of the host that sent the message
- The identity of the host that received the message
- The date and time that the message was received

The timestamps in the message header provide invaluable debugging information when there are mail delivery problems. For example, they might reveal that an item was stalled at some intermediate host for a day or two.

Timestamp formats vary, and diverse information is included by different vendors. Newer implementations provide timestamps which report the local time, followed by the offset from *Universal Time* (formerly called Greenwich Mean Time).

Computer clocks are sometimes inaccurately set, so some timestamp sequences don't seem to make very good sense. (For example, occasionally messages appear to arrive before they were sent.) Since network administrators usually are the only people who need to pay close attention to timestamps, the anomalies are tolerated.

When the mail arrives at its final destination, the recipient User Agent optionally may add a line that identifies the original sender.

The example that follows illustrates how lines are added to a message. The top line was added by the receiving UA and shows the identifier of the originator and the final arrival time at the mailbox.

The Message-Id on the bottom line is added to the message by the first Message Transfer Agent to handle the message.

The timestamps, which must be read from the bottom up, report the path followed by the mail—from *dial131.mbnet.mb.ca* to *access.mbnet.mb.ca,* then to *bulldog.cs.yale.edu,* and finally, to *pascal.math.yale.edu.*

```
From vsankar@ForeTell.CA Thu Aug 17 14:36:19 1995
Received: from BULLDOG.CS.YALE.EDU by PASCAL.MATH.YALE.EDU via SMTP; Thu, 17
Aug 1995 14:36:19 -0400
Received: from access.mbnet.mb.ca by bulldog.CS.YALE.EDU via SMTP; Thu, 17
Aug 1995 14:31:47 -0400
Received: from ftl6 (dial131.mbnet.mb.ca) by access.mbnet.mb.ca with SMTP id
AA02060
(5.67b/IDA-1.4.4); Thu, 17 Aug 1995 14:31:33 -0500
Date: Thu, 17 Aug 1995 14:31:33 -0500
Message-Id: <199508171831.AA02060@access.mbnet.mb.ca>
```

16.8 BOUNCED MAIL

Occasionally, it will be impossible to deliver mail to its destination. Most often, this is because the originator has provided an incorrect recipient identifier.

Mail that cannot be delivered is sent back to the originator and is called *bounced* mail.

16.9 SMTP COMMANDS

The scenario in Section 16.6.1 contained the most frequently used SMTP commands. The complete set of SMTP commands is described in Table 16.1.

A command is transmitted as a four-character mnemonic. Many commands are followed by a parameter.

A session between SMTP partners employs *telnet* NVT conventions such as sending 7-bit ASCII characters in 8-bit bytes and ending a line with carriage return and line feed.

TABLE 16.1   SMTP Commands

Command	Description
HELO	Identifies the sender to the receiver.
MAIL FROM	Starts a mail transaction and identifies the mail originator.
RCPT TO	Identifies an individual recipient. The command is repeated in order to identify multiple recipients. If possible, the receiver checks the validity of the recipient name and indicates the result in the reply message. An immediate check is not practical at a relay host. If it later turns out that some recipient was not valid, a brief mail item reporting the error will be sent back to the originator.
DATA	The sender is ready to transmit a series of lines of text. Each line must be terminated with <CR><LF>. The maximum length of a line, including <CR><LF>, is 1000 characters. An SMTP implementation must be able to send and receive messages that are up to 64 kilobytes in length. A larger maximum size is desirable, since mail often is used to copy files between users.
RSET	Abort the current mail transaction, clearing out all originator and recipient information.
NOOP	Asks the partner to send a positive reply.
QUIT	Asks the partner to send a positive reply and close the connection.
VRFY	Asks the receiver to confirm that a name identifies a valid recipient.
EXPN	Asks the receiver to confirm that a name identifies a mailing list and, if so, to return the membership of that list. This command is purely informational and will not add to the current list of recipients.
HELP	Asks the partner for information about its implementation, such as the list of commands that are supported.
Defined, but rarely implemented or used	
TURN	Asks the partner to switch roles and become the sender. The partner is allowed to refuse.
SEND	If the recipient is logged in, deliver a mail item directly to the recipient's terminal.
SOML	Send or Mail—if the recipient is logged in, deliver direct to the terminal. Otherwise, deliver as mail.
SAML	Send and Mail—deliver to the recipient's mailbox. If the user is logged in, also deliver to the terminal.

16.10   REPLY CODES

The SMTP reply codes are structured in very much the same way as the FTP reply codes. The codes are made up of three digits. The first digit indicates the status of the command:

1yz   Positive Preliminary reply (currently not used in SMTP)

2yz   Positive Completion reply

3yz   Positive Intermediate reply

4yz   Transient Negative reply ("try again")

5yz   Permanent Negative reply

The second digit classifies the reply:

x0z   In reply to a problem, this indicates a syntax error or unknown command.

x1z   Reply to information request such as *help*.

x2z   Reply referring to the connection.

x3z   Unspecified as yet.

x4z   Unspecified as yet.

x5z   Reply that indicates the status of the receiver mail system.

The meaning of the third digit varies depending on the command and the first two digits.

16.11   MORE ABOUT THE INTERNET MESSAGE FORMAT

The standard for the format of Internet text messages, defined in RFC 822, is straightforward. It consists of the following, in the order listed:

- A set of header fields (most of which are optional)
- A blank line
- The text or *body* of the message

A header field has the form:

Field-name: Field-contents

Field names and contents are expressed using ASCII characters. There are many header fields. A representative sample includes:

```
Received
Date
From
To
cc
bcc (blind cc)
Message-Id
Reply-To
Sender (if not the message creator)
In-Reply-To
References (to earlier Message IDs)
```

```
Keywords
Subject
Comments
Encrypted
```

We expect every message header to include *Date, From,* and *To* fields. *Received* fields are constructed using the timestamp information gathered as mail is transferred between Message Transfer Agents. Most mail software can create a message identifier that is included in the message. For example:

```
Message-Id: <199508271201.AA02330@PASCAL.MATH.YALE.EDU>
```

The Message-Id is designed to be unique across the network. To achieve this, it usually includes the originating host's name, along with a unique alphanumeric identifier. Note that the identifier above contains the date (1995 08 27), the Universal Time (12 01), and an additional string that assures that the ID is unique for that host and time.

Resent fields are added if a message is forwarded. Examples are *Resent-To, Resent-From, Resent-cc, Resent-bcc, Resent-Date, Resent-Sender, Resent-Message-Id,* and *Resent-Reply-To.*

The blank line that follows the headers is important. It tells the User Agent that the introductory header information is complete and that the actual message follows.

16.12   MAIL EXTENSIONS AND MIME

The simplicity of SMTP and the mail format made Internet mail easy to implement and led to widespread use. However, users grew impatient with its limitation to simple text messages. It was clear that SMTP needed an overhaul, but how could this be done without disturbing the installed base of mail applications?

A very practical approach was taken. New MIME clients would be implemented with the ability to create and receive multipart messages containing many useful types of information. These messages could be exchanged:

- Efficiently, via new Extended SMTP Message Transfer Agents.

- Less efficiently, via the old standard SMTP. Before passing a nontext body part to an old SMTP agent, the item would be converted so that it "looked" like ordinary NVT text.

Figure 16.7 shows how the architecture works.

16.12.1   Extended MTA

An Extended Message Transfer Agent needs to support one additional command. It sends an *EHLO* greeting instead of HELO. If the reply is positive, the partner also is an Extended MTA. If the reply is an error message, the MTA can revert to SMTP and send a HELO command.

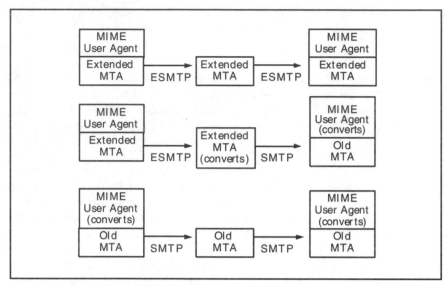

Figure 16.7   Delivering MIME messages.

Although the need to support MIME was the motive for extending the MTA, more services can be added at any time by defining new keywords for EHLO. For example, message sizes have been growing, and there is a new service that enables the sender to declare the size of a message before transmitting it. The receiver can indicate whether it is willing to accept a message of that size. The receiver also can indicate the biggest size that it is willing to accept.

Official extensions are registered with the Internet Assigned Numbers Authority. Some software is available that includes new, experimental extensions. These are assigned temporary names starting with X.

16.12.2 Extended SMTP Dialogue

The sample below illustrates how an Extended MTA sets up a transaction that will send a MIME message containing an 8-bit format:

- The receiver announces its extended capabilities, including 8BITMIME.

- The MAIL FROM command includes a BODY = 8BITMIME parameter.

```
EHLO MATH-GW.CS.YALE.EDU
250-Hello MATH-GW.CS.YALE.EDU, pleased to meet you
250-8BITMIME
250-HELP
250-SIZE
250-XONE
250-XVRB
250-XQUE
MAIL FROM:<feit-sidnie@math.yale.edu> BODY = 8BITMIME
250 <feit-sidnie@math.yale.edu>... Sender ok
```

```
RCPT TO:<Mary-Smith@jcn.com>
250 < Mary-Smith@jcn.com>... Recipient ok
DATA
354 Send 8BITMIME message, ending in CRLF.CRLF.
 ...
 .
250 OK
QUIT
250 Goodbye
```

16.13   FORMAT OF MIME MESSAGES

A MIME message contains a set of headers and one or more *body parts.* An ordinary Internet mail message starts with headers such as *From:, To:,* and *Date:.* A MIME message contains additional introductory headers that describe the structure and content of the message.

If there are multiple parts, one of the introductory headers defines a string that will be used to mark the *boundaries* between parts. Furthermore, after the boundary string that introduces a part, there will be additional headers that describe the body part that follows.

16.13.1   MIME Content-Type Headers

There are many different types of information that can be carried in a message. The overall structure of the message and the type of information in each part is announced by *Content-Type* headers. Sample headers include:

```
Content-Type: MULTIPART/MIXED; BOUNDARY = "xxxxxxxxx"
Content-Type: TEXT/PLAIN; charset = US-ASCII
Content-Type: image/gif
Content-Type: audio/basic
```

In general, a content-type header has the form:

Content-Type: *type / subtype; param = value; param = value;...*

Content-Types, subtypes, and parameter names are case-insensitive. They can be written in upper-, lower-, or mixed case. However, some parameter *values* will be case-sensitive.

Note that although MIME headers are written as English phrases, a *charset* = parameter could announce a part that is coded in ISO-8859-1 or in Japanese, Cyrillic, Hebrew, or Arabic.

16.13.2   A Sample MIME Message

The sample message that follows has multiple parts—it contains a text message and two attached text files.

The first Content-Type header:

```
Content-Type: MULTIPART/MIXED;
BOUNDARY = "plum.yale.edu:814898609:772210698:709846916:1916796928"
```

indicates that this is a multipart message. Its *BOUNDARY* parameter defines the delimiter that will mark the beginning and end of each part. The delimiter is selected by the User Agent. In this case, the delimiter is constructed from the host name and a string of digits generated by the User Agent. The actual boundary line will consist of two hyphens (--) followed by the boundary string.

MIME headers are displayed below in bold print so that they will stand out. Comments have been added to the right. Several lines in the message have been wrapped to make room for the comments.

	These are standard mail headers.
`Mime-Version: 1.0`	Announce MIME version.

`Content-Type: MULTIPART/MIXED;`
`BOUNDARY = "plum.yale.edu:814898609:`
`772210698: 709846916:1916796928"`
	Announce that it is multipart and provide a boundary delimiter.

	An empty line shows that there are no more introductory headers.
`-- plum.yale.edu:814898609:772210698:` `709846916:1916796928`	The boundary. Note the initial hyphens.
`Content-Type: TEXT/PLAIN; charset=` `US-ASCII`	Plain text content follows.
	Empty line to mark end of the headers for this part.
`Here are some attachments.`	The text contents.
`-- plum.yale.edu:814898609:772210698:` `709846916:1916796928`	Another boundary.

`Content-Type: TEXT/plain; SizeOnDisk`
`=28; name="ATT.TXT"; CHARSET=`
`US-ASCII`
	Plain text again. Parameters add some info.

`Content-Description: ATT.TXT`	This header provides extra information—the filename.
	End of headers for this part.
`** The first attachment **`	The text contents.
`-- plum.yale.edu:814898609:772210698:` `709846916:1916796928`	Another boundary.

`Content-Type: TEXT/plain; SizeOnDisk`
`=58368; name="NFSCAP.TXT"; CHARSET`
`=US-ASCII`
	More plain text.

`Content-Description: NFSCAP.TXT`	
	End of headers for this part.
`This is the second attachment. Some` `captured text follows:`	The text contents.
`. . .`	. . .
`. . .`	. . .
`-- plum.yale.edu:814898609:772210698:` `709846916:1916796928--`	Final boundary.

16.13.3 MIME Content Types

Table 16.2 shows the major content types and subtypes defined at the time of this writing. As might be expected, the *Assigned Numbers* document should be consulted for more current information.

TABLE 16.2   MIME Content Types

Type	Subtype	Description
text		
	plain	A standard mail text message.
	richtext	Portable word processor format.
	tab-separated values	
multipart		Consists of several parts, separated by boundary lines.
	mixed	
	alternative	A user can choose from several renditions. For example, ASCII text or Postscript.
	digest	Each part is itself a message.
	parallel	Parts that go together, such as video and its matching sound.
	appledouble	
	header-set	
message		An encapsulated message.
	rfc822	Classic electronic mail message.
	partial	Part of a message. Supports sending a very large message in pieces.
	external-body	Contains a pointer to a remote document, not the document itself.
	news	Contains Usenet News format.
application		Either uninterpreted binary, or data formatted for a specific application.
	octet-stream	
	postscript	Formatted for Postscript display of printing.
	oda	Office document architecture.
	atomicmail	
	andrew-inset	
	slate	
	wita	Wang information transfer.

TABLE 16.2   MIME Content Types (Continued)

Type	Subtype	Description
	dec-dx	DEC document format.
	dca-rft	IBM's Document Content Architecture, Revisable Format, for word-processing documents.
	activemessage	
	rtf	Rich text document format.
	applefile	
	mac-binhex40	Mac file converted for transfer.
	news-message-id	
	news-transmission	
	wordperfect5.1	
	pdf	Adobe Acrobat postscript.
	zip	Compressed.
	macwriteii	
	msword	
	remote-printing	
image		Image data.
	jpeg	Joint Photographic Experts Group, which specifies an image compression scheme.
	gif	Graphics Interchange Format, used for graphic files.
	ief	Image exchange format.
	tiff	Tag image file format.
audio		Audio data.
	basic	
video		"Movies."
	mpeg	
	quicktime	

16.13.4 Encoding Contents

RFC 822 defined the original format for Internet text messages. Mail contents consisted of a sequence of lines terminated by <CR><LF>. The maximum length of each line (including <CR><LF>) was 1000 characters.

How should the various types of contents of a MIME message be encoded for transmission? Wisely, the method of encoding is defined separately. This allows, for example, for SMTP to use:

- Inefficient encoding that makes binary data look like text, if that's all that the receiving MTA supports

- Efficient encodings when the receiver supports them

Encoding methods are listed in Table 16.3. If an encoding method other than ordinary NVT USASCII is used, it is announced in a Content-Transfer-Encoding header. For example:

```
Content-Transfer-Encoding: base64
Content-Transfer-Encoding: Quoted-printable
```

16.13.5 Quoted Printable Encoding Method

The quoted-printable encoding method is used for messages that just contain a few characters that do not belong to the basic ASCII set. These characters are mapped to special sequences, while the bulk of the item remains in its natural form. The encoding has the form:

= hex code for character

For example, a form feed, which is X' 0C, would be coded as *= 0C*.

TABLE 16.3   Encoding Methods

Method	Description
7bit	Ordinary NVT USASCII lines of text.
quoted-printable	Content is mostly ASCII text, but a few special characters need to be included. Each of these characters is mapped to special sequences of text characters.
base64	The entire content is mapped to a representation that looks like ordinary characters.
8bit	The message is still organized as a sequence of lines ending in <CR><LF> and at most 1000 characters long. However, 8-bit characters can be included.
binary	True binary data.
x-*token-name*	Any experimental encoding must be given a name starting with x.

16.13.6   Base64 Encoding Method

Base64 encoding converts any type of data to a one-third larger amount of textual characters. Data is broken up into sets of three 8-bit bytes. For example:

```
10001000 00110011 11110001
```

To convert, we first break this into four 6-bit groups:

```
100010 000011 001111 110001
```

We then interpret each group as a number:

```
34 3 15 49
```

And finally, we replace each number by the corresponding character from Table 16.4.

TABLE 16.4   Base64 Encoding

Value	Code	Value	Code	Value	Code	Value	Code
0	A	17	R	34	i	51	z
1	B	18	S	35	j	52	0
2	C	19	T	36	k	53	1
3	D	20	U	37	l	54	2
4	E	21	V	38	m	55	3
5	F	22	W	39	n	56	4
6	G	23	X	40	o	57	5
7	H	24	Y	41	p	58	6
8	I	25	Z	42	q	59	7
9	J	26	a	43	r	60	8
10	K	27	b	44	s	61	9
11	L	28	c	45	t	62	+
12	M	29	d	46	u	63	/
13	N	30	e	47	v		
14	O	31	f	48	w		
15	P	32	g	49	x		
16	Q	33	h	50	y		

If the total number of octets is not a multiple of 3, there will be 1 or 2 octets at the end. The "leftover" is padded with zero bits and encoded. One octet is then translated to two characters, followed by = =, while two octets are translated to three characters, followed by =.

16.14   POST OFFICE PROTOCOL

The Post Office Protocol (POP) is used to transfer mail from a mail server to a desktop station or a laptop portable.

The POP specification defines a lot of extra functions, such as the ability to view a listing of incoming mail items and their sizes and selectively retrieve and delete mail items. However, implementations usually simply download all waiting mail. The user may have the option either to leave copies of all mail at the server or else to delete items from the server after they have been downloaded.

A desktop system uses POP to download its mail and SMTP to send its mail. In most cases, the download server for incoming mail will be the same as the outgoing gateway for outgoing mail, as shown in Figure 16.8. However, your client application may allow you to use different systems as your POP server and outgoing gateway, if you want to.

16.15   OTHER MAIL APPLICATIONS

There are many Internet *mailing lists,* which enable participants to exchange questions and answers and receive the latest news for a specific topic—such as vacation spots, new CD-ROMs, or computer security problems.

Users subscribe to a list by sending a request to an advertised mailbox. Messages sent to a second mailbox are relayed to all subscribers, as shown in Figure 16.9. Free mailing list software, including a very popular program called *Majordomo,* is available for many platforms.

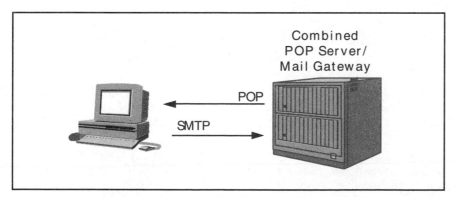

Figure 16.8   A combined POP server and Mail Gateway system.

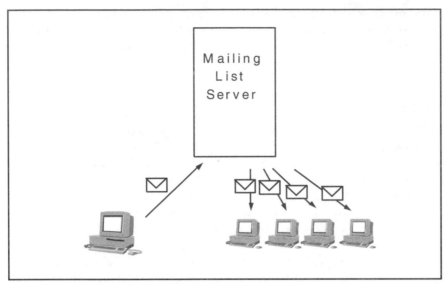

Figure 16.9   A mailing list server.

16.16   PERFORMANCE

Message Transfer Agent services use memory, disk, processing, and transmission bandwidth resources. A mail service is enormously useful, and traffic can be expected to build steadily.

Messages must be saved while they await transmission or relay. Mail sometimes is kept at a server until users login and access their mailboxes. It can be hard to predict the amount of storage that will be consumed in supporting a mail application.

Since mail handling is automated, items could conceivably sit in a Message Transfer Agent's queue forever. It is important for an administrator to define timeouts for every mail activity, to prevent black holes from swallowing up computer resources.

16.17   SECURITY

16.17.1   Sendmail Problems

The most commonly used message transfer program is an application called *sendmail*. *Sendmail* is a big, complicated program that performs many functions, including translation of mail alias names and expansion of mailing lists.

Since *sendmail* speaks SMTP, which runs on top of NVT *telnet,* it is easy for users to connect to *sendmail* at port 25 and try to break into a computer. Unfortunately, *sendmail* appears to be the source of an endless supply of security holes.

There is an easy solution. Very simple and robust programs have been written that can receive SMTP mail far more safely. One of these should be used. If the functionality of *sendmail* is needed, the simple program can queue incoming mail to *sendmail*.

16.17.2 Secure Electronic Mail

Sometimes crackers are able to eavesdrop on mail transmissions. Unfortunately, it also is fairly easy to forge electronic mail. Fortunately, the methods described in Chapter 3 have been applied to electronic mail, and products are available that authenticate and encrypt mail.

16.17.3 Secure MIME (S/MIME)

Secure MIME protects mail by means of message digests, public keys, and symmetric session keys. Public keys are reliably associated with their owners via a hierarchy of digital certificates whose format is defined by the X.509 standard (see Section 16.19.1).

16.18 MESSAGING VIA X.400

The International Telecommunications Union (ITU) is an agency responsible for supporting international communications and producing *Recommendations* that define how countries can promote telegraph, telex, and telephone connectivity.

The Telecommunication Standardization Sector of the International Telecommunications Union (ITU-T) presides over many study groups that produce specifications for new technologies. This sector formerly was called the International Telegraph and Telephone Consultative Committee (CCITT). As noted in Chapter 4, the CCITT was responsible for the X.25 data communications standards.

The CCITT established a study group that met during the 1981 to 1984 period and developed the X.400 set of recommendations for an international electronic message handling service. These recommendations were comprehensive and were subsequently adopted by ISO. The standards were updated in 1988, but some current implementations still are based on the 1984 specification. Some of the characteristics of X.400 are:

- Definition of a general store-and-forward service that can be used for electronic mail (which they call *interpersonal messaging*) as well as for other applications.

- Global, international scope for message delivery and support for international alphabets.

- The ability to send many information types besides text, such as binary, image, or digitized voice.

- If the sender wishes, notification of delivery to a recipient system and detailed nondelivery notices. There is an optional feature that can signal that the recipient end user has received[1] the mail.

- Support for mail priority.

- The ability to convert a message to a different medium—for example, deliver via fax or convert to hardcopy and use postal delivery.

- Definition of user-friendly identifiers for originators and recipients.

- Use of a formal envelope that contains fields that can be used to trace messages and gather other mail management information.

X.400 defines a standard for the exchange of mail between national administrations. It also can be seen as a gateway standard. Vendors of many proprietary mail systems provide software that converts their messages to and from X.400 format, enabling users of these products to exchange messages.

X.400 has won support in Europe and also has been mandated for use by some U.S. government agencies.

16.18.1 Sample X.400 Message

Unlike the Internet standards, X.400 did not rely on 7-bit ASCII and NVT conventions. Fields are formatted using the ISO Basic Encoding Rules (BERs) that will be described in Chapter 20. This encoding introduces each field with a hex identifier code and length value. Figure 16.10 shows an outline for a sample message that illustrates general features of the X.400 format.

16.18.2 Naming X.400 Recipients

How do you identify people when you refer to them in conversation? You might say "Mary Jones, who is a Technical Consultant at the Milwaukee unit of MCI Telecommunications Corporation." Or you might say "Jacques Brun, who lives at 10 Rue Centrale in Paris, France." The drafters of X.400 wanted to define a universal naming system that would correspond to the natural way that people are identified.

An X.400 originator or recipient name is a list of attributes. The standard defines many optional attributes which may be used in various combinations. The attributes that are expected to be most prevalent in electronic mail systems are:

- Country name
- Administration domain name
- Personal name (e.g., John H. Jones III)
- Organization name

---

[1]What it means to receive a message is deliberately left undefined. This could mean observing a summary of the contents of the mailbox, reading the item, or pressing a function key to acknowledge receipt.

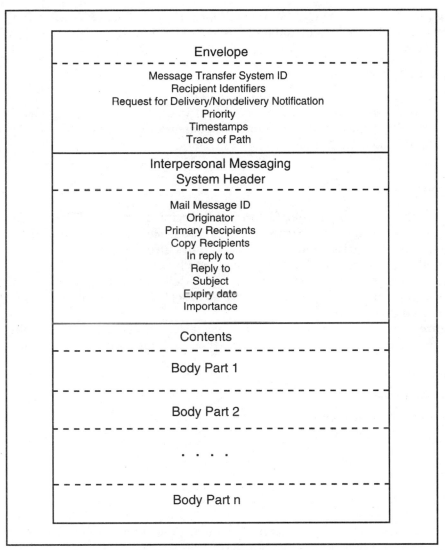

Figure 16.10   Format of an X.400 Interpersonal Message.

- Organizational unit names
- Private domain name
- Domain-defined attributes

Private domains include facilities such as commercial electronic mail services and corporate electronic mail systems based on proprietary mail products.

Domain-defined attributes allow names used by existing mail systems to be embedded in an X.400 identifier. This is an important feature. It allows an

X.400 gateway to switch mail between proprietary mail systems as well as between a proprietary system and an X.400-compliant system.

16.18.3   Interworking between X.400 and Internet Mail

Since both X.400 and Internet mail are store-and-forward services, mail can be passed between these services by means of mail gateways. Several RFCs have been written that deal with mapping between the Internet message format and X.400 message format.

16.19   ISO/ITU-T DIRECTORY

Producing the correct identifier for an X.400 recipient can be difficult. The naming attributes that are selected will vary from user to user. When X.400 was completed, it was realized immediately that a directory service was needed if X.400 was to succeed. The X.500 recommendations, which were prepared during the 1985 to 1988 CCITT study period, define directory services and protocols. Several research and commercial associations have created pilot implementations of X.500.

The directory standard is very broad in scope. The X.500 directory is a distributed database intended to include many types of information. For example:

- Names of people
- Postal addresses
- User identifiers for X.400 mail
- Internet-style mail identifiers
- Telex and fax numbers
- Telephone numbers
- Names and locations of printers

X.500 databases eventually are expected to include information that helps users to locate any type of network resource.

16.19.1   Directory Model

The Directory Information Base is distributed across a community of databases controlled by *Directory Service Agents* (DSAs). Users access directory information by means of a *Directory User Agent* (DUA). A DUA provides the user interface for interactive queries and updates and passes user requests to a DSA.

The X.500 standards define a complex formal protocol that governs the interaction between a DUA and a DSA. The Internet lightweight directory access protocol (LDAP) simplifies access to directory services. There also is a

DSA-to-DSA protocol that enables DSAs to relay user queries or download copies of parts of the Directory Information Base.

There are a number of structural similarities between the X.500 directory system and the Domain Name System. Both are distributed directory systems, and both have an overall structure organized as a hierarchical tree. Users of each interact with a local client to reach a designated server, and that server can initiate distributed queries on behalf of the user.

The X.500 standards include a method of validating the authenticity of a directory entry. An entry is validated by an encrypted certificate from a trusted source. The format of a certificate is defined in standard X.509.

16.20   RECOMMENDED READING

RFC 821 defines the Simple Mail Transfer Protocol, and RFC 822 describes the format of Internet messages. RFC 1939 describes the Post Office Protocol used to transfer mail between desktop workstations and a mail server.

RFCs 1521 and 1522 describe MIME. There have been many additions and updates defined in other RFCs. MIME types are published in the *Assigned Numbers* document, and registration procedures are described in RFC 1590. RFC 1848 defines a framework for secure MIME. The S/MIME specification can be obtained from RSA Data Security, Inc.

The Service Extension framework is described in RFC 1869, and RFC 1652 specifies the SMTP Service Extension for 8BITMIME transport.

X.400 was initially published as part of the 1984 CCITT recommendations and was updated in the 1988 recommendations. ISO published its version of X.400 in ISO 10021, which is made up of several parts. X.500 was an 1988 CCITT recommendation.

RFCs 1327 and 1495 currently define the mapping between X.400 elements and the classic RFC 822 format. This mapping has been updated frequently, so the RFC index should be consulted for the current version. RFC 1496 discusses MIME translation. RFC 1506 is a tutorial on gatewaying between X.400 and Internet mail. Several other RFCs discuss mail recipient address translations. See RFC 1777 and RFC 1798 for a description of the lightweight directory access protocol.

17

Network News

17.1 INTRODUCTION

Every day, up-to-date information on science, technology, computers, economics, travel, sports, music, education, and more is contributed to the Internet *Usenet News*. A *news group* is like a bulletin-board service. News is made available in the form of *articles* that are *posted* (sent) to the group.

Currently, there are thousands of public and private news groups, and many provide information not easily found elsewhere. Often, postings consist of questions and answers relating to a specific topic. Sometimes the flow of information is one way: the news group is used by an individual or organization as a way of publishing information.

Each news group is maintained by an administrator at a primary news server. If the news group is private, the news might reside exclusively at that server, and users would retrieve news items from that server. However, postings for a public Usenet News group are propagated from its primary news server to hundreds of other news servers all over the world.

The news application has proved to be useful beyond its original Internet bulletin-board role. News software is reused by organizations to post their own internal information. News software also has spawned a new publishing business. Publishers feed regular news stories from wire services, such as AP, UPI, and Reuters, to subscribing sites using the Internet news protocol.

17.2 HIERARCHY OF INTERNET NEWS GROUPS

Thousands of Internet news groups have been created. Each group is assigned a name that indicates its purpose. The names are arranged in a tree. Figure 17.1 shows part of the tree structure.

Unlike the other hierarchical names that we have met in this book, these names are read *from top to bottom*. For example:

rec.sport.basketball.college

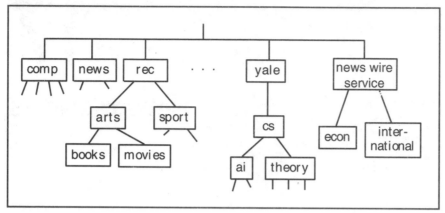

Figure 17.1 News group hierarchy.

17.3 NEWS AGENTS

Just as there are User Agent programs that enable end users to send and receive mail, there are programs—which we will call *news agents*—that enable users to subscribe to news groups, read news articles, and *post* their own articles to groups.

17.4 NEWS MODEL

A news client process interacts with a network news server via the *Network News Transfer Protocol* (NNTP). A client process can reside in an end user's news agent or in a peer news server. NNTP enables:

- A news server to obtain news from another news server
- A client's news agent to obtain news from a news server
- A client's news agent to post a new article to a news server

Figure 17.2 shows a client retrieving news from a server via NNTP and servers exchanging news via NNTP.

17.5 NNTP SCENARIO

Like the Simple Mail Transport Protocol (SMTP) NNTP runs over a *telnet* Network Virtual Terminal (NVT) session. The dialogue below shows a news transfer interaction. In the dialogue, the client:

- Connects to the server
- Asks what news commands are supported by the server
- Requests a list of news groups that have been created since October 23, 1995

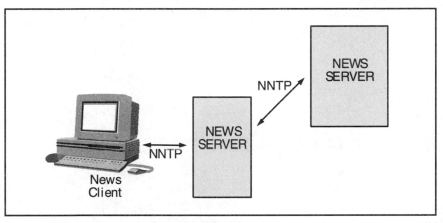

Figure 17.2 Requesting and transmitting news.

- Accesses a news group called *news.answers*
- Reads an article in *news.answers*

200 yale InterNetNews NNRP server   The server identifies itself and indicates
INN 1.4 22-Dec-93 ready (posting ok). that it accepts articles posted by users.

help

100 Legal commands   These are the commands supported by
the news server.

  authinfo user Name|pass Password

  article [MessageID|Number]

  body [MessageID|Number]

  date

  group newsgroup

  head [MessageID|Number]

  help

  ihave

  last

  list

[active|newsgroups|distributions
   schema]

  listgroup newsgroup

  mode reader

  newgroups yymmdd hhmmss ["GMT"]

    [<distributions>]

  newnews newsgroups yymmdd hhmmss
  ["GMT"]

    [<distributions>]

```
    next
    post
    slave
    stat [MessageID|Number]
    xgtitle [group_pattern]
    xhdr header [range|MessageID]
    xover [range]
    xpat header range|MessageID pat
[morepat...]
    xpath xpath MessageID

Report problems to
<usenet@cs.yale.edu>
    .
```

newgroups 951023 010000

This *newgroups* command requests a list of new groups that have been created since October 23, 1995 (1 A.M.).

```
231 New newsgroups follow.
rec.music.iranian 14 1 y
soc.atheism 0 1 m
soc.culture.jewish.parenting 1 1 m
soc.culture.rep-of-georgia 3 1 y
    .
```

newnews news.answers 951020 110101

Frequently Asked Questions (FAQ) documents on many topics are published in *news.answers*. This command asks for a list of new FAQs published since October 20, 1995 (11:01 A.M.).

```
230 New news follows
<unix-faq/faq/part2_814199602
  @rtfm.mit.edu>
```

```
<unix-faq/faq/part3_814199602
  @rtfm.mit.edu>
```

A very long list is displayed.

```
<unix-faq/faq/part4_814199602
  @rtfm.mit.edu>
```

```
. . .
<skydiving-faq_814424705
  @frc2.frc.ri.cmu.edu>
```

Here we just show a subset.

```
. . .
<civil-war-usa/faq/part1_814453424
  @rtfm.mit.edu>
<civil-war-usa/faq/part2_814453424
  @rtfm.mit.edu>
```

```
 . . .

<461fkk$lt2@cst715.iac.honeywell.com>

<461flf$lt2@cst715.iac.honeywell.com>

 . . .

 .
```

group news.answers   We move into the *news.answers* group.

`211 321 52807 53147 news.answers`

article   We ask to see an article.

<461fkk$lt2@cst715.iac.honeywell.com>   There is a long header.

`220 0 article`   The home host for the group is *iac.hon-eywell.com.*

```
<461fkk$lt2@cst715.iac.honeywell.com>

Path:

yale!yale.edu!spool.mu.edu!
  howland.reston.ans.net!newsfeed.
  internetmci.com

!in2.uu.net!news.iac.honeywell.
  com!dwe

From: dwe@eng.iac.honeywell.com
  (Dave Eaton)

 . . .

Archive-name:
  sw-config-mgmt/cm-tools

Last-modified: 1995/10/25
```

`Version: 2.5`   We finally reach the start of the article.

```
Posting-Frequency: monthly

-=-=-=-=-=-=-=-=-=-=-=-=

Configuration Management Tools Summary

-=-=-=-=-=-=-=-=-=-=-=-=

This is the newsgroup comp.
  software.config-mgmt
  "Frequently Asked Questions"

(FAQ) posting of a Software
  Configuration Management tools
  summary. This is part 2 of the
  3 part FAQ.

(. . . etc.)
```

`                                        .`   The end of the article is indicated by a line containing only a period.

quit   End the session.

```
205

Connection closed by foreign host.
```

17.6   USING DESKTOP NEWS AGENTS

Let's take a look at how a similar dialogue would look with a desktop news agent. Figure 17.3 shows a *Chameleon* news display. A list of new news groups can be requested by clicking a menu selection.

Figure 17.4 shows that the desktop news agent keeps track of a set of news groups selected (*subscribed to*) by the user.

A list of unread articles in the popular group *news.answers* is requested by double-clicking on the *news.answers* line. The result is shown in Figure 17.5. An article is displayed in Figure 17.6. The article's long header need not be viewed unless the user wishes to see it.

Finally, Figure 17.7 shows a news article displayed by a *Netscape Navigator* World Wide Web (WWW) browser that is being used to read news. The article was written by the Reuters news service and is published electronically by the *Clarinet* news service.

17.7   NNTP PROTOCOL

17.7.1   NNTP Commands

To access news articles, a client process connects to port 119 at a news server. The client sends a series of commands and receives responses. Commands are not case-sensitive.

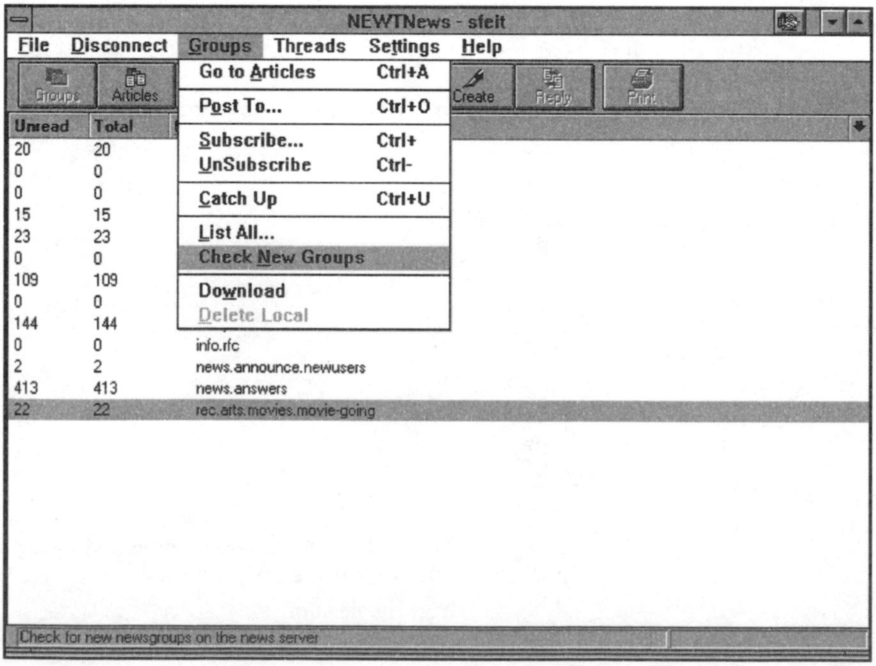

Figure 17.3   News group menu options.

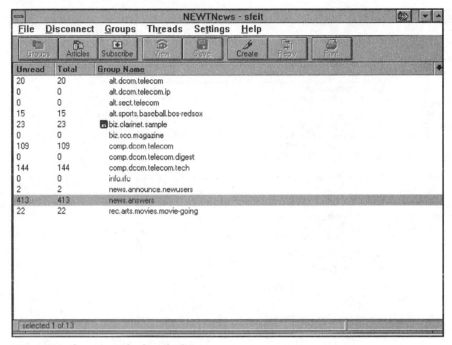

Figure 17.4 Overview of subscribed groups.

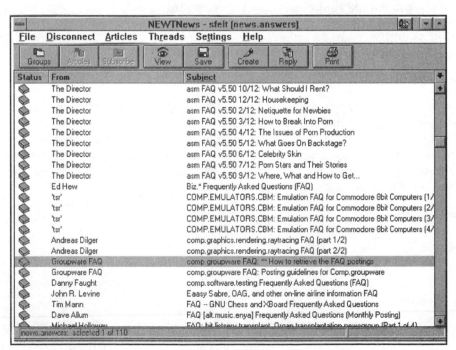

Figure 17.5 List of unread articles in *news.answers*.

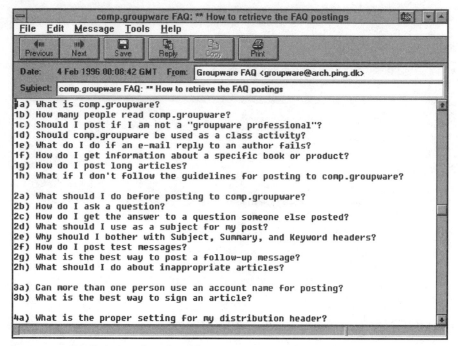

Figure 17.6 Display of a selected article.

There are commands that enable the requester to:

- List all groups
- Choose a group
- Select specific articles

A *current article pointer* at the server keeps track of the requester's position. The commands are summarized in Table 17.1.

The optional *distributions* parameter enables the user to select a list of top-level categories, such as *comp* or *news*. The list must be enclosed in angle brackets, and items must be separated by commas. For example, below we ask for a list of new news groups under *sci*.

```
newgroups 950601 010000 <sci>
231 New newsgroups follow.
sci.physics.cond-matter 552 1 y
sci.techniques.mass-spec 279 1 m
sci.psychology.consciousness 164 1 m
. . .
```

17.7.2 NNTP Status Codes

The dialogue displayed in Section 17.5 showed that each response from an NNTP server starts with a numeric status code. This is the same convention that was used for SMTP and FTP servers. The codes are:

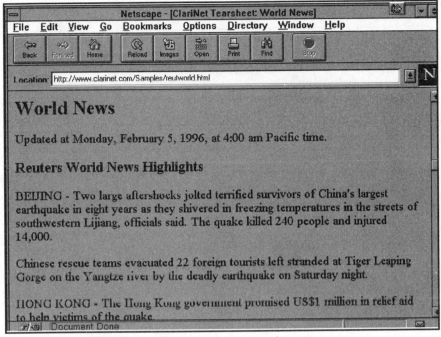

Figure 17.7   A news item.

1xx	Informative message
2xx	Command successful
3xx	Command successful so far; send the rest of it
4xx	Command was correct but couldn't be performed for some reason
5xx	Command unimplemented or incorrect or a serious program error occurred

As before, the second digit in the code provides more specific information about the response:

x0x	Connection, setup, and miscellaneous messages
x1x	News group selection
x2x	Article selection
x3x	Distribution functions
x4x	Posting
x8x	Nonstandard extensions
x9x	Debugging output

17.8   DIFFERENCES BETWEEN NEWS AND MAILING LISTS

A news application is more efficient than a mailing list in several ways. News

TABLE 17.1   NNTP Commands and Parameters

Command	Parameters	Description
article	"\<Messageid\>" or message number or none.	Retrieve the article identified by the ID or number or get the current article.
body		Retrieve the current article's body.
group	Groupname.	Move to the selected news group.
head		Displays the current article's header.
help		Asks for a list of commands supported by the server.
ihave	\<Messageid\>.	A server tells another server that it has an article. The other server can request a copy if it wishes.
last		The current article pointer is moved back one article in the current group.
list		Requests a list of news groups and the range of articles that are available.
newgroups	Date, time, and optionally, \<distribution\>.	Requests a list of news groups (optionally, within a category) created since the given date and time.
newnews	Newsgroup, date, time, and optionally, \<distribuition\>.	Requests a list of new articles for the groups since the given date and time.
next		The current article pointer is moved forward one article in the current group.
post		Send a new article to the news group.
quit		
slave		Indicates that the requester is a mail server rather than an individual client.
stat	Message number.	Selects an article.

is stored at centralized servers that can be accessed by many users. Many users can read news out of a single shared database.

Mailing list distributions can clutter up your mailbox with extraneous information, making it difficult to weed out the really important mail. In contrast, you access news at your own convenience, and sophisticated news screening capabilities are being built into news agents, making them even more convenient to use.

You do not have to subscribe to a news group in order to read its news or post to it. In fact, subscription is just a news agent function that is used to help you track what groups you wish to follow and which items you already have read.

There are many mailing lists that automatically feed their items to a news group.

17.9   RECOMMENDED READING

The Network News Transfer Protocol is defined in RFC 977.

Gopher and WAIS

18.1 INTRODUCTION

Gopher was developed in 1991 at the University of Minnesota Microcomputer, Workstation, and Networks Center. Personnel there had to support thousands of users who needed help with computer problems and wanted to know about campus information resources.

There was a ton of information, and it was all over the place. How could it be put on-line so that students could find what they needed easily? The answer was gopher—a hierarchy of simple menus and a clever client/server architecture.

Gopher provides an easy way to navigate your way through lots of information resources. You can:

- Find information provided at your local site
- Access remote sites transparently
- Retrieve what you need

Gopher's value in organizing and dispensing information was first recognized on college and university campuses around the world. Then gopher quickly spread to government organizations.

Gopher has been eclipsed by its more powerful sibling, the *World Wide Web* (WWW), which will be discussed in Chapter 19. However, there still are sites that provide information via gopher. World Wide Web browsers routinely access information at gopher servers, although users may not even be aware that this is happening.

18.2 USING GOPHER

The best way to get introduced to gopher is to use it. If you are logged into a multiuser host and are using a text-based user interface, you can type *gopher* to start a local gopher client. In Figure 18.1 we start a text-based client at *tig-*

```
> gopher

      Root gopher server: gopher.jvnc.net

->    1. About this gopher.
      2. Search GES Gopher Tree <?>
      3. GES/
      4. Educational Services/
      5. Internet Resources/
      6. Medical Resources/
      7. Gophers Hosted by GES/
      8. Other Interesting Gophers/
      9. Publishers Online/
     10. WAIS Based Information/
     11. InterNIC/


Press ? for Help, q to Quit, u to go up a menu
```

Figure 18.1   Text client accessing a gopher server.

ger and access the default local gopher server, which is operated by Global Enterprise Services.

As you can see, gopher displays menus. A menu item can lead to:

- A text document
- An image
- Another menu
- A search application
- A *telnet* session with an application on a remote host
- Some other application (such as FTP)

Any item can lead to a gopher server—or some other application—that resides on a different computer.

A gopher client is included in World Wide Web browser products. Today, this is the most popular way to access gopher servers. In Figure 18.2 a *Netscape Navigator* client accesses the same menu that was shown in Figure 18.1.

18.3   GOPHER INFORMATION TYPES

Entries in a gopher menu can include a variety of information types. Each type is assigned an identifier code. A text-based client announces the item type by displaying a tag at the end of the menu line. Types, their codes, and their tags are listed in Table 18.1. A graphical client displays a different icon for each type of information.

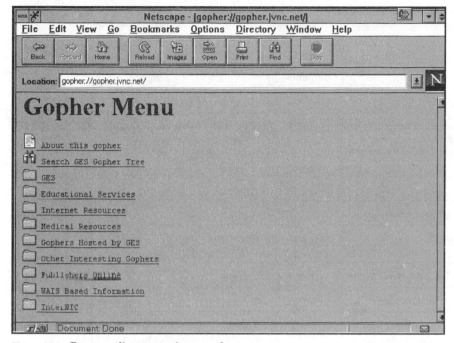

Figure 18.2 Browser client accessing a gopher server.

18.4 GOPHER MENU HIERARCHY

Gopher menus are arranged in a hierarchical tree. A menu item can point to another menu. The other menu may in fact be located at a completely different site. The "leaves" of a menu tree are documents or applications.

As we shall see later, gopher menus actually correspond to directories, so the use of / to denote a menu is not accidental. A gopher server's home directory is part of its initial configuration information. The initial default menu lists titles assigned to files and subdirectories in the home directory.

18.5 GOPHER ARCHITECTURE

Behind the scenes, gopher's architecture is very simple. As illustrated in Figure 18.3, a client connects to a gopher server, retrieves a menu or file, and disconnects. Gopher displays the item to the user. The user is not connected to the server while reading the menu or file.

A gopher server is *stateless*—that is, it does not have to remember anything about its clients. A client connects and asks for something. The server answers the request and then can forget about it. This makes gopher easy to run and pretty much foolproof. It also enables a gopher server to handle many more clients per hour than earlier services (such as *telnet* or file trans-

TABLE 18.1   Gopher Datatypes, Codes, and Tags

Identifier code	Type	Tag	Comments
0	A file.	.or blank	
1	A menu.	/	
2	A phone-book service (named for the Computer Services Organization of the University of Illinois).	<CSO>	A simple database application for phone numbers, electronic mail addresses, office address, etc.
3	Error.		
4	A BinHexed Macintosh file.		
5	A PC binary.	<PC Bin>	Client *must* read until the TCP connection closes.
6	A UNIX uuencoded file.		
7	An Index-Search service.	<?>	
8	A text-based *telnet* session.	<TEL>	*Telnet* session will be launched if selected.
9	A binary file.	<Bin>	Client *must* read until the TCP connection closes.
s	A sound file.	<)>	
e	An event.		
c	A calendar application.		
T	A text-based 3270 session.	<3270>	Session will be launched if selected.
g	A graphics file (in "GIF" standard format).	<Picture>	
I	Some other kind of image file.	<Picture>	Client decides how to display this.
M	A MIME message.	Blank or <MIME>	
h	A World Wide Web hypertext document.	Blank or <MIME>	

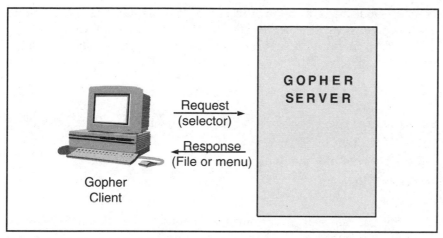

Figure 18.3 A client retrieving information from a gopher server.

for) could. The same principle was reused to make World Wide Web servers perform efficiently.

18.6 RELATIONSHIP TO FILE TRANSFER

A large portion of gopher's design was aimed at providing friendlier and more efficient access to a file transfer archive. Each gopher menu actually corresponds to a directory at the server. The directory includes special files that:

- Assign menu titles to files and subdirectories
- Describe links to files and directories at remote hosts
- Describe links to applications

We'll take a look at some examples a little later.

18.7 GOPHER PROTOCOL

A gopher session runs on top of a TCP connection. The usual server port is 70. Some NVT *telnet* conventions are used for the session. To get information from a gopher server, a client needs to:

- Connect to the gopher host at the appropriate port
- Send a *selector string* terminated with <CR><LF> to the server

The selector string identifies an item, such as a text document or menu.[1] A null selector string (consisting of just <CR><LF>) causes the server's default "root menu" to be sent back.

---

[1]Selector strings can identify many other things, such as a script or program to be executed or a query to be applied to a database.

If the server sends back a menu, the client will display the list of menu titles to the user. However, the server sends back a lot more information than just the titles. Each menu item sent back by the server consists of a sequence of fields separated by tabs. The fields include:

- The item type and title
- Selector string to be sent back to get this item (usually the item type along with a file or directory name)
- Name of the host holding the item
- Port for accessing the host

You can see this for yourself. A raw interaction with the Global Enterprise Systems (GES) gopher is shown below. We *telnet* to port 70 at the server and simply press the ENTER key after we have been connected:

```
> telnet gopher.jvnc.net 70
Trying 128.121.50.10 ...
Connected to nicol.jvnc.net.
Escape character is '^]'.
        (Pressing ENTER sends <CR><LF>)
0About this gopher          0/0about                        nicol.jvnc.net 70
7Search GES Gopher Tree     7/ts                            nicol.jvnc.net 70
1GES                        1/GES                           nicol.jvnc.net 70
1Educational Services       1/Educational_Services          nicol.jvnc.net 70
1Internet Resources         1/Internet_Resources            nicol.jvnc.net 70
1Medical Resources          1/Medical_Resources             nicol.jvnc.net 70
1Gophers Hosted by GES      1/Hosted                        nicol.jvnc.net 70
1Other Interesting Gophers  1/Other_Interesting_Gophers     nicol.jvnc.net 70
1Publishers Online          1/Publishers_Online             nicol.jvnc.net 70
1WAIS Based Information      1/WAIS_Based_Information         nicol.jvnc.net 70
1InterNIC                   /                               internic.net 70
.
Connection closed by foreign host.
```

Let's take a close look at the first item. "0About this gopher" indicates that the item is a text file and that the title, *About this gopher,* should be displayed. The selector string, "0/0about," repeats the type (0) and points to a file called *0about* in the server's home directory. If the user chooses this item, the gopher client will send the selector string to the server.

The next column identifies the host at which the item is located. We connected to *gopher.jvnc.net,* which is an alias name for *nicol.jvnc.net.* Finally, the last column indicates that the standard gopher port (70) should be used. The item is terminated with <CR><LF>.

The next few items identify subdirectories of *nicol*'s gopher home directory. The last item points to the default menu at the InterNIC gopher.

Note that the gopher server signals the end of the menu transmission by sending a line that contains only a period. When a text file is transmitted, a period also is used to signal the end of the file.

18.8   THE .NAMES FILE

A crude gopher server could be set up simply by configuring the gopher server program with the location of its home directory and starting it up. The server's main menu would list the names of the files and subdirectories in the home directory. If you chose a subdirectory item, its menu would list that directory's file and subdirectory names.

To replace the bald names with descriptive titles, the server's administrator simply creates a helper file called *.names* in each gopher directory. Below, we show some of the entries in the *.names* file for the GES gopher's home directory:

```
# The Top Level Directory
Path = 0/0about
Name = About this gopher
Numb = 1

Path = 1/GES
Name = GES
Numb = 3

Path = 1/Educational_Services
Name = Educational Services
Numb = 4

Path = 1/Internet_Resources
Name = Internet Resources
Numb = 5
```

Menu items that connect to remote gophers or start applications are listed in a *.Links* file. These entries include additional information—a formal information type identification, the host name, and the access port to be used. Some sample entries are:

```
Type = 7
Name = Search GES Gopher Tree
Path = 7/ts
Host = nicol.jvnc.net
Port = 70
Numb = 2

Type = 1
Name = InterNIC
Path = /
Host = internic.net
Port = 70
Numb = 11
```

As shown in Figure 18.4, the *Internet Resources* menu has many links to *telnet* sessions. A typical *telnet* entry in *.Links* is:

```
Type = 8
Name = CARL System
Path = CARL
Host = pac.carl.org
Port = 23
Numb = 2
```

```
                    Internet Resources


  -> 1. Area Code Info/
     2. CARL System <TEL>
     3. FreeNet (USA Today) <TEL>
     4. Ftp/
     5. Geographic Server <TEL>
     6. Libraries/
     7. Netfind (Internet White Pages) <TEL>
     8. News/
     9. Pilot Weather Service. [Airplane Pilot] <TEL>
     10. RFC/
     11. Sun Managers/
     12. Sunergy/
     13. Weather By State/
     14. Weather Service <TEL>
     15. World Wide Web  <TEL>



  Press ? for Help, q to Quit, u to go up a menu
```

Figure 18.4   Internet Resources menu.

Type = 8 means *telnet,* and in this case, the *Path* parameter identifies the userid that will be used for the *telnet* login.

18.9   WAIS

Gopher made many files available to users, but users needed a tool that would enable them to search an archive of text documents to find the ones that were useful. Most gopher servers are packaged with a full-text indexing and search tool called the *Wide Area Information Service,* or *WAIS*. There are free and commercial versions of WAIS.[2]

Other indexing and search tools have been developed. Searching is a very important application, and there is active competition in developing the most efficient indexing methodology, the most functional searches, and the fastest retrieval.

18.10   RECOMMENDED READING

The gopher protocol is described in RFC 1436. Free information and gopher software is available from the gopher server at the University of Minnesota (*gopher.tc.umn.edu*).

---

[2]WAIS is now a trademark of WAIS, Inc.

The World Wide Web

19.1 INTRODUCTION

19.1.1 Hypertext

Hypertext is an idea that has been around for several years. The basic idea is that:

- An underlined phrase in a document is associated with a pointer to another document.
- A user can *link* to the other document by clicking on the phrase.

Users of Microsoft Windows or Macintosh help screens depend on Hypertext routinely, although the users might not ever have heard of the term. For example, suppose that we are presented with a help menu like:

> Saving Files
> Finding and Replacing
> Cutting and Pasting
> Page Formats

It is intuitively clear that we can get more information on an underlined topic by clicking on that topic. In this case, phrases that provide hypertext links to other documents are underlined. Other user interfaces might present links in a different color or highlighted.

19.1.2 Hypermedia

This idea has been extended to *Hypermedia*—you can have an underlined phrase that points to a picture, a sound file, a film clip, or some other type of binary data. Or a picture can contain clickable elements that point to documents, pictures, sound data, or a film clip. This kind of display is commonplace on CD-ROMs.

19.1.3 Hypermedia and the WWW

The use of Hypermedia is extended to *networked* information via the *World Wide Web* (WWW). An underlined phrase can point to a local item—or to an item that actually is stored at a remote computer. This simple idea has led to attractive user interfaces that make it easy to navigate the Internet.

19.2   WHERE WWW CAME FROM

The idea for the World Wide Web came out of the physics research community. It was the brainchild of Tim Berners-Lee at CERN, the European Laboratory for Particle Physics, located in Switzerland.

19.3   WWW BROWSERS

The Web was given a tremendous boost by Marc Andreessen, who conceived the powerful *Mosaic* WWW client in 1992, when he was an undergraduate student at the University of Illinois and a staff member at the university's National Center for Supercomputing Applications (NCSA). Mosaic is an Internet *browser*—a program that can access data from multiple sources—including hypertext archives, gopher servers, search databases, file transfer sites, and news sites.

As illustrated in Figure 19.1, a browser can run the multiple protocols required to reach this information. Mosaic spawned a popular commercial browser called *Netscape Navigator,* marketed by Netscape Communications Corporation. Figure 19.2 shows the Netscape home page being retrieved by a Netscape browser.

Use of Web servers and browsers has grown explosively, and the protocols and technology have advanced rapidly.

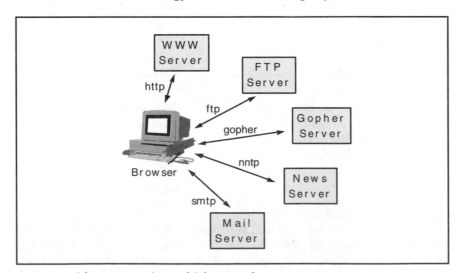

Figure 19.1 A browser running multiple protocols.

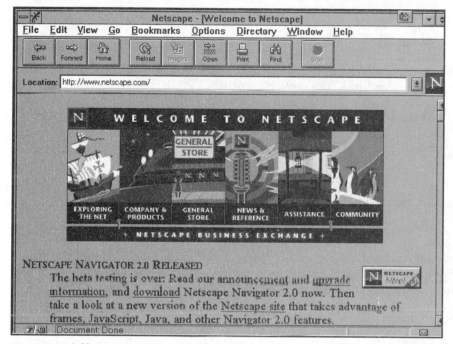

Figure 19.2 A *Netscape* browser retrieving a *Netscape* home page.

19.4   UNIFORM RESOURCE LOCATORS

A very important unifying concept came out of the World Wide Web effort. Every Web information resource is identified by its *Uniform Resource Locator* (sometimes called Universal Resource Locator), or *URL.* The URL:

■ Names the item

■ Tells where it is

■ Indicates the protocol to be used to get it

URLs are a special case of *Universal Resource Identifiers* (URIs). URI syntax provides a general way of expressing the names of information resources.

19.4.1   Hypertext URL

If you give a Web browser the URL for a hypertext document, the browser will go out and get the document using a protocol called the *Hypertext Transfer Protocol,* or *HTTP.* A hypertext URL has the form:

> *http://system-name/filename*

For example:

> http://www.ibm.com/index.html

If we provide just:

http://system-name/

the Web server will return its default *home page,* which often is called *home.html* or *index.html.* The most general HTTP URL is:

http://host:port/path?searchpart

URLs easily extend to other protocols.

19.4.2 Gopher URL

From a browser client, if we type the URL,

gopher://gopher.jvnc.net/

the browser will act as a gopher client and connect to the gopher server *gopher.jvnc.net.* If the gopher server is not listening at the usual port (70) but at some other, say 3333, the URL has the form:

gopher://gopher.somewhere.edu:3333/

19.4.3 File Transfer URL

We plug into file transfer with a URL such as:

ftp://ds.internic.net/

or identify a specific file with:

file://ds.internic.net/rfc/rfc1738.txt

To FTP to a site where you must enter a userid and password, use:

ftp://username:password@hostid/

By the way, a host can be identified by its IP address rather than its domain name.
When we access a file with a URL such as:

```
file:// ds.internic.net/rfc/rfc1738.txt
```

note that a protocol is not identified. However, implementations default to FTP.

19.4.4 Telnet URL

We can tie into *telnet,* for example, with

telnet://ds.internic.net/

More generally, it is:

telnet:///username:password@hostid/

19.4.5   News URL

The URL for a news group has the form *news:newsgroup-name,* for example:

news:rec.airplane

The news server is not identified in the URL. Instead, the user's news server name (or address) is entered into a browser's configuration information.

19.4.6   Mail URL

There even is a URL for sending electronic mail:

mailto:user@mail-location

As was the case for news, the name or address of a mail gateway is entered into a browser's configuration information.

19.4.7   WAIS URL

Although rarely (if ever) used, a URL has been defined to access WAIS databases via a protocol called Z39.50. For example, an interface to a directory of public WAIS servers has the form:

wais://cnidr.org/directory-of-servers

More generally, WAIS URLs have the form:

wais:///host:port/database
wais:///host:port/database?search
wais:///host:port/database/wtype/wpath

At the time of this writing, few (if any) browsers support the WAIS access protocol. Database searches are carried out by entering query information into a form and sending it to a World Wide Web server, which then invokes an appropriate search engine.

19.5   GENERAL URL FORMAT

To summarize, note that:

- A URL starts with the access protocol to be used.
- For applications other than news and electronic mail, this is followed by the delimiter ://.

- Next, the name of the server host is shown.

- Finally, the resource to be accessed is identified—or else a default file is retrieved.

Recall that in the case of news and electronic mail, the location of a preferred news server and mail gateway are part of the browser's configuration information. Only the : part of the delimiter is used, and no host server is identified in the URL.

19.5.1 Special Characters

Sometimes a resource identifier contains a space or some other character—such as a slash or a colon—that normally is used as a delimiter or special character in a URL. For example, Macintosh and Windows 95 filenames can contain spaces and other unusual characters.

Special characters are identified as belonging to the actual resource name by mapping them to strings that start with a % symbol. Table 19.1 shows the mappings.

19.6 AN INTRODUCTION TO HTML

WWW documents containing hypertext links are written using *Hypertext Markup Language,* or *HTML*. Hypertext files compatible with versions 1 and 2 usually have names of the form:

filename.html

TABLE 19.1   Mappings of Special Characters

Special character	Mapped representation
Space	%20
/	%2F
#	%23
=	%3D
;	%3B
?	%3F
:	%3A
~	%7E

A file that includes advanced features introduced in version 3 would be called:

filename.html3

On DOS Windows computers, a suffix of *.htm* or *ht3* is used.

HTML is based on a text markup standard called *Standard Generalized Markup Language* (SGML). The main idea is that an author places *tags* into a document to identify elements such as its title, section headings, paragraph boundaries, bulleted lists, figures, and so forth.

HTML is intended to be platform-independent, enabling hypertext documents to be viewed by client devices ranging from dumb text terminals to sophisticated workstations. Clients can display documents on a screen of any size and can use locally selected fonts.

We'll learn some HTML basics in the next sections. We will follow the conventions of HTML version 3. HTML is becoming a large language, and there are many capabilities that we will not cover.

For example, an important capability that has been omitted is the construction of sophisticated data entry forms that can be sent from the server to the client. These forms can be used to enter database queries or to order products from an on-line store.

Another important capability that is not included is the construction of clickable maps. An end user can click on an area in a picture in order to choose a linked document.

19.6.1 Writing Hypertext Markup Language

Recall that some presentation details are left to the client. A desktop browser usually lets its end user choose the text fonts that will be used. Text in an HTML document will be rearranged on a user's screen according to the size of the screen window and the font that has been chosen.

An HTML author defines document elements such as:

- Document title
- Section headers
- Paragraph definition
- Links via URLs
- Lists
- Preformatted text
- Character formatting
- Special characters
- In-line images
- External images
- Data entry forms

- Clickable maps
- Tables and formulas

Tags included in an HTML document state what each element is. For example, the tag <TITLE> introduces the document's title.

You can write a hypertext document using an ordinary text editor. However, popular word processor programs provide add-ons that automate the creation of tags and let you work in "What-You-See-Is-What-You-Get" mode. There are several products dedicated to the creation of hypertext documents. These automate the creation of the various elements and hide the underlying tags if you wish.

Another good way to create HTML documents is to create the document using some other formatting software and then use a converter that automatically translates the document to HTML.

A basic understanding of how HTML works can be helpful in learning how to use any of these tools effectively. Also, new capabilities are outstripping releases of the tools, and therefore some information may have to be entered manually. Fortunately HTML is fairly easy to understand.

19.6.2   HTML Tags

Tags consist of element names and attribute parameters enclosed in angle brackets (<...>). Below we list some commonly used tags. Tags are not case-sensitive, but for consistency, we will write all tags in uppercase.

Most tags come in pairs, showing where an element begins and where it ends. The closing tag name mirrors the opening tag name but is enclosed by </...>. For example:

<TITLE>Welcome To The Web</TITLE>

19.6.3   Overall Format

A few tags are used to delimit the beginning and end of an HTML document and divide it into a head and body. This is illustrated in the example that follows:

`<HTML>`	Start of hypertext doc.
`<HEAD>`	Start of header items.
`<!--Last Modified on October 21, 1995-->`	A comment.
`<BASE HREF = "http://www.abc.com/ind.html3">`	Useful. Identifies where this document is stored.
`<TITLE>Welcome to the Web</TITLE>`	The title usually is displayed at the top of the client's screen.
`</HEAD>`	End of header items.
`<BODY>`	Start of document body.
`...`	
`</BODY>`	End of document body.
`</HTML>`	End of hypertext.

19.6.4 HTML Headers

Chapters, sections, and subsections of a document are introduced by header elements. Six levels of headers are available for use. Each will be displayed with a different format. For example, level 1 headers usually are presented in large and boldface type:

<H1>This is a level 1 header—which is a major header.</H1>

<H2>A level 2 header might be used for sections.</H2>

<H3>You also can insert level 3, 4, 5, or 6 headers. </H3>

19.6.5 Paragraphs and Breaks

An author must identify paragraph boundaries. Otherwise, all text will just be run together when it is displayed. A client program will collapse multiple spaces and multiple blank lines into a single space unless told to do otherwise.

Older versions of HTML delimited a paragraph by putting <P> at the start of a new paragraph:

```
<P>This is a paragraph.
<P>This is another paragraph.
```

This still can be done for version 3, or a pair of tags can show where a paragraph begins and ends:

```
<P> This is a paragraph.</P>
```

By default, most browsers insert a blank line between paragraphs.[1] If you do not want to start a new paragraph, but just want to move to the next line, use a break:

```
Roses are red,<BR>
Violets are blue.<BR>
```

19.6.6 Unordered Lists

An unordered list will be presented as a bulleted series of items. For example:

```
<UL>
<LI>Apple
<LI>Pear
</UL>
```

---

[1]There are HTML version 3 instructions that can be used to define a different paragraph style, such as indenting the first line.

Version 3 added an optional list header and end of list item tag:

```
<UL>
<LH>Types of Fruit</LH>
<LI>Apple</LI>
<LI>Pear</LI>
</UL>
```

19.6.7   Ordered Lists

Ordered lists have the same structure, but items will be numbered:

```
<OL>
<LH>This is an ordered list.</LH>
<LI>The first item.
<LI>Another item.
</OL>
```

As before, the end of list item tag, </LI>, and the list header, <LH>...
</LH>, are optional.

19.6.8   Definition Lists

A definition list is a sequence of terms and their definitions:

```
<DL>
<LH>World Wide Web Terminology</LH>
<DT>Hypertext Markup Language (HTML)
<DD>A markup language used to write hypertext documents. Tags in the document
identify elements such as headers, paragraphs, and lists.
<DT>Hypertext Transfer Protocol (HTTP)
<DD>A protocol used to request and transmit hypertext documents.
</DL>
```

Here is a rendition of these definitions:

```
World Wide Web Terminology
Hypertext Markup Language (HTML)
   A markup language used to write hypertext documents. Tags in the document
   identify elements such as headers, paragraphs, and lists.
Hypertext Transfer Protocol (HTTP)
   A protocol used to request and transmit hypertext documents.
```

Lists of all kinds can be nested.

19.6.9   Miscellaneous Tags

A horizontal rule can be used to separate parts of a presentation. A horizontal
line stretching across the page will be inserted:

```
<P><HR></P>
```

Sometimes, you will have text that already is laid out exactly the way you
want it. A preformatted (<PRE>) tag tells the browser to present it as is:

```
<PRE>
    This text will be displayed
    just as it is written—including blank spaces.
</PRE>
```

A block quote is a way to set off text. Usually it is indented. In version 2, the <BLOCKQUOTE> tag is used.

```
<BLOCKQUOTE>
    This is a block quote.
    It will probably be indented when displayed to the user.
</BLOCKQUOTE>
```

In version 3, the tag is shortened to <BQ>.

19.6.10 Emphasizing Text

There are times when you want some text to appear in a special form—for example, in boldface or italics. This can be done in two ways:

1. Leave the details to the browser

```
<EM> Typically displayed as italics. </EM>
<STRONG> Typically boldface. </STRONG>
<CODE> Typically a fixed-width font. </CODE>
```

2. Specify exactly how the text should look:

```
<I> Put this in italics. </I>
<B> Put this in bold. </B>
<U> Underline this text. </U>
<S> Strike through. </S>
<TT> Typewriter text—use a fixed-width font. </TT>
<SUB> Subscript. </SUB>
<SUP> Superscript. </SUP>
```

Version 3 adds many features, giving an author even more control over the way that items will be displayed to a version 3 client.

19.6.11 Links

To embed a link in a document you need to:

- Use start and end link tags
- Provide the URL parameter that identifies the linked document
- Provide a clickable label that will be displayed and underlined

A sample link is shown below. The *A* is the name of the tag and is called an *anchor*. The *HREF* parameter identifies the item to be linked. The text before the </A> delimiter becomes the clickable label for the link:

```
<A HREF = "http://www.abc.com/wwwdocs/showme.html">Click here to see some-
thing good</A>
```

You don't always have to write a complete URL for a linked item. Suppose that document *showme.html* contains a link to a file named *more.html.* in the same directory. Then the following would work:

```
<A HREF = "more.html">more here</A>
```

This is called a *relative pathname*. You also can use a relative pathname for documents in a subdirectory of the current directory.

19.6.12   Links to Local Documents

You can reference a document in your local host. For example, here is a link to a local DOS document:

```
<A HREF = "file:///c:\webdocs\home.htm">My Home Document</A>
```

You do not use the hypertext transfer protocol to retrieve a local file. Note that the host name is null—no host is listed between the slashes (///).

You even can link to a location within the same document. First, mark the spot. For version 2, this is done by inserting an anchor at the location and using a NAME parameter:

```
<A NAME = "Section3"> 3. Airplanes </A>
```

Then you can refer to that location by prefacing the name with a pound sign:

```
See <A HREF = "#Section3">section three</A> for more information.
```

If the user clicks on the underlined phrase, (<u>section three</u>), the client will jump to the marked location.

For version 3, instead of marking a location by creating a separate anchor tag, you can add an ID parameter to an existing tag for an item. For example, below we add an ID parameter to an H2 tag:

```
<H2 ID = "Section3">3. Airplanes</H2>
```

19.6.13   Images

The IMG tag is used to insert a picture into a document. The tag contains an *SRC* parameter that identifies the URL for the file that contains the picture. Image URLs look just like any other URL. An image reference looks like:

```
<IMG SRC = "http://www.abc.com/wwwdocs/ourlogo.gif">
<IMG SRC = "bigpic.jpeg">
<IMG SRC = "file:///c:\webdocs\building.gif">
```

Graphics Interchange Format (GIF) files often are used to hold images for a WWW page. Portable Network Graphics (PNG) is a standard for compressed bitmapped image files. Another popular type is a *Joint Photographic Experts*

Group (JPEG) compressed picture. JPEG was designed for photographic images but sometimes is used for other types of pictures.

A browser that cannot display images will ignore an IMG element unless it contains an ALT parameter. For example:

```
<IMG SRC = "bigpic.jpeg" ALT = "Washington's Monument">
```

A text-only browser would display the string "Washington's Monument" instead of the picture.

19.6.14 Viewing HTML Sources

A good way to learn HTML is to peek at the source of the documents that you retrieve. Usually, your browser will let you do this—or else, you can save the document to disk and then read it with an ordinary text editor.

19.7 HTTP ARCHITECTURE

Like gopher, hypertext retrieval is simple. As shown in Figure 19.3, a client connects to a WWW server, retrieves an item, and disconnects. The browser displays the item to the user, and then the user can think and choose the next step.

A Web server that primarily dispenses text documents will work very efficiently and can support many concurrent users. However, information is greatly enhanced by still and moving pictures and by sound. These objects are large, and transferring them consumes far more CPU resources and bandwidth than text. Furthermore, some requests trigger programs which create the information for the response. This uses still more system resources.

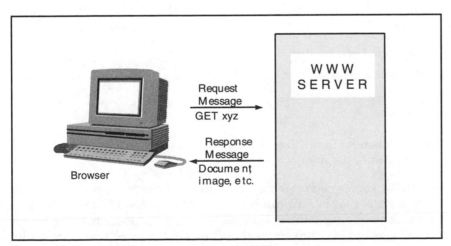

Figure 19.3   A browser retrieving an item from a WWW server.

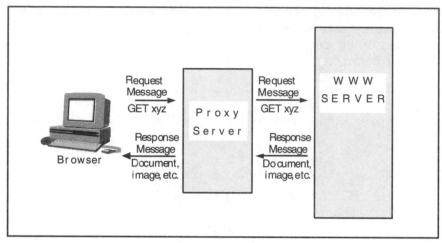

Figure 19.4 Retrieving information via a proxy WWW server.

19.7.1 Proxy Server

A *proxy* WWW server sometimes is used when clients that reside within a secure environment wish to access external WWW servers. The client browser would be configured to forward all requests to the proxy server. The proxy interacts with the actual server and relays results back to the client. Figure 19.4 shows a client accessing a WWW server via a proxy.

Some proxy systems cache documents so that they can respond to repeated requests directly.

19.8 HYPERTEXT TRANSFER PROTOCOL

The Web grew up on the Internet and was implemented on top of TCP connections, but other reliable transports could be used. A WWW server operates in a simple way:

- A client connects to the server.
- The client sends a request—for example:

```
GET /home.html HTTP/1.0
ACCEPT: text/html
```

- The server responds, indicating the type of information that will be sent and then transmits the item.

One reason that a server can interwork with many different types of clients is that the information that is sent can be tailored to the client's capabilities. A client can announce its capabilities by sending *Accept:* statements along with its requests. One client may announce that it can only accept text/html, while another might state that it can handle text, images, and sound.

Normally, a WWW server operates at well-known TCP port 80. Occasionally servers are set up to operate out of other ports.

In keeping with object-oriented language, the term *method* is used in HTTP documents instead of *command* or *request*. A client can send three standard methods:

GET   Retrieve an item

HEAD   Ask to see the headers that would be sent with a particular item

POST   Send an item to the server—such as data entered into a long form

A GET retrieves a *page*. A page is a document, along with whatever images and sounds may accompany it. A page may consist of a single screen or may include an entire book.

The HEAD command enables a client to find out about the length and datatype of an item, as well as the last-modified date associated with the current version, before acquiring it. If a browser already has cached the most recent version on the local disk, the document does not have to be retrieved.

19.8.1 Sample HTTP Dialogue

One reason that the WWW protocol has evolved quickly is that its designers did not take the time to reinvent the wheel. Header formats and datatypes have been borrowed from classic electronic mail and MIME standards.

The dialogue below shows how simple an *HTTP* interaction can be. GET / HTTP/1.0 requests the server's default document and announces that the client is running http version 1.0. The client also announces that it will accept only text HTML documents.

The server's response announces its HTTP version (1.0) and a status code; 200 means that the request has succeeded. A series of MIME-like headers follow. An empty line (<CR><LF>) signals the end of the header section and then the document follows.

```
GET / HTTP/1.0
ACCEPT: text/html

HTTP/1.0 200 Document follows
Date: Sat, 28 Oct 1995 14:07:25 GMT
Server: NCSA/1.5.1
Content-type: text/html
Last-modified: Tue, 09 May 1995 01:22:41 GMT
Content-length: 1563

<TITLE>InterNIC Directory and Database Services Home Page</TITLE>
<IMG src = "/Pics/logo.gif" alt = "">
<a href = ds/dspg01.html>
<H1>InterNIC Directory and Database Services</H1></a>
<p>
Welcome to InterNIC Directory and Database Services provided by AT&amp;T.
These services are partially supported through a cooperative agreement with
the National Science Foundation.
 .  .  .
```

The server closes the connection when it has completed its transmission.

TABLE 19.2   HTTP General Headers

General headers	Description
Date: *date*	Date is in Universal Time. *Date: Sun, 29 Oct 1995 15:15:23 GMT*
MIME-Version: *version*	The MIME version for the headers. *MIME-Version: 1.0*
Pragma: *directive*	An implementation-specific directive. For example, *Pragma: no-cache* tells a proxy to get a fresh version of the item even if it is cached.

19.8.2   Message Headers

Tables 19.2 through 19.5 provide brief definitions of the headers that can accompany a request or a response.

- *General* headers appear first in a message and may be found in requests or responses. These are listed in Table 19.2.

- Next are headers specific to either requests (Table 19.3) or responses (Table 19.4).

- Finally, *entity* headers appear last and provide detailed information about an item. These are listed in Table 19.5.

Keep in mind that a POST request transfers an item—such as the information from a form—from a client to a server. Therefore entity headers can appear in requests as well as responses.

TABLE 19.3   HTTP Request Headers

Request headers	Description
Authorization: *credentials*	Includes client authentication information for accessing a protected resource.
From: *mailid*	Like the electronic mail field.
If-Modified-Since: *date*	Used to make a GET conditional. If the item has not been modified, the response will have a 304 return code and no body.
Referer: *URL*	The identity of the item from which this link was obtained. *Referrer: http://www.abc.com/index.html*
User-Agent: *product*	Identifies the client software.

TABLE 19.4   HTTP Response Headers

Response headers	Description
Location: *URL*	The server's preferred location for the item
Server: *product*	Identifies the server software
WWW-Authenticate: *challenge*	Provides parameters that identify an authentication scheme and challenge the client to authenticate itself

19.8.3   Status Codes

Status codes are used much as they are in electronic mail and file transfer. General code assignments are:

1xx	Informational. Not used, but reserved for future use
2xx	Success. The action was successfully received, understood, and accepted
3xx	Redirection. Further action must be taken in order to complete the request
4xx	Client Error. The request contains bad syntax or cannot be fulfilled
5xx	Server Error. The server failed to fulfill an apparently valid request

More detailed information is provided by the specific codes.

19.9   ONGOING WORK

HTTP and HTML are expanding in response to customer demands for greater functionality. At the time of this writing, creation of standards for secure

TABLE 19.5   http Entity Headers

Entity headers	Description
Allow: *method*	Lists methods supported by a resource. *Allow: GET, HEAD*
Content-Encoding: *content-coding*	For a compressed or encrypted body, states the algorithm that was used. *Content-Encoding: x-gzip*
Content-Length: *length*	Indicates the size of the body to be transferred. *Content-Length: 2048*
Content-Type: *media-type*	Types are listed by the IANA. *Content-Type: text/html*
Expires: *date*	The item will no longer be valid after this date.
Last-Modified: *date*	Time that the item was last modified.

client/server interactions and secure commerce are top priorities. Other efforts are aimed at defining and implementing a location-independent naming scheme—*Uniform Resource Names* (URNs). One problem with linking to documents across a network is that if the document is moved to a different directory or a different computer, the link is useless.

Given a URN, the location might be retrieved from a directory system. Several locations could be listed for replicated documents, and the nearest location could be selected.

19.10   RECOMMENDED READING

RFC 1738 contains descriptions of URLs. RFC 1630 is a technical description of Universal Resource Identifiers.

The HTTP 1.0 specification was published in RFC 1945. Several documents relating to HTML also were in Internet draft form. (Internet drafts are located at *ftp://ftp.internic.net/internet-drafts.*)

Information about HTTP, HTML, and WWW security was available at a site established by the W3 Consortium, *http://www.w3.org/.*

20

Simple Network Management Protocol

20.1 INTRODUCTION

Network management has been a slow runner, lagging far behind other network facilities. Very large TCP/IP networks have operated and functioned quite well, but administration and management of these networks has been a labor-intensive task, requiring experienced personnel with a high level of technical skill.

This especially has been the case for the Internet, with its ever-expanding size and complexity. In the late 1980s, the Internet Architecture Board (IAB), which is charged with setting technical policy for the Internet, concluded that there was a critical need to define a network management framework and set of protocol standards and to turn these into working tools as quickly as possible.

Although quite a lot of work had been carried out by OSI committees responsible for producing network management standards, there was no prospect for quickly translating their draft documents into tools that would fit TCP/IP management needs.

An Internet working group created the Simple Network Management Protocol (SNMP) to meet immediate TCP/IP needs. SNMP's architecture was designed with OSI's model in mind. It was believed that the OSI network management standards—Common Management Information Services/Common Management Information Protocol (CMIS/CMIP)—would be the long-term solution. However, within a few months, it was clear that SNMP needed to evolve independently and be shaped by the experience of implementers and the needs of network managers.

20.1.1 Results of IAB Adoption of SNMP

The initial SNMP specifications established a starting point. The IAB expected that changes and enhancements would evolve rapidly. As stated in RFC

1052, IAB Recommendations for the Development of Internet Network Management Standards:

> We will learn what [Inter]Network Management is by doing it.
>   (a) in as large a scale as is possible
>   (b) with as much diversity of implementation as possible
>   (c) over as wide a range of protocol layers as possible
>   (d) with as much administrative diversity as we can stand.

The results of the IAB policy have exceeded their expectations. Since the time that the SNMP specifications and sample source code were made available on the Internet, the protocol has been incorporated into hundreds of products ranging from complex mainframe hosts to the simplest communications devices. The scope of SNMP has steadily been enlarged and strengthened.

Vendors have been able to create network management stations that use a well-defined protocol to communicate with a vast array of different devices. A thriving new market has been created in which vendors compete to enhance their management stations with features such as Graphical User Interfaces (GUIs), history databases, and report generation capabilities.

The continuing production of a stream of RFCs devoted to network management is evidence of the dynamic expansion of the scope of the protocol.

Version 2 of SNMP was published early in 1996. We will include some of the important new features available in version 2 within this chapter.

20.2 SNMP MODEL

20.2.1 Logical Database

SNMP follows a database model. Every network system contains configuration, status, error, and performance information that network administrators would like to access. This information is viewed as being stored in a *logical database* at the system.

20.2.2 Agents

In order to make the information accessible, a managed system must contain a software component called an *agent*. The agent responds to queries, performs updates, and reports problems. One or more *management stations* send query and update requests to agents and receive responses and problem messages.

20.2.3 Managers

As shown in Figure 20.1, a management station contains *manager* software that sends and receives SNMP messages and has a variety of *management applications* that communicate with network systems via the manager.

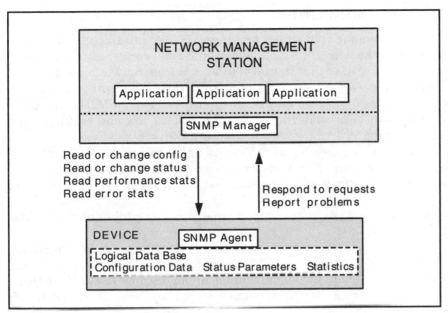

Figure 20.1   The SNMP model.

20.2.4   Management Information Base

The *Management Information Base* (MIB) is the logical description of all network management data. There are many RFC documents that describe MIB variables. Each document describes a MIB *Module*, which is a set of related variables. There also are additional MIB Modules written by vendors. Vendor MIB documents define product-specific variables.

A MIB variable definition does not deal with how the variable is stored; it includes:

- A definition of what a variable is
- A description of how its value is measured
- A name to be used when reading or updating the variable's database value

Although the network management MIB formally is a set of definitions, it is convenient to call the specific data that is stored at a device its *MIB database*—or simply, its *MIB*. A typical MIB database contains:

- System and status information
- Performance statistics
- Configuration parameters

A system's MIB contains only those variables that make sense for the system. For example, a simple LAN bridge does not need variables that count TCP statistics.

20.3   ROLES OF MANAGERS AND AGENTS

A network management application provides the user interface that enables an operator to invoke network management functions, view the status of components, and analyze data that has been extracted from network nodes.

A manager *supervises* a system by asking the system's agent to send back data values that are in its MIB database. Typical values in a MIB include the types of physical network interfaces that the system has and traffic counts through each interface.

A manager *controls* a system by asking its agent to update MIB status or configuration parameters. A parameter change can be tied to an action. For example, a network interface can be disabled by setting a status variable to *down*.

New supervision and control functions are defined by adding new variables to the MIB database.

An ever-increasing range of devices can be monitored from today's management stations. There are management station products that run on platforms ranging from PCs to mainframes. The sections that follow will include screens from an *HP OpenView for Windows Workgroup Node Manager,* which runs on a Windows PC.

20.3.1   Proxy Agents

In the basic SNMP model, an agent and MIB database coreside within the device that is being monitored or controlled. A *proxy* management agent extends the model, enabling indirect access to devices. A management station interacts with the proxy, requesting and receiving information. The proxy exchanges information with the device by means of a separate interaction (see Figure 20.2).

In version 2 of SNMP, proxies are used to relay information between version 1 and version 2 environments.

20.4   NATURE OF MANAGEMENT INFORMATION

The definition of management *variables* was completely separated from the specification of the *protocol* used between managers and agents. This is an important feature of the management architecture.

The definition of variables is delegated to committees of experts for each technology. Separate groups have designed MIBs for bridges, hosts, telephony interfaces, and so on.

The first MIB document focused on information that would be useful in managing a TCP/IP network. It included data such as:

- What kind of system is this?
- What is its name and location?
- What types of network interfaces does the system have?

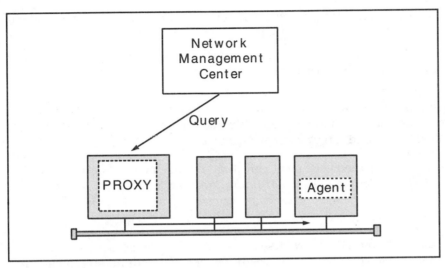

Figure 20.2   A Proxy agent.

■ How many frames, datagrams, and TCP segments has it sent and received?

For example, Figure 20.3 shows system information retrieved from a router using HP *Openview.*

20.5   STRUCTURE OF MANAGEMENT INFORMATION

A framework for defining network management variables needs to include:

■ *An administrative structure.*   The work of defining MIB variables for different types of network components is delegated to experts in the field. An

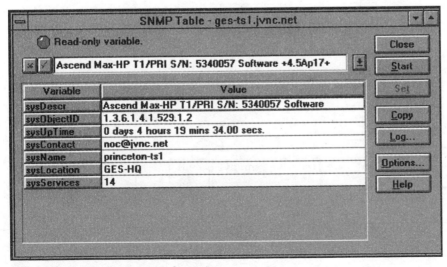

Figure 20.3   Polling for system information.

administrative structure is needed to describe and track the partitioning of the work and delegation of authority.

- *An information structure.*   Network information will not remain static. Information must be structured so that it is easy to extend or revise old technologies and add new ones.

- *A naming structure.*   There will be hundreds of variables that will be defined for network management. We need a consistent method of defining, describing, and naming these variables.

A tree-structured framework meets all three requirements. The framework is called the *Structure of Management Information, or SMI.*

20.5.1   SMI Tree

Recall that initially, SNMP was supposed to be a temporary stopgap until ISO management standards were ready for use. The administrative/naming tree, shown in Figure 20.4, reflects the initial effort to fit into the ISO world.

The higher nodes in the tree were supposed to represent the administrative authorities responsible for the lower parts of the tree, as shown in Table 20.1. The tree is administratively very out of date. There no longer is an effort to coordinate SNMP standards with ISO. The Department of Defense (dod) no longer runs the Internet.

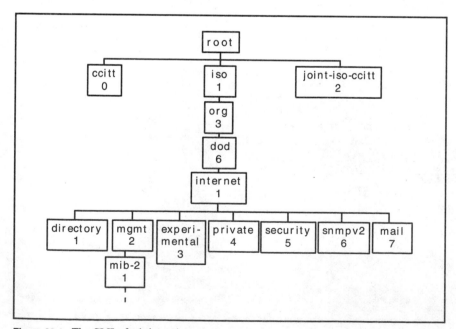

Figure 20.4   The SMI administrative naming tree.

TABLE 20.1   Nodes in the SMI Tree

Label	Description
iso (1)	International Standards Organization
org (3)	National and international organizations
dod (6)	Department of Defense
internet (1)	Internet Architecture Board

However, the tree still serves its primary function of defining MIB variable names. The tree is immensely useful. Whenever a new technology is added to the network management environment, a committee is created and is assigned a new node in the tree. The committee then creates whatever variables are needed within its own subtree.

20.6   OBJECT IDENTIFIER NAMES

Figure 20.5 focuses on the important parts of the tree. The tree is used to assign names called *OBJECT IDENTIFIER*s to management variables.

OBJECT IDENTIFIERs are formed by starting at the top of the tree and concatenating the numeric identifiers for each node. Each node also is

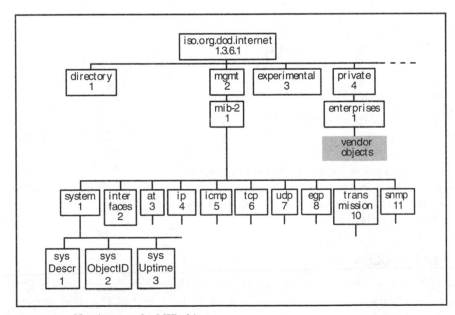

Figure 20.5   Naming tree for MIB objects.

assigned a text label, intended to help users and designers to understand what the variable is. For example:

OBJECT IDENTIFIER   **1.3.6.1.2.1.1.1**
Text Name   **iso.org.dod.internet.mgmt.system.sysDescr**

20.6.1   Identifying MIB Database Values

In order to identify an actual value at a device, a "which one" index is added at the end of the variable's OBJECT IDENTIFIER. For example, information about all of a device's interfaces is stored in a table. The OBJECT IDENTIFI-ER for the *ifType* table variable is 1.3.6.1.2.1.2.2.1.3. If we want to identify the *ifType* for the fourth interface on a router, its ID is:

1.3.6.1.2.1.2.2.1.3.*4*

The convention of adding an index at the end is extended to one-of-a-kind variables, such as *sysDescr* or *sysUpTime*. A 0 is added at the end of one-of-a-kind variables. For example, the full identifier for a *sysDescr* variable is 1.3.6.1.2.1.1.1.*0*.

20.6.2   Lexicographic Order

The variables in a device's MIB are ordered *lexicographically* (like a dictionary). To compare two values:

1.3.6.1.2.1.2.2.1.19.3
1.3.6.1.2.1.2.2.1.21.2

- Start at the left.
- Compare until you find the first value that is different.
- The item with the bigger number in this position is the bigger item.

The second item above is bigger. What about the next example?

1.3.6.1.2.1.2.2.1
1.3.6.1.2.1.2.2.1.21.2

For this kind of tie, the longer identifier is bigger.

By the way, when traversing a table in lexicographic order, you would march all the way down one column and then jump to the top of the next column, as shown in Figure 20.6.

20.7   IMPORTANT MIB MODULES

Dozens of MIB modules have been written, covering everything from RS-232 interfaces to electronic mail servers. We'll describe a few of the most important ones in the sections that follow.

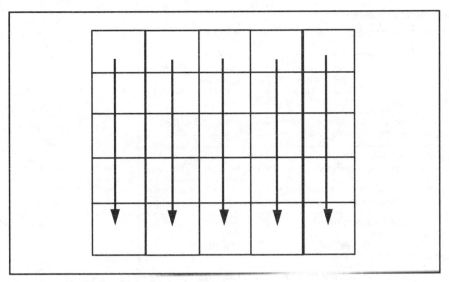

Figure 20.6   Lexicographic order in a table.

20.7.1   MIB-II

The groups of variables shown in Figure 20.5 (*system, interfaces,* etc.) were defined in the first MIB document, which described variables relating to TCP/IP networking. After some testing and experience, the module was updated and called *MIB-II*. MIB-II provided the basic variable definitions for version 1 SNMP implementations. An update that includes some additional variables and a few modifications has been published for version 2 of SNMP.

20.7.2   Transmission Modules

Many modules that describe variables relating to local and wide area technologies have been written. Some of the subtrees that have been created under the *transmission* node are shown in Figure 20.7. See the *Assigned Numbers* document for a complete list.

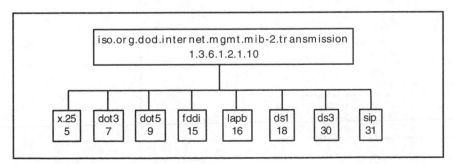

Figure 20.7   Transmission MIB modules.

TABLE 20.2   Groups of RMON MIB Variables

Variable	Description
statistics	Statistics specific to a given type of interface, such as Ethernet (collisions, jabbers) or Token-Ring (beacons, lost tokens).
history	Compiles statistics for a configured polling interval.
alarm	Generates an event if values for an interval exceed a configured threshold.
host	Reports hosts that have been detected by the monitor as well as related statistics, such as how many frames each sent.
hostTopN	Reports statistics for the hosts that top a list sorted on values for a selected performance or error statistic.
matrix	Reports statistics for conversations between pairs of addresses.
filter	Defines criteria for selecting a particular set of frames for closer examination.
packet capture	Allows frames that match filter criteria to be captured.
event	Controls the generation and notification of events. An event may be caused by a local occurrence such as exceeding a threshold. An event can trigger a local activity such as writing a message to a log or initiating packet capture or can cause a *trap* message to be sent to a management station.

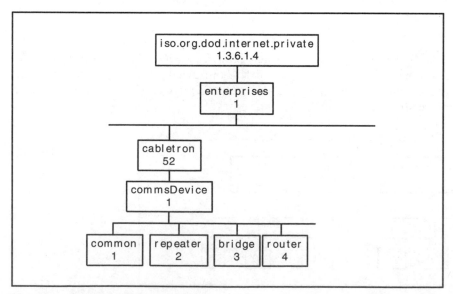

Figure 20.8   Part of the Cabletron MIB.

20.7.3   RMON MIB

A network *monitor* or *probe* is a device that passively watches link traffic and can be configured to gather data about the traffic, revealing usage patterns and providing network performance statistics. Monitors are configured with error count thresholds that can be used to spot problems before they become critical.

The *Remote Network Monitoring MIB* (RMON MIB) integrates the valuable information collected by monitors into the SNMP framework. This gives a significant boost to the power of SNMP management stations.

A remote monitor can independently collect local data, carry out diagnostics, and detect alarm situations. Since problems will be reported as they arise, a network management station can cut back on the frequency of its requests to read MIB data from individual devices.

The nine data groups defined for the RMON MIB are shown in Table 20.2.

20.7.4   How Vendor MIBs Fit In

From the beginning, there was room in the object tree for vendor[1] MIB objects. To obtain a branch in the tree, a vendor simply registers with the Internet Assigned Numbers Authority (IANA).

Figure 20.8 shows part of the Cabletron MIB. Cabletron has been assigned OBJECT IDENTIFIER

1 . 3 . 6 . 1 . 4 . 1 . 52.

20.8   SNMP MESSAGE PROTOCOL

Now let's look at the message protocol that enables managers to communicate with agents. There were a few principles that guided the design of SNMP:

- Choose a very undemanding preferred transport but do not make the choice of transport an absolute—leave the choice open so that SNMP can be used on non-TCP/IP networks.
- Use a very small number of message types.

20.8.1   SNMP Version 1 Message Types

Managers and agents communicate with each other by sending SNMP messages. As shown in Figure 20.9, there are only five message types for version 1 of SNMP:

get-request   Requests one or more values from a managed system's MIB

get-next-request   Enables the manager to retrieve values sequentially; used to read through the rows of a table or to "walk" through an entire MIB

---

[1]"Enterprises," such as companies, organizations, or government agencies, also can obtain tree identifiers and define their own MIB variables.

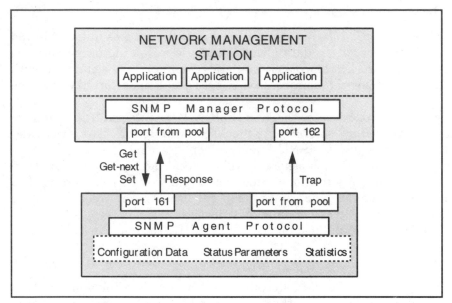

Figure 20.9 SNMP version 1 message types.

set-request	Enables the manager to update variables
get-response	Returns the results of a *get, get-next,* or *set* request operation (just called a *response* in version 2)
trap	Enables an agent to report important events or problems

Limiting exchanges to these five message types kept the implementation simple, while still providing for plenty of functionality.

Typically, network administrators configure a management station to read statistics at regular periods—such as every 15 minutes. These values can be saved and analyzed in order to discover normal baseline behavior, detect peak hour bottlenecks, and spot unusual activities.

A *trap* is used to report general events such as:

- Reinitializing self
- Local link failure
- Link functioning again

MIB standards committees have defined additional trap messages for specific communications technologies. In addition, vendors define traps that deliver critical problem information that relates to their products.

It is part of the SNMP philosophy that the number of *trap* messages that are transmitted should be kept relatively small. Network managers are familiar with the phenomenon that when one thing goes wrong, a lot of other problems get triggered. A flood of problem messages can clog the network, slowing down recovery procedures.

20.8.2 Transports

UDP was selected as the preferred transport because it is simple and can be implemented with very little code. It also is the choice most likely to work, even when a device is stressed or damaged. However, other transports can be used. For example, SNMP can be run over IPX in the NetWare environment.

When UDP is used, each SNMP message is enclosed within a single UDP datagram and is delivered via IP. As shown in Figure 20.9, requests are sent from any convenient UDP port to port 161. *Responses* are sent back to the requesting port. *Traps* are sent from any convenient UDP port to port 162.

Every version 1 implementation must be able to handle messages of at least 484 octets.

20.9 SNMP MESSAGE FORMATS

An SNMP version 1 message consists of some introductory "wrapper" material followed by a message Protocol Data Unit, which is one of the five types, *get-request, get-next-request, get-response, set-request,* or *trap.* The introductory material includes the:

Protocol version   0 for version 1 and 1 for version 2

Community name   Used like a password

An agent is configured to restrict what information can be accessed, and whether it can be read or written, based on the community name. Agents also are configured with the IP addresses of the management stations that are permitted to read or write MIB information.

Unfortunately, the community name in a message can be read by LAN eavesdroppers, and IP source addresses can sometimes be forged. One solution is to access important devices, such as routers, via a separate, secured communications link when performing updates to a system's configuration or status.

20.9.1 Format of Version 1 *Gets, Sets,* and *Responses*

The main information content in all of these messages is the same. It consists of a list:

Variable name	Value[2]
Variable name	Value
. . .	. . .

---

[2]The pairing of a variable name and a value is formally called a *variable binding.*

OBJECT IDENTIFIERs are used as variable names. In a *get* or *get-next,* the value field is a null place holder. The agent just has to fill in the missing values.

A complete *get-request, get-next-request, set-request,* or *response* Protocol Data Unit consists of:

A request-id	Used to correlate requests with responses.
An error-status field	0 in requests. A non-0 value in a response means something went wrong.
An error-index field	0 in requests. In responses, it indicates which variable caused the problem.
List of object identifiers and values	In a *get* or *get-next,* the values are null. In a *set* or a *response,* the values are filled in.

20.9.2   Get Request and Response

Figure 20.10 shows a Network General *Sniffer* analysis of a *get-request* and its *response.* The request includes a list of five variables whose values are desired. A "null" placeholder is included after each variable identifier. To create a response, the agent just has to "fill in the blanks" and replace the nulls with actual values.

20.9.3   Get-Next Request and Response

A *get-next* works differently. When you send an OBJECT IDENTIFIER, the *next* OBJECT IDENTIFIER and matching value are returned. For example, if you sent a *get-next* request with:

```
SNMP: Object = {1.3.6.1.2.1.5.1.0} (icmpInMsgs.0)
SNMP: Value = NULL
```

The response would be the name and value for the next variable:

```
SNMP: Object = {1.3.6.1.2.1.5.2.0} (icmpInErrors.0)
SNMP: Value = 0 messages
```

This enables you to "walk" through a MIB or move from row to row of a table.

20.9.4   Set Request

A *set* request writes data into an agent's database. The message format is very simple—it looks just like a *get-request,* except that the actual update values are filled in. Figure 20.11 shows a trace of a *set* request.

All updates must succeed, or else the whole request will fail. This makes sense because there often will be several variables that ought to be changed together or not at all. This all-or-nothing rule for *sets* is kept in version 2.

The response to a *set* looks just like the request, except that the error-status and error-index fields will be given nonzero values if there was a problem.

```
SNMP: Version = 0
SNMP: Community = public
SNMP: Command = Get request
SNMP: Request ID = 112
SNMP: Error status = 0 (No error)
SNMP: Error index = 0
SNMP:
SNMP: Object = {1.3.6.1.2.1.1.3.0} (sysUpTime.0)
SNMP: Value  = NULL
SNMP:
SNMP: Object = {1.3.6.1.2.1.5.1.0} (icmpInMsgs.0)
SNMP: Value  = NULL
SNMP:
SNMP: Object = {1.3.6.1.2.1.5.2.0} (icmpInErrors.0)
SNMP: Value  = NULL
SNMP:
SNMP: Object = {1.3.6.1.2.1.5.3.0} (icmpInDestUnreachs.0)
SNMP: Value  = NULL
SNMP:
SNMP: Object = {1.3.6.1.2.1.5.4.0} (icmpInTimeExcds.0)
SNMP: Value  = NULL
```

```
SNMP: Version = 0
SNMP: Community = public
SNMP: Command = Get response
SNMP: Request ID = 112
SNMP: Error status = 0 (No error)
SNMP: Error index = 0
SNMP:
SNMP: Object = {1.3.6.1.2.1.1.3.0} (sysUpTime.0)
SNMP: Value  = 1037388 hundredths of a second
SNMP:
SNMP: Object = {1.3.6.1.2.1.5.1.0} (icmpInMsgs.0)
SNMP: Value  = 1 messages
SNMP:
SNMP: Object = {1.3.6.1.2.1.5.2.0} (icmpInErrors.0)
SNMP: Value  = 0 messages
SNMP:
SNMP: Object = {1.3.6.1.2.1.5.3.0} (icmpInDestUnreachs.0)
SNMP: Value  = 0 messages
SNMP:
SNMP: Object = {1.3.6.1.2.1.5.4.0} (icmpInTimeExcds.0)
SNMP: Value  = 0 messages
```

Figure 20.10 Sample *get-request* and *response*.

20.9.5 Trap Messages

An agent uses *trap* messages to report serious problems to a manager.

Very few traps were defined in the SNMP standard. The definition of traps is left to technology standards committees and vendors—with a warning to keep the number down. When the network is stressed, you do not want to get

```
SNMP: Version = 0
SNMP: Community = xyz
SNMP: Command = Set request
SNMP: Request ID = 0
SNMP: Error status = 0 (No error)
SNMP: Error index = 0
SNMP:
SNMP: Object = {1.3.6.1.2.1.4.1.0} {ipForwarding.0}
SNMP: Value  = 2
SNMP: Object = {1.3.6.1.2.1.4.2.0} (ipDefaultTTL.0)
SNMP: Value  = 70
```

Figure 20.11 Updating MIB values with a *set*.

dozens of messages from every device on the network, complaining about its problems.

Version 1 traps were slightly more complicated than they needed to be. Version 2 got it right. Let's look at the version 1 *trap* message first. There is a field called *generic trap* whose value identifies the type of trap as one of the following:

coldStart(0)	The sender is reinitializing, and its configuration may change.
warmStart(1)	The sender is reinitializing, but its configuration will not change.
linkDown(2)	An adjacent link has failed.
linkUp(3)	An adjacent link has come up.
authentication Failure(4)	Someone has sent the agent a request that was not properly authenticated (i.e., the message had an inappropriate community name).
egpNeighbor Loss(5)	An Exterior Gateway Protocol neighbor is down.
enterprise Specific(6)	Other. This is a trap defined by a standards committee, a vendor, or some other enterprise.

Figure 20.12 shows a very simple *trap* message which reports a cold start.

- The *enterprise* field indicates that this trap was sent by a system running FTP Software's TCP/IP product.

- Since the *generic trap* value is 0, this message reports a cold start.

- The *time ticks* field contains the *sysUpTime,* which is 0 because this system just made a cold start initialization.

Any trap that was defined by a MIB committee or vendors will have *generic trap* = 6. In this case, the *enterprise* field combined with the *specific trap* field tells you what the trap is all about.

If this seems too complicated, you are right. Traps are simplified in version 2.

```
SNMP: Version = 0
SNMP: Community = public
SNMP: Command = Trap
SNMP: Enterprise = {1.3.6.1.4.1.121.1.1}
SNMP: Network address = [198.207.177.10]
SNMP: Generic trap = 0 (Cold start)
SNMP: Specific trap = 0
SNMP: Time ticks = 0
```

Figure 20.12 A version 1 *trap* message.

20.9.6 Version 1 Problems and Version 2 Corrections

There were some features of SNMP version 1 that were irritating:

- If just one variable in a *get* or *get-next* request was not in the agent's database, the entire operation failed.
- If a *request* asked for several variables, and the agent could not fit the entire answer into the biggest message that it was able to send, the entire operation failed.
- *Traps* performed a simple function but were hard to describe.

Version 2 solves these problems. An agent can put an error code into the *value* field for a variable that cannot be retrieved. There is a new *get-bulk* request that asks the agent to return as much of the requested information as it can. And *trap* messages are given the same simple format as all of the other messages.

Version 2 also expands the list of error codes that are supported, which gives managers a better idea of what has gone wrong when a request fails.

20.9.7 Version 2 Get-Bulk Message

Get-bulk behaves like a *get-next*. The agent will return variables whose OBJECT IDENTIFIERs follow the OBJECT IDENTIFIERs in the request.

A *get-bulk* request has parameters that indicate:

- The number of initial stand-alone (nonrepeater) variables requested
- For the remaining (repeater) variables, the number of repeats requested

For example, you might ask for the two nonrepeater stand-alone variables:

sysDescr

sysUpTime

and then ask for 10 rows of the table variables: *ifIndex, ifDescr, ifType, ifMTU,* and *ifSpeed.* In this case:

- There would be seven variables in the variable list.
- Nonrepeaters = 2
- Max-repetitions = 10

The response will pack in as much of the requested information as it can. If it can't hold everything, it is easy for an application to send another *get-bulk* to ask for more.

Since the error-status and error-index fields really serve no function in requests, they are commandeered in the *get-bulk* request to hold the *nonrepeaters* and *max-repetitions* parameters. This means that the basic message format does not have to be changed at all to accommodate *get-bulk*.

20.9.8   Version 2 Traps

In version 2, a *trap* has the same format as a response. It starts with the standard header information, followed by a variable list:

```
OBJECT IDENTIFIER    Value
      . . .           . . .
      . . .           . . .
```

The *sysUpTime* and a unique trap identifier are moved to the head of the variable list. Additional variables that shed light on the problem can be included.

20.9.9   Version 2 Inform Messages

Version 2 also added the idea of *inform* messages, which simply are acknowledged *traps*. These are very handy manager-to-manager communications, in situations where the sender really wants to know that the destination manager received the message. An ordinary response message is used as the acknowledgment.

20.9.10   Other Version 2 Enhancements

How precisely does the implementation of a module have to match its MIB definition in order for a vendor to claim compliance? And how can a vendor declare departures from the specification that may have been necessary because of some product limitation?

Version 2 provides the mechanisms for expressing:

- Compliance statements: The actual minimum requirements for a module
- Capability statements: Vendor-provided statements that describe the actual capabilities of an agent

These statements enable a customer to look further than the claim "we support SNMP" when evaluating a product.

20.10   READING MIB DOCUMENTS

The documents that define MIB variables contain a lot of extremely useful information. They describe exactly how each variable is defined and measured. Often there is extensive additional material that describes the technology, error conditions, and typical configurations.

In the sections that follow, we'll discuss some concepts that should be helpful in reading MIBs.

20.10.1   Managed Objects

Up to this point, we have used the informal term, *MIB variable*. But MIB standards actually define *Managed Objects*. A variable just has a name and a value, but the definition of a Managed Object includes:

- A name—the OBJECT IDENTIFIER
- A set of attributes, including:

  1. A datatype
  2. A description providing implementation details
  3. Status information

- A set of operations that can be performed on the object

Let's look at a typical MIB definition:

```
sysDescr OBJECT-TYPE
  SYNTAX DisplayString (SIZE (0..255))
  ACCESS read-only
  STATUS mandatory
  DESCRIPTION
    "A textual description of the entity. This value
    should include the full name and version
    identification of the system's hardware type,
    software operating-system, and networking
    software. It is mandatory that this only contain
    printable ASCII characters."
  ::= { system 1 }
```

The definition starts with the text label for the node, *sysDescr*. It ends with { *system 1* }, which means "put this node under *system* and assign the number 1 to the node." This allows us to construct the complete OBJECT IDENTIFIER, which is:

$$1.3.6.1.2.1.1.1$$

The rest of the definition consists of a series of *clauses*—SYNTAX, ACCESS, STATUS, and DESCRIPTION.

In this case, the *SYNTAX* (datatype) is a display string, that is, a string of printable characters, at most 255 characters in length.

The *ACCESS* identifies the operation(s) that may be performed. In this case, the ACCESS is read-write, so a manager can read or update the value.

In early MIB documents, the *STATUS* could be *mandatory, optional, obsolete,* or *deprecated.* However, values of mandatory and optional were not useful. Newer MIBs do not include variables unimportant enough to be labeled optional. STATUS now is used to indicate whether a variable is *current, deprecated* (on the way out), or *obsolete.*

20.10.2   Abstract Syntax Notation 1

MIB definitions like the one shown above are written in an ISO standard language called *Abstract Syntax Notation 1* (ASN.1). ASN.1 resembles a computer language. There also are standard ISO *Basic Encoding Rules* (BER) that define the transmission format for values defined using ASN.1.

A management station learns MIB variables by *compiling* ASN.1 MIB definitions. Good management stations allow you to compile as many MIBs as you need.

A management station then is ready to send and receive SNMP messages that include any of the compiled variables. A well-designed station also can display variable descriptions. Figure 20.13 shows how HP *OpenView* displays the DESCRIPTION clause in the *sysDescr* definition.

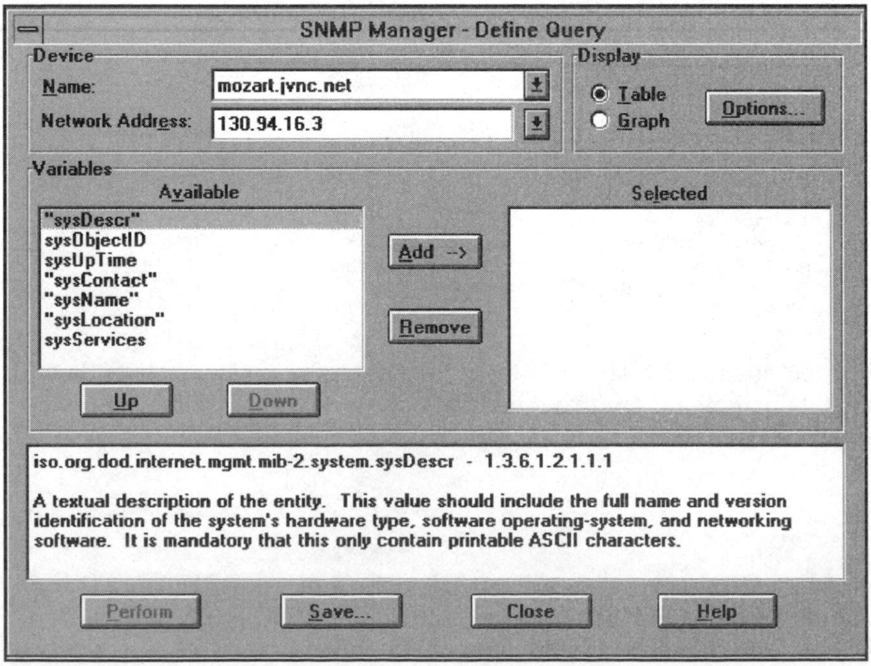

Figure 20.13   Manager display of a variables description.

20.10.3 MIB Datatypes

One reason that SNMP has been implemented widely is that the designers stuck to a "Keep It Simple" rule:

- All MIB data consists of simple scalar variables, although parts of a MIB may be *logically* organized into a table.

- Only a few datatypes—such as integers and octet strings—were used to express the values of MIB variables.

In fact, the underlying datatypes are INTEGER, OCTET STRING, and OBJECT IDENTIFIER.

20.10.4 Integers

Integers are used in two ways:

- To state "how many" of some item
- To enumerate a list of possibilities, such as 1 = up, 2 = down, 3 = testing

The definitions that follow illustrate the use of these datatypes. Note that in the first definition, the SYNTAX statement restricts the range of values.

```
tcpConnLocalPort OBJECT-TYPE
  SYNTAX INTEGER (0..65535)
  ACCESS read-only
  STATUS mandatory
  DESCRIPTION
    "The local port number for this TCP
    Connection."
  ::= { tcpConnEntry 3 }
```

```
ifAdminStatus OBJECT-TYPE
  SYNTAX INTEGER {
      up(1), — ready to pass packets
      down(2),
      testing(3) — in some test mode
      }
  ACCESS read-write
  STATUS mandatory
  DESCRIPTION
    "The desired state of the interface. The
    testing(3) state indicates that no
    operational packets can be passed."
  ::= { ifEntry 7 }
```

20.10.5 Counters

A counter is a nonnegative integer that increases to a maximum value and then wraps around. Specifically, a 32-bit counter can increase to $2^{32}-1$ (4,294,967,295) and then wraps around to 0. Version 2 adds a 64-bit counter that can increase to 18,446,744,073,709,551,615 before wrapping.

A single counter value has no intrinsic value. Counters are polled, and their current values are compared with their previous values. *Differences* in the readings are what matter. An example of a counter variable is:

```
ifInOctets OBJECT-TYPE
  SYNTAX Counter
  ACCESS read-only
  STATUS mandatory
  DESCRIPTION
   "The total number of octets received on
    the interface, including framing
    characters."
  :: = { ifEntry 10 }
```

20.10.6   Gauges

A *gauge* is an integer that behaves in quite a different manner. Gauge values go up and down. Gauges are used for quantities such as queue lengths. Sometimes the value builds up and sometimes it decreases.

A 32-bit gauge can increase to $2^{32}-1$ (4,294,967,295). If the quantity that it is measuring goes higher than this, the gauge just has to "latch" at the maximum until the value comes down again, as shown in Figure 20.14. An example of a gauge variable is:

```
ifOutQLen OBJECT-TYPE
  SYNTAX Gauge
  ACCESS read-only
  STATUS mandatory
  DESCRIPTION
    "The length of the output packet queue
     (in packets)."
  :: = { ifEntry 21 }
```

20.10.7   TimeTicks

Time intervals are measured in *TimeTicks,* which measure time in hundredths of a second. A TimeTick value is a nonnegative integer ranging from 1 to $2^{32}-1$ (4,294,967,295). It takes over 497 days to exhaust a TimeTick counter.

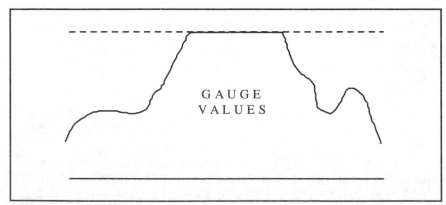

Figure 20.14   Behavior of gauge values.

The *sysUptime,* which measures the time elapsed since the agent software initialized, is the most frequently used TimeTick variable.

```
sysUpTime OBJECT-TYPE
    SYNTAX  TimeTicks
    ACCESS  read-only
    STATUS  mandatory
    DESCRIPTION
      "The time (in hundredths of a second)
      since the network management portion
      of the system was last re-initialized."
  ::= { system 3 }
```

20.10.8   OCTET STRINGs

An OCTET STRING is just a sequence of bytes. Just about anything can be represented by an OCTET STRING.

20.10.9   Textual Conventions

Rather than define a lot more datatypes, *Textual Conventions* are used in MIB definitions to indicate what kind of information is packaged in an OCTET STRING value and how the information should be displayed to users.

A type that is defined using a Textual Convention is encoded for transmission as a plain OCTET STRING. However, its actual meaning is determined from the Textual Convention definition. There is a MIB template that is used to define Textual Conventions. Here is the definition of *DisplayString:*

```
DisplayString ::= TEXTUAL-CONVENTION
    DISPLAY-HINT "255a"
    STATUS    current
    DESCRIPTION
      "Represents textual information taken from the NVT ASCII
        character set, as defined in pages 4, 10-11 of RFC 854."
```

Keep in mind that in a message, a value always is introduced by an OBJECT IDENTIFIER. A management station application could use the MIB definition that corresponds to that identifier and the Textual Convention definition to decide how to display, store, and use an OCTET STRING value.

20.10.10   BER Encoding of Datatypes

Along with the ASN.1 datatype definition language, ISO defined a set of *Basic Encoding Rules (BERs)* that are used to encode SNMP data values for transmission. The BER encoding for a data value has the form:

[*identifier*] [*length (of contents)*] [*contents*]

For example, the identifier X'02 is used for an INTEGER, X'04 is used for an OCTET STRING, and X'06 is used for an OBJECT IDENTIFIER.

An entire SNMP message actually is a sequence of ASN.1 values, and each message is entirely encoded using BER.

20.11 WHAT NEXT?

The most important piece of work that was left undone in the currently proposed version 2 update of SNMP was the definition of a new administrative structure and provision of authentication and encryption standards so that devices could safely be configured remotely. However, as we shall see in Chapter 24, mechanisms for authentication and encryption of traffic at the IP level have been proposed.

There has been bustling activity among vendors and standards writers to expand and refine MIB definitions. The result is that an abundance of raw information is available to network management stations.

Good applications are needed in order to use SNMP information effectively for problem detection and long-term capacity planning. Developers want standardized application development toolkits so that they can port new tools across management station products.

Intelligent systems, such as routers and hosts, are capable of monitoring themselves. Some interesting work has been done on embedding management applications into systems and accessing that information with Web browsers and HTTP.

20.12 RECOMMENDED READING

There is a long and growing list of RFCs dealing with SNMP and MIBs. The InterNIC RFC archive should be consulted for the most recent versions of these documents.

The author's book, *SNMP: A Guide to Network Management,* contains a description of SNMP concepts and structure and discusses several key MIB modules in detail.

21

The Socket
Programming Interface

21.1 INTRODUCTION

Communications standards define all of the rules needed to exchange data across a network. However, until recently, the need to standardize Application Programming Interfaces (APIs) for communication has been ignored. How can a programmer write a client/server application if the programs are completely different on every computer?

21.1.1 Berkeley Programming Interface

Fortunately, most TCP/IP implementations offer a programming interface that follows a single model, the *socket programming interface*. The socket programming interface was first introduced with the 1982 4.1c Berkeley Software Distribution (BSD) version of the Unix operating system. A number of improvements have been incorporated into the original interface over time.

The socket programming interface was designed for use with several communications protocols, not for TCP/IP alone. However, when the OSI transport layer specification was completed, it was clear that the socket interface was not general enough to satisfy OSI requirements.

In 1986, AT&T introduced the Transport Layer Interface (TLI) for Unix System V. The TLI can be used to interface to the OSI transport layer, TCP, and other protocols.

Another important addition to the socket family was the Windows Socket programming interface, or WinSock. It has enabled Windows applications to run on top of TCP/IP stacks implemented by many different vendors. Windows 95 provides a multiprotocol programming interface.

The socket interface is a de facto standard because it is almost universally available and is in widespread use. This chapter is intended to provide a general understanding of how the socket interface works. There will be minor differences in the APIs offered on various computers, due to the way that each vendor has implemented communications services within its operating system. The manual for the appropriate system should be consulted for programming details.

21.1.2 Unix Orientation

The original socket interface was written for a Unix operating system. The Unix architecture provides a framework in which standard file, terminal, and communications I/O all operate in a similar fashion. Operations are performed on a file by means of calls such as:

```
descriptor = open(filename, readwritemode)
read(descriptor, buffer, length)
write(descriptor, buffer, length)
close(descriptor)
```

When a program opens a file, the call creates an area in memory called a *file control block*. Information about the file, such as its name, attributes, and location, are stored in the file control block.

The call returns a small integer called a *file descriptor*. The program uses this descriptor to identify the file in any subsequent operations. As the user reads from or writes to the file, a pointer in the file descriptor keeps track of the current location in the file.

A very similar framework is used for TCP/IP socket communications. The primary difference between the socket programming interface and the Unix file I/O interface is that a couple of preliminary calls are required in order to assemble all of the information that is needed to carry out communications. Apart from the extra work at setup time, ordinary read and write calls can be used to receive and send data across a network.

21.2 SOCKET SERVICES

The socket programming interface provides three TCP/IP services. It can be used for TCP *stream* communication, UDP *datagram* communication, and *raw* datagram submission to the IP layer. Figure 21.1 illustrates these services.

Recall that the socket API was not designed for use exclusively with TCP/IP. The original idea was that the same interface could also be used for other communications protocol families, such as Xerox Network Systems (XNS).

The result can be slightly confusing. For example, we shall see later that some socket calls contain optional parameters that are not relevant for TCP/IP communications—they are needed for some other protocol. Also, occasionally a programmer will be required to spell out the length of a fixed quantity such as a version 4 IP address. The reason for this is that although it is obvious that a version 4 IP address contains 4 bytes, the programming interface can be used for other protocols with different address lengths.

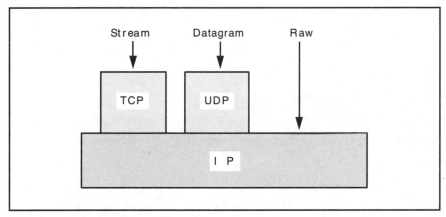

Figure 21.1   Socket Application Programming Interfaces.

21.3   BLOCKING AND NONBLOCKING CALLS

When a program reads data from a network connection, it is hard to predict how long it will take before some data arrives and the call can complete. One issue that a programmer must decide is whether to wait for the outcome of a read or return immediately and get the data either by checking a status variable periodically or by responding to an interrupt.

- Calls that wait are called blocking, or synchronous.
- Calls that return right away are called nonblocking, or asynchronous.

Are socket programming interfaces blocking or nonblocking? The answer is Yes. The programmer usually can take control of how the calls will behave.

21.4   SOCKET CALL

The *socket* call prepares for communication by creating a *Transmission Control Block* (TCB). Some manuals call this process "creating a socket." The *socket* call returns a small integer called a *socket descriptor* that is used to identify the communication in any subsequent operations.

There are many parameters that can be included in a TCB. We list a few to give an idea of the type of information that is included in the TCB for a TCP session:

- Local IP address
- Local port
- Protocol (e.g., TCP or UDP)
- Remote IP address
- Remote port

- Send buffer size
- Receive buffer size
- Current TCP state
- Smoothed round-trip time
- Smoothed round-trip deviation
- Current retransmission timeout value
- Number of retransmissions that have been sent
- Current send window size
- Maximum send segment size
- Sequence number of last byte that was ACKed
- Maximum receive segment size
- Sequence number of next byte to be sent
- Enable or disable tracing

21.5 TCP SOCKET PROGRAMMING

In this section, we will examine the socket programming calls that are used to interface with TCP. For simplicity, we will omit the I/O parameters for the calls at this stage and will concentrate on their major functions and their relation to one another. The details will be provided later.

21.5.1 TCP Server Model

A typical scenario for a TCP server is that there is a master process that spends most of its time listening for clients. When a client connects, usually the server creates a new "child" process that will do the actual work for the client. The server passes the client over to the new child process and then goes back to listening.

Sometimes clients arrive faster than the master process can get to them. What should be done with them? The standard mechanism is that when the master starts up, it tells TCP to create a queue that can hold a certain number of connection requests. Clients that can't be served immediately are put on the queue and served in turn. Suppose that the queue fills up and another client arrives? The new client's connection request will not be accepted.

21.5.2 TCP Server Passive Open

A server gets ready to communicate and then waits passively for clients. To get ready, the server makes a series of calls:

socket()   The server identifies the type of communication (TCP in this case). The local system creates an appropriate TCB data structure for the communication and returns a *socket descriptor*.

bind() The server establishes the local IP address and port that it wants to use. Recall that a host may have multiple IP addresses. The server may specify one IP address or else indicate that it is willing to accept connections arriving at any local IP address. The server may ask for a specific port or else let the bind call obtain a free port that it can use.

listen() The server sets the length of the client queue.

accept() The server is ready to accept client connections. If the queue is not empty, the first client connection request is accepted. The *accept()* call creates a *new TCB* that will be used for this client's connection and returns a *new descriptor* to the server.

Usually a synchronous form of *accept* is used so that if the queue is empty, *accept()* will wait for the next client to show up before returning.

21.5.3 TCP Client Active OPEN

A client actively requests a connection via two calls:

socket() The client identifies the type of communication (TCP in this case). The local system creates an appropriate TCB data structure for the communication and returns a local socket descriptor.

connect() The client identifies a server's IP address and port. TCP will attempt to establish a connection with the server.

If the client wishes to specify exactly which local port it wants to use, the client must call *bind()* before calling *connect()*. If the port is available, the *bind* will assign it to the client.

If the client does not call *bind()* to ask for a port, the *connect* call will assign an unused port to the client. The port number will be entered into the TCB.

21.5.4 Other Calls

The remaining calls are used in exactly the same way by both the client and the server. Data can be transmitted and received using ordinary *write* and *read* calls. The connection can be terminated by calling *close*. There also are *send* and *recv* calls that are specific to communications. They support sending and receiving urgent data as well as ordinary data:

send() Writes a buffer of data to the socket. Alternatively, write() may be used.

sendv() Passes a sequence of buffers to the socket. Alternatively, writev() may be used.

recv() Receives a buffer of data from a socket. Alternatively, read() may be used.

recvmsg() Receives a sequence of buffers from a socket. Alternatively, readv() may be used.

Sometimes a program needs information that is stored in the TCB:

getsockopt() Reads selected information out of the TCB. Optionally, a system may provide additional I/O system calls that can be used to read various parts of the TCB.

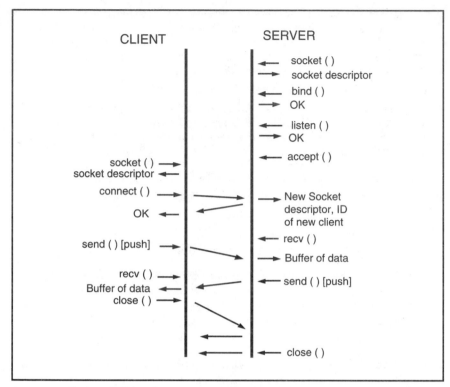

Figure 21.2   Series of TCP socket program calls.

Later, when we examine the input parameters for the opening calls, sends, and receives, we shall see that very few parameters are included in these calls. The reason is that a set of default values normally are used for most TCB parameters. For example, default values are preset for important environment information such as the receive buffer size, whether event logging is enabled, and the use of blocking or nonblocking processing for calls such as *recv*. Some defaults can be changed by using the functions:

setsockopt()	Sets a number of TCB parameters such as input and output buffer sizes, use of logging, whether urgent data should be received in the normal sequence order, and whether a close should block until all outstanding data has been safely sent.
iocntl() *or fcntl*()	Sets socket I/O to blocking or nonblocking.

Figure 21.2 shows the sequence of calls in a typical TCP session. *Socket*(), *bind*(), and *listen*() calls complete quickly and return immediately.

The *accept*(), *send*(), and *recv*() calls shown are assumed to be blocking,[1] which is their normal default. *Write*() and *read*() could have been used instead of *send*() and *recv*().

---
[1]A *send* call blocks when the TCP send buffer already is full.

21.6   A TCP SERVER PROGRAM

Now we are ready to take a close look at a sample server program. The server is designed to run forever. The server will:

1. Call *socket* to create a master TCB and return an integer socket descriptor that will identify this TCB in future calls.

2. Enter the server's local socket address information into a program data structure.

3. Call *bind,* which will copy the local socket address into the TCB.

4. Set up a queue that can hold up to five clients.

The remaining steps are repeated over and over:

5. Wait for clients. When a client arrives, create a new TCB for the client. The new TCB is constructed by making a copy of the master TCB and writing the client's socket address and other client parameters into the new TCB.

6. Create a child process to serve the client. The child will inherit the new TCB and handle all further communication with the client. The child will wait for a message from the client, write the message, and exit.

Each step in the program will be explained in the section that follows.

```
/*  tcpserv.c
 *  To run the program, enter "tcpserv".
 */

/* First we include a bunch of standard header files.
 */
#include <sys/types.h>
#include <sys/socket.h>
#include <stdio.h>
#include <netinet/in.h>
#include <netdb.h>
#include <errno.h>

main()
{
int sockMain, sockClient, length, child;
struct sockaddr_in servAddr;

/*  1. Create the master transmission control block.
 */

if ( (sockMain = socket(AF_INET, SOCK_STREAM, 0)) < 0)
    {perror("Server cannot open main socket.");
exit(1);
    }

/*  2. Create a data structure to hold the local IP address
 *      and port that we will use. We are willing to accept
 *      clients connecting to any local IP address (INADDR_ANY).
 *      Since this server will not use a well-known port,
 *      set the port = 0. The bind call will assign a port to the
 *      server and write the port into the TCB.
 */
```

```
bzero( (char *) &servAddr, sizeof(servAddr));
servAddr.sin_family = AF_INET;
servAddr.sin_addr.s_addr = htonl(INADDR_ANY);
servAddr.sin_port = 0;

/*  3. Call bind. Bind will pick a port number and write it
 * into the TCB.
 */
if ( bind(sockMain, &servAddr, sizeof(servAddr)) )
    { perror("Server's bind failed.");
      exit(1);
    }

/* We want to look at the port number. We use the
 * getsockname() function to copy the port into servAddr.
 */
length = sizeof(servAddr);
if ( getsockname(sockMain, &servAddr, &length) )
    { perror("getsockname call failed.");
      exit(1);
    }

printf("SERVER: Port number is %d\n", ntohs(servAddr.sin_port) );

/*  4. Set up a queue that can hold up to five clients.
 */
listen(sockMain,5);

/*  5. Wait for an incoming client. Accept will return
 * a NEW socket descriptor that will be used for this client.
 */
for ( ; ; ) {

if ( (sockClient = accept(sockMain, 0, 0)) < 0)
    { perror("Bad client socket.");
      exit(1);
    }

/*  6. Create a child process to handle the client.
 */
if ( (child = fork()) < 0)
    {perror("Failed to create child.");
     exit(1);
    }
else if (child == 0) /* This is code for the child to execute */
    { close(sockMain); /* The child is not interested in sockMain.*/
      childWork(sockClient);
      close(sockClient);
      exit(0);
    }

/*  7. This is the parent. It is no longer interested in
 *      the client socket, since the child is taking care
 *      of the client. The parent closes its entry to
 *      the client socket and loops back to issue a new accept().
 */
  close(sockClient);
    }
}

/* The child reads one incoming buffer, prints a message and quits.
 */
#define BUFLEN 81
int childWork(sockClient)
int sockClient;
{
char buf[BUFLEN];
int msgLength;
```

```
/* 8. Zero out the buffer. Then issue a recv to get a message
 *       from the client.
 */
bzero(buf, BUFLEN);
if ( (msgLength = recv(sockClient,buf, BUFLEN, 0)) < 0)
   { perror("Bad receive by child.");
     exit(1);
   }
printf("SERVER: Socket used for this client is %d\n", sockClient);
printf("SERVER: Message length was %d\n", msgLength);
printf("SERVER: Message was: %s\n\n", buf);
 }
```

21.6.1 Calls Used in the TCP Server Program

1. *sockMain = socket(AF_INET, SOCK_STREAM, 0)*;

The socket call has the form:

```
socket_descriptor = socket(address_domain, communications_type, protocol)
```

Recall that the socket interface can be used for other kinds of communications, such as XNS. *AF_INET* selects the Internet Address Family. *SOCK_STREAM* requests a TCP socket. This variable would be set to *SOCK_DGRAM* to create a UDP socket and *SOCK_RAW* to interface directly to IP.

We do not have to specify any other protocol information for TCP (or for UDP). However, the *protocol* parameter is needed for the raw interface, and for some of the other families that use sockets.

2. *struct sockaddr_in servAddr*;

   ...
   *bzero((char *) &servAddr, sizeof(servAddr))*;
   servAddr.sin_family = AF_INET;
   servAddr.sin_addr.s_addr = htonl(INADDR_ANY);
   servAddr.sin_port = 0;

The *servAddr* program structure is used to hold server address information. The *bzero()* call just initializes *servAddr* by putting 0s into all parameters. The first variable in the *servAddr* structure indicates that the rest of the values contain Internet Address Family data.

The next variable holds the local IP address at which this server can be reached. For example, if the server is attached to an Ethernet LAN and to an X.25 network, it might want to restrict access to clients reached through the Ethernet interface. In this program, we don't care. *INADDR_ANY* means that clients can connect through any interface.

The *htonl()*, or host-to-network-long, function is used to translate a 32-bit integer stored in the local computer to the Internet format for a 32-bit IP address. Internet standards represent integers with the most significant byte first. This is called the Big Endian style of data representation. Some computers store data with the least significant byte first, in a Little Endian style. If the local computer is Big Endian, *htonl()* will have no work to do.

If this server were operating at a well-known port, we would fill that port number into the next variable. Since we want the operating system to assign us a port for this test program, we just enter a zero value.

3. *bind(sockMain, &servAddr, sizeof(servAddr))*;
   getsockname(sockMain, &servAddr, &length);

The *bind* call has the form:

```
return_code = bind(socket_descriptor, address_structure,
length_of_address_structure)
```

If the address structure identifies a desired port, *bind* will try to get it for the server. If there is a 0 in the port variable, *bind* will obtain an unused port. *Bind* will enter the port number and IP address into the TCB.

The *getsockname* call has the form:

```
return_code = getsockname(socket_descriptor, address_structure,
length_of_address_structure)
```

We asked *bind* to get us a port, but *bind* does not tell us what port it got. If we want to find out, we have to read it out of the TCB. The *getsockname()* function retrieves information from the TCB and copies it into the address structure where we can read it. The port number is extracted and printed in the statement:

```
printf("SERVER: Port number is %d\n", ntohs(servAddr.sin_port) );
```

The *ntohs()*, or network-to-host-short, function is used to convert the network byte order of the port number to local host byte order.

4. *listen(sockMain,5)*;

The *listen* call is used by connection-oriented servers and has the form:

```
return_code = listen(socket_descriptor, queue_size)
```

The *listen* call indicates that this will be a passive socket and creates a queue of the requested size that will hold incoming connection requests.

5. *sockClient = accept(sockMain, 0, 0)*;

The *accept* call has the form:

```
new_socket_descriptor = accept(socket_descriptor,
client_address_structure, length_of_client_address_structure)
```

By default, the call will block until a client connects to this server. If a *client_address_structure* variable is provided, the client's IP address and port will be entered into this variable when a client connects. In this sample program, since we are not checking up on the client's IP address and port number, we just fill 0s into the last two parameter fields.

6. *child = fork()*;

   . . .

   close(sockMain);

This is the C language *fork* command that creates a new child process. The child process will inherit all of the parent program's I/O descriptors and will have access to *sockMain* and *sockClient*. The operating system keeps track of the number of processes that have access to a socket.

A connection is closed when the last process accessing the socket calls *close()*. When the child closes *sockMain,* the parent will still have access to this socket.

7. *close(sockClient)*;

This call is made within the parent part of the program. When the parent closes *sockClient,* the child will still have access to this socket.

8. *msgLength = recv(sockClient,buf, BUFLEN, 0))*;

   . . .

   close(sockClient);

The *recv* call has the form:

```
message_length = recv(socket_descriptor, buffer, buffer_length, flags)
```

By default, the *recv* call blocks. (Either the *fcntl()* or *iocntl()* function could be used to change the status of a socket to nonblocking.)

After the child has received data and printed its messages, it closes its access to *sockClient*. This will cause the connection to enter its termination phase.

21.7   A TCP CLIENT PROGRAM

The client connects to the server, sends a single message, and terminates. The program steps will be explained in the next section. To run this program, an end user inputs the server host's name and port and a message to be sent to the server. For example:

```
tcpclient plum.cs.yale.edu 1356 hello

/* tcpclient.c
 * Start the server before starting a client. Find out
 * the server's port.
 * To run the client, enter:
 * tcpclient hostname port message
 */
#include <sys/types.h>
#include <sys/socket.h>
#include <netinet/in.h>
#include <netdb.h>
#include <stdio.h>
#include <errno.h>
```

```
main(argc, argv) /* The client program has input arguments. */
int argc;
char *argv[];
{
int sock;
struct sockaddr_in servAddr;
struct hostent *hp, *gethostbyname();

/* Args are 0:program-name, 1:hostname, 2:port, and 3:message */
if (argc < 4)
   {printf("ENTER tcpclient hostname port message\n");
    exit(1);
   }

/*  1. Create a transmission control block. */
if ( (sock = socket(AF_INET, SOCK_STREAM, 0)) < 0)
   {perror("Could not get a socket\n");
    exit(1);
   }

/*  2. We will fill the server's address and port into the servAddr.
 *      First we fill the address structure with 0s.
 *      Next we look up the IP address for this host name and
 *      fill it in.
 *      Finally, we fill in the port number, which is in argv[2].
 */
bzero( (char *) &servAddr, sizeof(servAddr) );
servAddr.sin_family = AF_INET;
hp = gethostbyname(argv[1]);
bcopy(hp->h_addr, &servAddr.sin_addr, hp->h_length);
servAddr.sin_port = htons(atoi(argv[2]) );

/*  3. Connect to the server. We do not have to call bind.
 *      The system will assign a free port while performing the
 *      connect function.
 */

if ( connect(sock, &servAddr, sizeof(servAddr) ) < 0)
    {perror("Client cannot connect.\n");
     exit(1);
    }

/*  4. The client announces that it is ready to send the message.
 *      It sends and prints a goodbye message.
 */

printf("CLIENT: Ready to send\n");

if (send(sock, argv[3], strlen(argv[3]), 0) < 0)
   {perror("problem with send.\n");
    exit(1);
   }

printf("CLIENT: Completed send. Goodbye.\n");
close(sock);
exit(0);
}
```

21.7.1 Calls Used in the TCP Client Program

1. *sock = socket(AF_INET, SOCK_STREAM, 0)*;

The client creates a Transmission Control Block ("socket"), just as the server did.

2. The server had to initialize an address structure to use in its *bind* call.

This structure included the server's local IP address and port number. The client also initializes an address structure—and again it contains information about the *server*'s IP address and port. This structure will be used by the *connect* call to identify the destination.

The *bzero()* call below just puts 0s into the server address structure, *servAddr*. Once again, we identify the Address Family as Internet.

Next we must convert the host name entered by the user to an IP address. The *gethostbyname* function does this, returning a pointer to a *hostent* structure. This structure contains the server's name and IP address.

The *bcopy* function is used to copy the IP address (which is in *hp->h_addr*) into *servAddr*.

The second argument entered by the end user was the server's port. This was read in as an ASCII text string, so it must first be converted to an integer via *atoi()* and then converted to network byte order by *htons()*. Finally, the port number is copied into the address variable in *servAddr*.

```
bzero( (char *) &servAddr, sizeof(servAddr) );
servAddr.sin_family = AF_INET;
hp = gethostbyname(argv[1]);
bcopy(hp->h_addr, &servAddr.sin_addr, hp >h_length);
servAddr.sin_port = htons(atoi(argv[2]) );
```

3. *connect(sock, &servAddr, sizeof(servAddr));*

The *connect* call has the form:

```
connect(socket_descriptor, address_structure, length_of_address_structure)
```

The client will open a connection with the server whose IP address and port are contained in the address structure.

4. *send(sock, argv[3], strlen(argvs[3]), 0);*

The *send* call has the form:

```
return_code = send(socket_descriptor, buffer, buffer_length, flags)
```

Recall that the third argument entered by the end user (which appears in the program as *argv[3]*) is a text message. A common use for the flags parameter is to signal urgent data. In this instance, the flags parameter is set to 0.

5. *close(sock);*

The client issues a *close* to terminate the connection.

21.8   A SIMPLER SERVER

Many servers have the form shown in the earlier example. However, a simpler model can be used when the server needs to perform only a simple task for a client, as was the case in the example above.

Instead of creating a child process for each client, the server can directly perform the task and then close the connection to the client. The server queue enables a few other clients to wait until the server is ready for them.

Code for a simpler server follows. This server also can be accessed by clients running the *tcpclient* program discussed above.

```
/* tcpsimp.c
 * To run the program, enter "tcpsimp".
 */

/* First we include a bunch of standard header files.
 */
#include <sys/types.h>
#include <sys/socket.h>
#include <stdio.h>
#include <netinet/in.h>
#include <netdb.h>
#include <errno.h>

main()
{
int sockMain, sockClient, length, child;
struct sockaddr_in servAddr;

/*  1. Create the master socket.
 */

if ( (sockMain = socket(AF_INET, SOCK_STREAM, 0)) < 0)
    {perror("Server cannot open main socket.");
     exit(1);
    }

/*  2. Enter information into a data structure used to hold the
 *     local IP address and port. The "sin" in the variable names is
 *     short for "socket internet."
 */
bzero( (char *) &servAddr, sizeof(servAddr) );
servAddr.sin_family = AF_INET;
servAddr.sin_addr.s_addr = htonl(INADDR_ANY);
servAddr.sin_port = 0;

/*  3. Call bind. Bind will write a usable port number into servAddr.
 */
if ( bind(sockMain, &servAddr, sizeof(servAddr)) )
    { perror("Server's bind failed.");
     exit(1);
    }

/*  4. We want to look at the port number. We use the
 *     getsockname() function to copy the port into servAddr.
 */

length = sizeof(servAddr);
if ( getsockname(sockMain, &servAddr, &length) )
    { perror("getsockname call failed.");
     exit(1);
    }
printf("SERVER: Port number is %d\n", ntohs(servAddr.sin_port) );

/*  5. Set up a queue that can hold up to five clients.
 */
listen(sockMain,5);

/*  6. Wait for an incoming client. Accept will return
 *     a new socket descriptor that will be used for this client.
 */
```

```
for ( ; ; ) {
if ( (sockClient = accept(sockMain, 0, 0)) < 0)
{ perror("Bad client socket.");
exit(1);
}

/* 7'. Serve the client and close the client's connection.
 */
  doTask(sockClient);
  close(sockClient);
  }
}

/* Read one incoming buffer, print some information and quit.
 */

#define BUFLEN 81
int doTask(sockClient)
int sockClient;
{
char buf[BUFLEN];
int msgLength;

/* 8'. Zero out the buffer and then issue a recv
 * to got a message from the client.
 */

bzero(buf, BUFLEN);
if ( (msgLength = recv(sockClient,buf, 80, 0)) < 0)
    { perror("Bad receive.");
      exit(1);
    }

printf("SERVER: Socket used for this client is %d\n", sockClient);
printf("SERVER: Message length was %d\n", msgLength);
printf("SERVER: Message was: %s\n\n", buf);
}
```

21.9   UDP SOCKET PROGRAMMING INTERFACE

We have tackled the TCP programming interface, which is the most complex, first. Now let's take a look at programming a UDP server and client. Figure 21.3 shows an outline of a UDP dialogue between a client and server. The *socket*() and *bind*() calls complete quickly and have an immediate return. The *recvfrom* call is assumed to be blocking, which is its normal default. It can be changed to nonblocking (i.e., asynchronous) mode.

21.10   A UDP SERVER PROGRAM

The program that follows creates a UDP socket, binds to a port, and then begins to receive and print messages that are sent to its port:

```
/* udpserv.c
 * To run the program, enter "udpserv".
 *
 * First we include a bunch of standard header files.
 */
#include <sys/types.h>
#include <sys/socket.h>
#include <stdio.h>
```

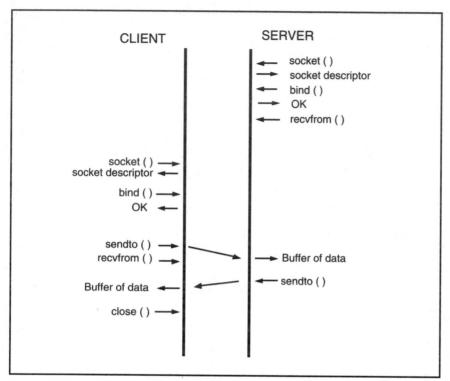

Figure 21.3   Typical UDP socket calls.

```
#include <netinet/in.h>
#include <netdb.h>
#include <errno.h>
#define BUFLEN 81

main()
{
int sockMain, addrLength, msgLength;
struct sockaddr_in servAddr, clientAddr;
char buf[BUFLEN];

/*  1. Create a UDP socket.
 */
if ( (sockMain = socket(AF_INET, SOCK_DGRAM, 0)) < 0)
    {perror("Server cannot open UDP socket.");
     exit(1);
    }

/*  2. Enter information into a data structure used to hold the
 *     local IP address and port. We will let bind get a free
 *     port for us.
 */
bzero( (char *) &servAddr, sizeof(servAddr) );
servAddr.sin_family = AF_INET;
servAddr.sin_addr.s_addr = htonl(INADDR_ANY);
servAddr.sin_port = 0;

/*  3. Call bind. Bind will write a usable port number into the TCB.
 */
```

```
if ( bind(sockMain, &servAddr, sizeof(servAddr)) )
   { perror("Server's bind failed.");
     exit(1);
   }
/*  4. We want to look at the port number. We use the
 *     getsockname() function to copy the port into servAddr.
 */
addrLength = sizeof(servAddr);
if ( getsockname(sockMain, &servAddr, &addrLength) )
   { perror("getsockname call failed.");
     exit(1);
   }
printf("SERVER: Port number is %d\n", ntohs(servAddr.sin_port) );
/*  5. Loop forever, waiting for messages from clients.
 */
for ( ; ; ) {

   addrLength = sizeof(clientAddr);
   bzero(buf, BUFLEN);
   if ( (msgLength = recvfrom(sockMain, buf, BUFLEN, 0,
                     &clientAddr, &addrLength)) < 0)
     { perror("Bad client socket.");
       exit(1);
     }

/*  6. Print the client's IP address and port, and the message.
 */
   printf("SERVER: Client's IP address was: %s\n",
                   inet_ntoa( clientAddr.sin_addr) );

   printf("SERVER: Client's port was: %d\n",
                   ntohs(clientAddr.sin_port));
   printf("SERVER: Message length was %d\n", msgLength);
   printf("SERVER: Message was: %s\n\n", buf);
   }
}
```

21.10.1   Calls Used in the UDP Server Program

1. *sockMain = socket(AF_INET, SOCK_DGRAM, 0)*;

The Address Family is again Internet.

2. *bzero((char *) &servAddr, sizeof(servAddr));*
   servAddr.sin_family = AF_INET;
   servAddr.sin_addr.s_addr = htonl(INADDR_ANY);
   servAddr.sin_port = 0;

The calls initializing the server address structure are the same as those used in the TCP programs.

3. *bind(sockMain, &servAddr, sizeof(servAddr))*;

As before, *bind* will get a port for the server and write values into a Transmission Control Block. Of course, UDP has a very small amount of control information compared to TCP.

4. *getsockname(sockMain, &servAddr, &length)*;

We use *getsockname* to extract the port assigned to the socket.

5. $msgLength = recvfrom(sockMain, buf, BUFLEN, 0, \&clientAddr,$ $\&length)$;

The *recvfrom* call has the form:

```
recvfrom(socket_descriptor, receive_buffer, buffer_length,
flags, source_address_structure,
pointer_to_length_of_source_address_structure)
```

The flags parameter can be set to allow the caller to peek at a message without actually receiving it.

On return, the source address structure will be filled with the client's IP address and port number. A pointer to the length of the source address is used because this length may be changed when the actual client address fields are received and filled in.

6. $inet\_ntoa(clientAddr.sin\_addr)$;

This call converts the client's 32-bit Internet address to the familiar dot notation for IP addresses.

21.11   A UDP CLIENT PROGRAM

The client connects to the server, sends a single message, and terminates. To run this program, an end user inputs the host name, the port of the server, and a message to be sent to the server. For example:

udpclient plum.cs.yale.edu 2315 "This is a message."

```
/* udpclient.c
 * Start the server before starting a client.
 * Find out the server's port.
 * To run the client, enter:
 *    udpclient hostname port message
 */
#include <sys/types.h>
#include <sys/socket.h>
#include <netinet/in.h>
#include <netdb.h>
#include <stdio.h>
#include <errno.h>

main(argc, argv)
int argc;
char *argv[]; /* These are the arguments entered by the enduser. */
    /* argv[0] is the program name. argv[1] points to a hostname. */
            /* argv[2] points to a port, */
            /* and argv[3] points to a text message. */

{
int sock;
struct sockaddr_in servAddr, clientAddr;
struct hostent *hp, *gethostbyname();

/* Should be four args. */
if (argc < 4)
    {printf("ENTER udpclient hostname port message\n");
     exit(1);
    }
```

```
/*  1. Create a UDP socket. */
if ( (sock = socket(AF_INET, SOCK_DGRAM, 0)) < 0)
   {perror("Could not get a socket\n");
    exit(1);
   }

/*  2. We will fill the server's address and port into the servAddr.
 *      First we fill the address structure with 0s.
 *      We use the gethostbyname function to look up the host name
 *      and get its IP address. Then we copy the IP address
 *      into servAddr using the bcopy function.
 *      Finally, we fill in the port number, which is in argv[2].
 */
bzero( (char *) &servAddr, sizeof(servAddr) );
servAddr.sin_family = AF_INET;
hp = gethostbyname(argv[1]);
bcopy(hp->h_addr, &servAddr.sin_addr, hp->h_length);
servAddr.sin_port = htons(atoi(argv[2]) );

/*  3. We have to call bind to get a UDP port. The system
 *      will assign a free port.
 */
bzero( (char *) &clientAddr, sizeof(clientAddr) );
clientAddr.sin_family = AF_INET;
clientAddr.sin_addr.s_addr = htonl(INADDR_ANY);
clientAddr.sin_port = 0;
if ( bind(sock, &clientAddr, sizeof(clientAddr)) < 0)
    {perror("Client cannot get a port.\n");
     exit(1);
    }

/*  4. The client announces that it is ready to send the message.
 * It sends and prints a goodbye message.
 */
printf("CLIENT: Ready to send\n");
if (sendto(sock, argv[3], strlen(argv[3]), 0, &servAddr,
  sizeof(servAddr)) < 0)
    {perror("problem with sendto.\n");
     exit(1);
    }
printf("CLIENT: Completed send. Goodbye.\n");
/* Close the socket */
close(sock);
 }
```

21.11.1 Calls Used in the UDP Client Program

1. *sock = socket(AF_INET, SOCK_DGRAM, 0);*

The UDP client creates a UDP socket.

2. *bzero((char *) &servAddr, sizeof(servAddr));*
   servAddr.sin_family = AF_INET;
   hp = gethostbyname(argv[1]);
   bcopy(hp->h_addr, &servAddr.sin_addr, hp->h_length);
   servAddr.sin_port = htons(atoi(argv[2]));

The *servAddr* structure is filled in using values entered by the end user, just as was done for the TCP client.

3. *bind(sock, &clientAddr, sizeof(clientAddr);*

The client calls *bind* to get a port.

4. *sendto(sock, argv[3], strlen(argv[3]), 0, &servAddr, sizeof(servAddr))*;

The *sendto* call has the form:

```
sendto(socket_descriptor, buffer, buffer_length, flags,
    destination_address_structure,
    length_of_destination_address_structure)
```

Note that this call contains all of the destination information required to send a User Datagram.

21.12   RECOMMENDED READING

Any Unix programmer's manual contains descriptions of socket program calls. *Unix Network Programming* by W. Richard Stevens provides an indepth discussion of socket programming. TCP/IP programmer's manuals for other operating systems describe the socket calls and often include sample programs. The manual for the TCP/IP product in use should be consulted since there are small variations due to differences between operating systems.

IP Version 6

22.1 INTRODUCTION

Over a remarkably short span of years, PCs have been connected to LANs, LANs have been connected to one another and to WANs, and all of it often has been connected to the outside world. The result is an astonishingly diverse Internet with millions of users.

The design of the original IP address system was not at all appropriate for this environment. The number space was small. Unlike the telephone system, which uses country and area codes, numbering was not hierarchical. Blocks of numbers were assigned to organizations very inefficiently, and much of the number space was wasted.

The result is that the number space is being depleted. In addition, because numbers are not assigned hierarchically, routing tables are growing very quickly.

The expansion of the Internet is not expected to slow down. There is a steady growth in the spread of personal computers and of their connectivity into the global network. In addition, new challenges are presented by:

- The networking of the coming generations of mobile personal computers— the offsprings of today's pagers and personal digital assistants.

- The anticipated demand for real-time audio and video, which will push the current technology to its limits.

The world of serious commerce has moved onto the Internet, and it also is clear that it is time to build security into the network infrastructure.

There is another problem that plagues managers. Many organizations use IP backbones to tie their networks together and tunnel traffic from site to site wrapped in IP headers. Today this is done awkwardly and often without any mechanism for network congestion control.

The development of IP version 6 (IPv6; also called *IP next generation*) has been spurred on by the urgent need to solve these Internet addressing, routing, performance, security, and congestion problems. We will present the major features of IPv6 in this chapter. Implementers should consult the most recent RFCs for details.

22.2 OVERVIEW OF IPV6

IPv6 does the following:

- Introduces 128-bit (16-octet) addresses that can be structured hierarchically to simplify address delegation and routing.

- Simplifies the main IP header but defines many optional *extension headers.* This will enable new networking functions to be added as needed.

- Supports authentication, data integrity, and confidentiality at the IP level.

- Introduces *flows,* which can be used to support many new kinds of transmission requirements—such as real-time video.

- Makes it easy to encapsulate other protocols and provides a mechanism for congestion control when carrying "foreign" protocols.

- Provides new automatic address self-configuration methods and builds in a test for IP address uniqueness.

- Improves router discovery and the detection of dead routers or unreachable neighbors on a link.

At the time of this writing, many IPv6 design details are still being worked out. However, the major architectural elements are in place and will be described here.[1]

22.3 TERMINOLOGY

Version 6 makes some adjustments to version 4 nomenclature and introduces some new terms:

- A *packet* is an IPv6 header plus payload.

- A *node* is any system that implements IPv6.

- A *router* is a node that forwards IPv6 packets not explicitly addressed to itself.

- A *link* is a medium over which nodes can communicate at the link layer.

- *Neighbors* are nodes attached to the same link.

The term *packet* is one of the most abused in the networking world. People use it to describe Protocol Data Units (PDUs) from the link layer to the application layer.

---

[1]IPv6, ICMPv6, DNS extensions, and the IPv6 addressing architecture were proposed standards at the time of this writing.

Why did the version 6 authors switch from *datagram* to *packet?* One of the innovations in IPv6 is that it can be used to carry traffic for many other protocols—hence its payload might not be a PDU from the TCP/IP suite. When the PDU is native IP, the term *datagram* is still appropriate.

In this chapter, we will follow the usage in the current IPv6 documents and use the term *packet*.

22.4   IPV6 ADDRESSES

IPv6 addresses are 16 octets (128 bits) long. A fairly compact (if ugly) notation is used to write these addresses. They are represented as eight hexadecimal numbers separated by colons. Each hexadecimal clump represents 16 bits. For example:

41BC:0:0:0:5:DDE1:8006:2334

Note that leading zeros in a hex field can be dropped (e.g., 0 instead of 0000 and 5 instead of 0005). The format can be compressed further by replacing one contiguous string of 0 fields with ::— for example:

41BC::5:DDE1:8006:2334

Three clumps are missing, and so :: represents the string :0:0:0:.

Finally, version 4 IP addresses often will be embedded in the last 4 octets of a version 6 address. These can be written using a mixed address format that uses both the colon and the dot notations, for example:

0:0:0:0:0:FFFF:**128.1.35.201**

22.4.1   Address Allocations

A 128-bit address space allows room for many different types of addresses, including:

- Hierarchical Service Provider-based global unicast addresses
- Hierarchical geographical global unicast addresses
- Private site addresses, for use within an organization only
- Local and global multicast addresses

Version 6 does not use broadcasts but relies on multicasts for control functions such as address resolution and booting. The reasoning behind this is that broadcasting a message causes interrupts at every device on a link. In most cases, only a few devices really need to examine the message. In addition, restricting version 6 control messages to multicast addresses prevents version 6 nodes and version 4 nodes from interfering with one another when they share the same link.

22.4.2 Overall Address Allocations

The *Internet Assigned Numbers Authority* (IANA) has the job of delegating chunks of the IPv6 address space to regional registries around the world. Regional registries can, in turn, pass blocks of addresses to smaller regions, national registries, or Service Providers.

Table 22.1 shows the suggested overall plan for address space allocation:

- A large block is used for Service Provider-based addressing.

- There are blocks allocated to stand-alone LANs or entire sites that are not connected to the Internet, enabling them to assign addresses to themselves.

- Blocks have been allocated for IPX addresses and OSI Network Service Access Point (NSAP) addresses.

- A large block has been reserved for geographically based addresses.

Currently, almost three-quarters of the address space has not been assigned to any use.

22.4.3 Address Format Prefixes

The first few bits of an address, called the *Format Prefix,* identify the type of address. For example, prefix 010 introduces Service Provider-based unicast IP addresses. As one might expect, the format of the rest of the address depends on the Format Prefix.

22.4.4 Provider-Based Addresses

A simple hierarchical structure currently is proposed for Provider-Based addresses:

3 bits	n bits	m bits	o bits	125-n-m-o bits
010	Registry ID	Provider ID	Subscriber ID	Intrasubscriber

Note that it is easy to route to a provider by comparing the first part of the address with entries in a routing table. The provider then can route to its subscribers by comparing a larger chunk of the address with its table entries.

When using this format, a subscriber organization will own sufficient address space to build a convenient internal hierarchy. An organization could structure its address space into subnets and hosts, as is done now, or could add one or more extra levels of hierarchy. For example, a hierarchy consisting of area, subnet, and host could be used.

By the way, all-0 or all-1 fields are not forbidden in version 6 addresses.

TABLE 22.1   IPv6 Address Space Allocation

Allocation	Prefix (binary)	Fraction of address space
Reserved	0000 0000	1/256
Unassigned	0000 0001	1/256
Reserved for NSAP Allocation	0000 001	1/128
Reserved for IPX Allocation	0000 010	1/128
Unassigned	0000 011	1/128
Unassigned	0000 1	1/32
Unassigned	0001	1/16
Unassigned	001	1/8
Provider-Based Unicast Address	**010**	**1/8**
Unassigned	011	1/8
Reserved for Geographic-based Unicast Addresses	100	1/8
Unassigned	101	1/8
Unassigned	110	1/8
Unassigned	1110	1/16
Unassigned	1111 0	1/32
Unassigned	1111 10	1/64
Unassigned	1111 110	1/128
Unassigned	1111 1110 0	1/512
Link-Local Use Addresses	1111 1110 10	1/1024
Site-Local Use Addresses	1111 1110 11	1/1024
Multicast Addresses	1111 1111	1/256

22.4.5   Addresses for Stand-Alone Sites

Today, a version 4 LAN or network that is not connected to the Internet uses a special block of addresses, such as 10.0.0.0 or 172.16.0.0, that has been reserved for this purpose. But if the organization subsequently needs to connect to the outside world, it has a big reconfiguration job on its hands.

Version 6 takes care of this address reassignment problem far more gracefully, as we shall see in the sections that follow.

22.4.6 Link-Local Addresses

Recall that a *link* is a communication facility, such as an Ethernet,[2] Token-Ring, or Fiber Distributed Data Interface (FDDI) LAN, frame relay network, Asynchronous Transfer Mode (ATM) network, or point-to-point line. It is easy to automate addressing on an isolated link that is not connected to a router. *Link-Local* addresses have the form:

1111111010 (10 bits)	00...00	Unique address for link technology

For example, if the link is a LAN:

1111111010	00...00	LAN MAC address

A Link-Local address also is useful during initialization.

22.4.7 Site-Local Addresses

A site that has routers, but is not connected to a Service Provider, can automatically generate internal addresses of the form:

1111111011 (10 bits)	00...00	Subnet ID	Unique address for link technology (e.g., LAN MAC address)

Routers advertise the prefix (including the subnet ID) onto the link.

Note how easy it is to migrate to Service Provider connectivity. A router simply is configured with a new prefix which includes Registry, Service Provider, and Subscriber numbers along with the subnet number. The router advertises the new prefix, and hosts start to use it. None of the site-assigned part of an address would need to change.

22.4.8 Multicast Address Format

Version 6 multicast addresses have a clearer, more flexible definition than version 4 multicast addresses. There are many different types of multicast addresses. They have slightly different introducers, based on whether the

---

[2]A new Ethernet type code, X' 86-DD, has been defined for IPv6.

multicast address is permanent or transient, local, or global. Multicast addresses have the format:

8 bits	4	4	112 bits
11111111	000T	Scope	Group ID

T = 0 for a well-known, permanent multicast address.

T = 1 for a transient multicast address.

Scope codes indicate whether the scope is same-node, link-local, site-local, organizational-local, or global. A same-node scope covers the case where a client sends a multicast message to servers that are located at the same host. The specific scope codes are:

0	*reserved*
1	node-local scope
2	link-local scope
3	*unassigned*
4	*unassigned*
5	site-local scope
6	*unassigned*
7	*unassigned*
8	organization-local scope
9	*unassigned*
A	*unassigned*
B	*unassigned*
C	*unassigned*
D	*unassigned*
E	global scope
F	*reserved*

22.4.9 Anycast Addresses

A new (and experimental) kind of addressing is proposed—the *anycast* addresses. An anycast address is a unicast address that is assigned to more than one interface. Initially, only routers may be assigned anycast addresses. For example, an anycast address might identify:

- All routers owned by a particular Service Provider
- All routers on the boundary of a given Autonomous System
- All routers attached to a particular LAN

An anycast address can be included in a source route. It means "Use the nearest router that has this anycast address." For example, if the anycast address identified routers owned by a Service Provider, it would be used to say "Get to this Service Provider using the shortest path."

Of course, a router interface that has been assigned an anycast address also has its own real address.

22.5   SPECIAL ADDRESSES

There are several special IPv6 address formats.

22.5.1   Unspecified Address

The all-0s address

0:0:0:0:0:0:0:0

means "unspecified address." It sometimes is used as a source during initialization, when a system does not yet know its own address.

22.5.2   Version 6 Loopback

The version 6 *loopback* address is:

0:0:0:0:0:0:0:1

22.5.3   Version 4 Addresses

In a mixed version 4 and 6 environment, the addresses of IP version 4 systems that do *not* support version 6 are mapped to version 6 addresses of the form:

0:0:0:0:0:FFFF:a.b.c.d

where a.b.c.d is the original IP address.

22.5.4   Version 6 Addresses Interfacing to Version 4 Nets

Another special format is used by version 6 nodes that communicate with one another across an intermediate version 4 network. (This is called IPv4 tunneling.) As shown in Figure 22.1, the interfaces at the boundaries must be assigned version 4 addresses. These are mapped to the special *IPv4-compatible* IPv6 address format:

0:0:0:0:0:0:a.b.c.d

Thus, these addresses are easily mapped between their version 4 and version 6 representations.

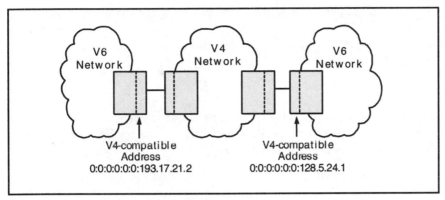

Figure 22.1 IPv4-compatible IPv6 addresses.

22.6 IPV6 HEADER FORMAT

The basic header is very simple, as shown in Figure 22.2. Note that there are very few fields:

Version	Equals 6 for IP next generation.
Priority	Differentiates interactive from bulk traffic or defines discard eligibility during congestion.
Payload length	(16 bits) If the length is less than or equal to 64 kilobits, this field reports the length of the part of the packet that follows the initial IPv6 header. If the length is greater than 64 kilobits, the payload length is set to zero, and the actual length will be reported in a *Jumbo Payload* option in a later header.
Hop limit	Is decremented by 1 at each router. The packet will be discarded if the value reaches 0 at a router.

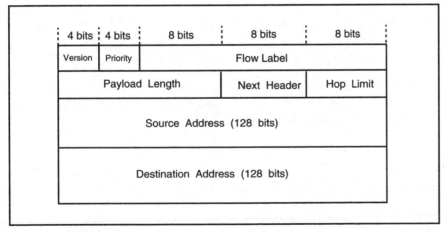

Figure 22.2 IPv6 header format.

Next header	Identifies the type of protocol header that follows (e.g., 6 for a TCP header).
Flow label	Identifies traffic that needs a special type of handling (e.g., real-time video).

22.6.1   Priority

The priority field has a dual function. For congestion-controlled TCP traffic, it assigns high numbers to control packets and interactive traffic and low numbers to bulk traffic. Specifically, the values are:

0	Uncharacterized traffic
1	"Filler" traffic (e.g., netnews)
2	Unattended data transfer (e.g., electronic mail)
3	Reserved
4	Attended bulk transfer (e.g., file transfer)
5	Reserved
6	Interactive traffic (e.g., *telnet*)
7	Internet control traffic (e.g., routing protocols)

IPv6 can be used to carry traffic for ISO, DECnet, and so forth. Priority values 0 through 7 can be used for any protocol that imposes its own flow control.

Values 8 through 15 are used as a tool for handling congestion when a protocol (e.g., UDP or IPX) does not impose its own congestion control. When a network is congested, traffic is discarded. It is more harmful to discard some types of application data than others. Low values, such as 8 or 9, mean that a packet is more eligible for discard.

22.6.2   Use of the Flow Label

A *flow* is a sequence of packets from a source to a destination that requires some kind of special treatment. For example, voice or real-time video require different handling from bulk data transfer.

The *flow label* is used to identify a stream of traffic that has a special handling mechanism—such as bandwidth reservation.

The fact that packets belong to a flow is indicated by a nonzero flow label. Packets belonging to a particular flow all have the same source address, destination address, priority, and flow label.

22.7   IPV6 EXTENSION HEADERS

The use of extension headers is a very innovative idea that allows functionality to be added incrementally to IP version 6.

Recall that in an IP version 4 header, the *Protocol* field is used to identify what type of header (e.g., TCP or UDP) follows the IP header. Version 6 uses a more general *Next Header* field. If the next header is a TCP or UDP header, the value in the Next Header field is 6 or 17, the protocol identifier for TCP or UDP.

But several extension headers can be sandwiched between the IPv6 header and a higher-layer header. These are used for options, such as source routing or security. Fragmentation also has been moved into an extension header.

As illustrated in Figure 22.3, each of the extension headers contains a Next Header field, and so headers are chained together. The next layer protocol finally is identified in the final extension header.

This scheme provides great flexibility. New options can be defined as needed at any time, and their overall length need not be restricted. Also note that the final extension header can point to a header that belongs to an entirely different protocol suite (such as ISO or DECnet).

The currently defined header identifiers are listed in Table 22.2. Some headers contain information that must be processed at every node along the way, while others contain information that only needs to be processed at the destination.

The order shown in Figure 22.3 reflects the recommended order for whatever headers may be included. Note that two *Destination Options* headers could appear. The first would be placed before a *Routing* header and would be applied to each hop listed in the *Routing* header. The second would appear as the last header and applies only to the final destination.

It is possible that someday there may be a use for sending a packet that consists of a header and no payload. In this case, the final Next Header identifier is 59, meaning "nothing follows."

22.7.1 Routing Header Use

The *Routing* header is a very important feature of version 6. When combined with anycast addresses, it can be used to control paths based on which

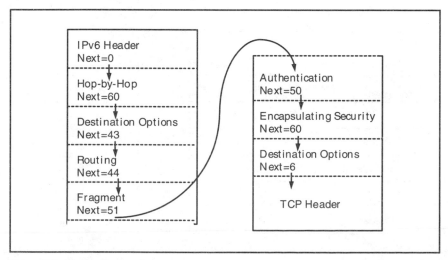

Figure 22.3 Extension headers.

TABLE 22.2 IPv6 Headers

Header	Number in previous "Next Header" field
Hop-by-Hop Options	0
Destination Options	60
Routing	43
Fragment	44
Authentication	51
Encapsulating Security Payload	50
No Next Header	59

providers are preferred or on the need to use a specific provider (e.g., to reach a mobile user). Recall that an anycast address can be used to say "Go to the nearest router belonging to Service Provider X."

When a Routing header is used, the destination must turn the route around and follow the reverse path back to the source.

22.7.2 Routing Header Operation

The Routing header includes a type field, enabling various types to be added in the future. Currently, only type 0 is defined. A type 0 Routing header is similar to an IPv4 Source Route.

The format of a type 0 Routing header is shown in Figure 22.4. The type 0 Routing header contains a list of nodes to be traversed on the way to a destination.

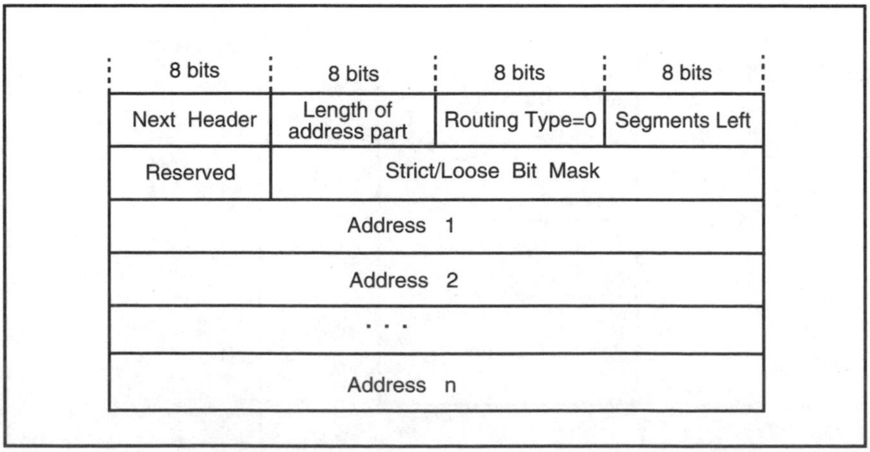

Figure 22.4   Type 0 Routing header.

As in IP version 4, the final destination is *Address n.* The packet is first forwarded to the address in the main IPv6 header. Then the *Routing* header is consulted. Address 1 is swapped into the IPv6 header's destination address field, the *Segments Left* counter is reduced by 1, and the packet is forwarded. The final address in the *Routing* header is the true destination. On arrival, the address list contains the addresses of the nodes that have been visited.

The strict/loose bit mask indicates whether the corresponding hop must be to a neighbor (strict) or not (loose).

22.7.3 Hop-by-Hop Extensions Header

The Hop-by-Hop header carries option information that must be examined by every hop along the way. The format of the Hop-by-Hop header is shown in Figure 22.5.

The Hop-by-Hop header can carry a variable number of options. Each is self-identifying and is encoded in three fields:

Option Type 8 bits	Option Length 8 bits	Option Value n bits

The *Jumbo Payload* option is an example of a Hop-by-Hop option. It is used to declare the length of a payload that is bigger than 64 kilobits. The payload length (in octets) is described in a 4-byte value. The reported payload length includes all of the packet except for the IPv6 header.

22.7.4 Fragmentation

Unlike version 4, *fragmentation is never performed by routers* but only by a source node. Fragmentation should be avoided whenever possible, but it will be needed occasionally. It is up to the source node to fragment packets, and the destination node must reassemble them.

If a router receives a packet that is too large to forward, it will discard the packet and send back an ICMP message that announces the next-hop Maximum Transmission Unit (MTU).

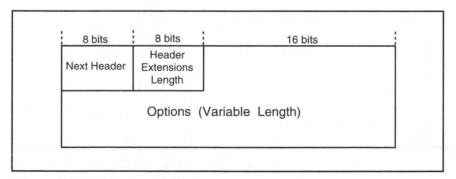

Figure 22.5 The Hop-by-Hop header.

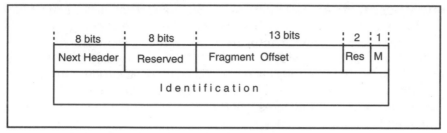

Figure 22.6 Format of a Fragment header.

When a source node creates a fragment, it must include a *Fragment* header. The Fragment header has the form shown in Figure 22.6.

As in version 4, the *Fragment Offset* field is 13 bits long and measures offsets in 8-octet blocks. The *more* bit indicates whether this is the last fragment or not. The *Identification* field has been expanded to 32 bits.

22.7.5 Destination Options

The *Destination Options* header contains options that need to be processed by the packet's destination (or destinations, for multicasts). Currently, no options (other than padding fields) have been defined for this header. The *Destination Options* header format is shown in Figure 22.7.

Recall that if a *Routing* header is included, the protocol allows two *Destination Options* headers to be included. The first, right before the *Routing* header, contains options that apply to every node listed in the header. The second, placed after all other headers, applies only to the final destination.

22.8 VERSION 6 AUTOCONFIGURATION

In the past, network managers paid for the benefits of having an IP network with a hefty amount of configuration and maintenance. One of the goals of

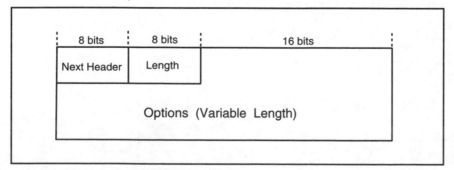

Figure 22.7 The Destination Options header.

version 6 is to provide a lean, effective automatic initialization procedure. This is important in helping a site migrate to the new address format. It also is vital to automate address changes that may come about because of a change in the choice of Service Provider.

On a stand-alone LAN, an IPv6 host can build an IP address automatically, using a network interface card address or other unique link level identifier known to the system.

When an organization has a routed network or is connected to a Service Provider, routers provide hosts with the information that the hosts need in order to autoconfigure their addresses and start to operate.

22.8.1 Role of Routers

Each router supplies hosts with information such as:

- The address of the router
- A list of all address prefixes that are used on the link
- Which prefixes hosts should use to create their own addresses
- A suggested maximum hop limit to be used
- Whether the host should retrieve additional configuration data from a Dynamic Host Configuration Protocol (DHCP) server
- The MTU for a link that has variable MTU
- Values for various timers

22.8.2 Address Prefix List

Many network managers will be happy to hear of the demise of the loathed subnet mask. Instead, routing decisions are based on comparing address prefixes.

A router advertises the list of address prefixes that are used on the local link. A prefix is expressed as all or part of an IPv6 address, along with a number that tells how many bits actually belong to the prefix. Hosts store these prefix lists.

When a host needs to decide whether a destination is on or off the link, the host runs through its list of on-link prefixes and compares the relevant number of bits with the corresponding destination address bits.

22.8.3 IPv6 Interface Addresses

Every version 6 interface has a *list* of addresses associated with it. At minimum, the list includes the unique *link local address,* which has the form:

1111111010 (10 bits)	00...00	Unique address for link technology

Each node needs a way to generate its own unique interface link address. For example, for a LAN interface the MAC address is the unique part of the address and occupies the right-most 48 bits. A system can communicate with other systems on the link using its link local address.

How can a host generate site-local or global addresses automatically? Recall that a router advertises a list of prefixes. Some of these are flagged for use in constructing host addresses. A new site-local or global address is constructed by placing an advertised prefix in front of the unique link address. This address is then added to the host's list.

The router's advertisement also tells hosts whether they should pick up additional address information from a DHCP server (which can assign administrator-configured addresses). And the advertisement indicates whether additional configuration information should be obtained from a server.

By the way, manual configuration still will be supported for version 6, if anyone wants to use it.

22.8.4 Changing Addresses

The ability to use more than one global prefix can be used to ease a transition from one Service Provider to another.

The router's advertisements associate an invalidation timer with each provider's prefix. When switching from one provider to another, the old prefix will simply be allowed to age out. Of course, the timeout values for a new, active prefix are refreshed periodically so that it will not age out.

Timeouts also make it possible to pick up a host and plug it into a different link at the site. Recall that the prefix includes the subnet identifier as well as the provider and region information. Old prefixes will age out, and new ones will be learned.

22.8.5 Testing Address Uniqueness

Before using the link local address, a host will multicast a query that checks that the address is not already in use on the link. This assures that the link local IP address—and every address formed from it by using a different prefix—all are unique. Addresses that are manually configured, or are learned from a DHCP server, also are tested for uniqueness before they are used.

22.9 CONFIGURATION VIA DHCPV6

Systems can obtain a full set of configuration parameters from a DHCP server. Some changes are needed to migrate DHCP to version 6.

Obviously, the updated protocol must support version 6 addresses. In addition, the old lease timeout is replaced by deprecation and invalidation lifetimes.

It is intended that DHCPv6 will not only autoconfigure hosts but also will automatically autoregister host names and addresses in the Domain Name

System. A host that is being initialized can request the use of a specific host name or else can be assigned a name by the DHCPv6 server.

If the client's invalidation lifetime expires, the DHCPv6 server should then delete the client's DNS records.

22.10   IPV6 TRANSITION

Given the universality of IP around the world, it will *not* be possible to say "On such-and-such a day, everybody must cut over to version 6." Clearly transition needs to be gradual:

- Version 6 nodes need to interwork with version 4 nodes.
- Organizations must not be forced to give up their current addresses.
- Organizations should be able to upgrade some nodes, while leaving others unchanged.
- The transition should be easy to understand and easy to do.

22.10.1   Why Change?

Service Providers need IPv6 in order to perform more efficient backbone routing and provide future subscribers with numbers. But why should an organization with a properly functioning stand-alone IP network switch to version 6? If it is not having problems in managing its IP addresses, and if some of the new services (such as flows) are not needed, the answer is—don't bother.

It is likely that Internet servers will run with dual stacks and dual addresses for a long time. However, at some point, it will become more convenient to use version 6 than not to use it.

22.10.2   How to Change

The first step in moving toward version 6 is to upgrade a site's Domain Name Server software so that its DNS servers can respond to queries that use the new address format.

It is very likely that the first systems to be converted to dual versions 4 and 6 protocol stacks will be routers that interface to external networks. Little by little, important servers will add a version 6 stack. In a mixed environment, version 6 traffic sometimes will have to be tunneled across a version 4 network.

During an interim period, IPv6 site-local addresses can be used. When sites attach to a Service Provider, the addresses will be augmented with the appropriate Region, Provider, and Subscriber prefixes.

22.10.3   DNS Changes

A new address resource record type, *AAAA,* maps domain names to IP version 6 addresses. A sample entry is:

MICKEY IN AAAA 4321:0:1:2:3:4:567:89AB

Reverse lookups also have to be supported. A new domain has been added in order to handle IPv6 *address-to-name* mappings. The reverse lookup domain is rooted at *IP6.INT*.

Recall that version 4 IP addresses are reversed to obtain their labels in the *in-addr.arpa* domain. A version 6 address also is reversed and is rewritten as a series of hexadecimal digits separated by dots. For example, a reverse lookup entry for:

4321:0:1:2:3:4:567:89AB

appears in the domain tree as:

B.A.9.8.7.6.5.0.4.0.0.0.3.0.0.0.2.0.0.0.1.0.0.0.0.0.0.0.1.2.3.4.IP6.INT

22.10.4   Tunneling through a Version 4 Network

During the transition period, datagrams sometimes will traverse a path like the one shown in Figure 22.8. In the figure, Service Providers A and C support version 6, but Service Provider B does not. The boundary router interfaces will be assigned IPv4-compatible IPv6 addresses, which can easily be converted to version 4 addresses by dropping their zero prefixes. Version 6 packets will be wrapped inside a version 4 header and tunneled across the intervening network.

Tunneling also can occur within a site that has converted some of its networks to version 6. Tunneling can be used anywhere that it is convenient to do so. It can be used between routers, between hosts, or on a host/router path.

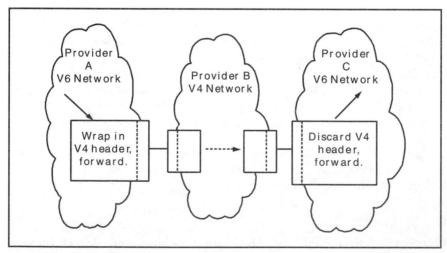

Figure 22.8   Tunneling traffic across a version 4 network.

22.11   SUMMARY

The IP next generation working groups have laid the foundation for a version that solves the Internet address space depletion problem and promotes more efficient routing. The new protocol provides attractive options for automatic configuration and enables coexistence and gradual migration. The chained headers allow for fairly painless future updates and also provide a graceful way for other protocols to ride across an IP network.

22.12   RECOMMENDED READING

RFC 1884 describes IPv6 addresses, and RFC 1883 outlines the version 6 protocol. RFC 1885 describes ICMPv6, and RFC 1886 deals with extensions to the Domain Name System. RFC 1887 discusses an architecture for address allocation. Several more RFCs were being prepared at the time of this writing.

23

ICMPv6 and Neighbor Discovery

23.1 INTRODUCTION

Version 6 of the Internet Control Message Protocol (ICMPv6) retains many of the version 4 functions, but there are some major changes:

- ICMPv6 messages assist in automatic address configuration.
- There are new ICMPv6 messages and procedures that replace the Address Resolution Protocol (ARP).
- Path Maximum Transmission Unit (MTU) discovery is automatic. Since routers no longer fragment packets, whenever a packet is dropped because it is too big, a new *Packet Too Big* message is sent to the source.
- ICMPv6 does not send *Source Quench* messages.
- ICMPv6 incorporates Internet Group Management Protocol multicast membership reporting functions.
- ICMPv6 assists in detecting that a router is not functioning or a communicating partner is no longer active.

ICMPv6 is sufficiently different that it has a new protocol number—it has been assigned Next Header value 58.

23.2 BASIC ICMP MESSAGES

Table 23.1 lists the basic ICMPv6 message types. Note that error messages have types in the range 0 to 127, while informational messages have types in the range 128 to 255. The general ICMP message format is shown in Figure 23.1. First, let's examine the ICMP messages that resemble version 4 counterparts.

23.2.1 Destination Unreachable

The reasons for sending a *Destination Unreachable* message are described by the codes:

TABLE 23.1   ICMP Message Types

Error message	Type
Destination Unreachable	1
Packet Too Big	2
Time Exceeded	3
Parameter Problem	4
Information message	Type
Echo Request	128
Echo Reply	129
Group Membership Query	130
Group Membership Report	131
Group Membership Reduction	132

0   No route to destination.

1   Communication with destination administratively prohibited.

2   Next destination in Routing header is not a neighbor, and the "strict" bit is set.

3   Address unreachable.

4   Port unreachable.

The format of a *Destination Unreachable* message is shown in Figure 23.2.

23.2.2   Packet Too Big

A router sends a *Packet Too Big* message when the packet is bigger than the MTU for the next-hop link. The next-hop link's usable MTU is reported in the message. Note that in version 4, this fact was reported in a *Destination Unreachable* message. Figure 23.3 shows the format of a *Packet Too Big* message.

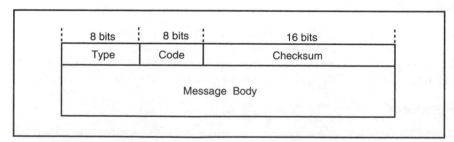

Figure 23.1   Format of an ICMPv6 message.

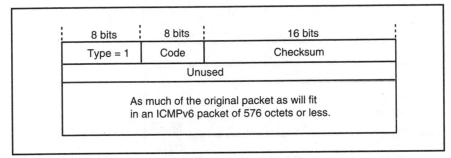

Figure 23.2 Format of a Destination Unreachable message.

23.2.3 Time Exceeded Message

A *Time Exceeded* message is sent by a router that has decremented the Hop Limit to 0 (code = 0) or by a system whose reassembly timeout has expired (code = 1). The message format is identical to the format of the *Destination Unreachable* message, except that the type is 3.

23.2.4 Parameter Problem Message

A *Parameter Problem* message is sent by a system that cannot process a packet because of a header field. The codes are:

0   Erroneous header field encountered

1   Unrecognized Next Header type encountered

2   Unrecognized IPv6 option encountered

The message format is similar to the format of the *Destination Unreachable* message, except that the "unused" field is replaced by a pointer that indicates the octet offset to the error, and the type is 4.

23.2.5 Echo Request and Reply

The *Echo Request* and *Reply* messages have formats identical to the version 4 messages, except for the fact that type = 128 is used for Echo Requests and type = 129 is used for Echo Replies.

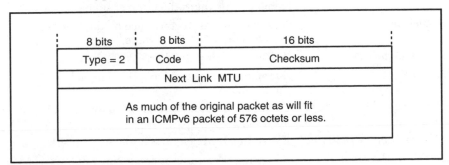

Figure 23.3 Format of a Packet Too Big message.

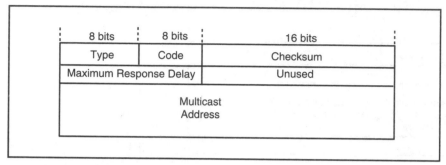

Figure 23.4 Group Membership message format.

23.2.6 Group Membership Messages

The format of multicast Group Membership messages (shown in Figure 23.4) has been changed so that it is consistent with the general ICMPv6 format. The *Maximum Response Delay* is nonzero only in query messages. It reports the maximum time that responding *Report* messages may be delayed. Message types include:

130	Group Membership Query
131	Group Membership Report
132	Group Membership Reduction

23.3 NEIGHBOR DISCOVERY

At the time of this writing, work was still proceeding on a very important set of protocols aimed at automating many important link functions. They include:

Router Discovery	Finding routers on the local link.
Prefix Discovery	Discovering and using prefixes that indicate which destinations are on the link and which are remote. This takes the place of subnet masks.
Parameter Discovery	Discovering parameters such as the link MTU and a default hop limit value.
Address Autoconfiguration	Self-configuring IP addresses for link interfaces.
Address Resolution	Mapping a link neighbor's IP address to its link-layer address.
Next-hop Determination	Mapping an IP address to the address of the next hop.
Neighbor Unreachability Detection	Detecting dead neighbor hosts and routers.
Duplicate Address Detection	Checking that an IP address that you have been assigned is not already in use.

Redirect   Getting notification that there is a better router
                                        to use for a given destination or that a destina-
                                        tion is on the local link.

Table 23.2 lists the ICMPv6 messages that have been proposed to implement
the Neighbor Discovery functions.

23.3.1 Autoconfiguration via Routers

Routers provide hosts with:

- Router address information
- A list of all prefixes that are used on the link
- Prefixes that hosts should use to create their own addresses
- The maximum hop limit that should be used for routes through that router
- Indication whether hosts should use a boot server to get additional configu-
  ration data
- The MTU for a link that has variable MTU
- Values for various timers

This is done via ICMPv6 *Router Advertisement* messages, which have type
134. Hosts listen for *Router Advertisement* messages on the all-nodes link-
local multicast address.

When a host boots, it may not want to wait for a *Router Advertisement*. The
host may send out a *Router Solicitation* message (type 133) in order to trigger
an advertisement. The router responds by sending an advertisement to the
host's link-local address.

23.3.2 Neighbor Solicitation and Advertisement

Current work proposes replacing the old Address Resolution Protocol (ARP)
requests with the new ICMP *Neighbor Solicitation* and *Neighbor
Advertisement* multicast messages. A *Neighbor Advertisement* is the response

TABLE 23.2 ICMP Neighbor Discovery Messages

Information Message	Type
Router Solicitation Message	133
Router Advertisement Message	134
Neighbor Solicitation Message	135
Neighbor Advertisement Message	136
Redirect	137

to a *Neighbor Solicitation*. In addition to discovering neighbor link-layer addresses, *Neighbor Solicitation* messages also are used to:

- Detect duplicate IP addresses
- Test whether a router is dead
- Test whether a neighbor to whom you are sending packets is dead

23.3.3   Address Resolution

To discover the link level address for a neighbor, a *Neighbor Solicitation* is sent to a special address called the *solicited-node multicast address of the target address*. The solicited-node multicast address is formed by taking the low-order 32 bits of the target IP address and prepending the 96-bit prefix FF02:0:0:0:0:1. This produces a link-local scope multicast address. The sender includes its own link layer address in the message.

Note that using this specialized multicast cuts down substantially on the number of systems likely to "hear" the request. In fact, it is very likely that only the targeted system will examine the request.

23.3.4   Detecting Duplicate IP Addresses

Before using its link-local IP address or any other address that is *not* constructed by adding a prefix to the link-local address, a node will send a *Neighbor Solicitation* message asking whether any neighbor has that IP address. The node uses the unspecified source address as the source address for the message. If the IP address already is in use, the address holder will multicast a response.

23.3.5   Neighbor Unreachability Detection

The detection of a dead router was a chancy business in IPv4. In version 6, if a timeout indicates that a router might be inactive, a system checks by sending a *Neighbor Solicitation* unicast message to the router.

The same procedure is used to check whether a host neighbor has become unreachable.

23.3.6   Redirect Messages

Just as in version 4, when a host has forwarded a datagram to the wrong local router, the router sends back a *Redirect* message indicating the correct first-hop node. A *Redirect* message also can be used to notify a sender that the destination is actually on the local link. Perhaps this is why *Redirect* messages are defined in the Neighbor Discovery specification.

Figure 23.5 shows the proposed format for ICMPv6 *Redirect* messages. The target address is the IP address of the next hop that should be used. The destination address is the desired destination. The options field includes the link

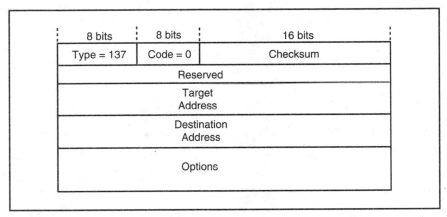

Figure 23.5   Format of a Redirect Message.

layer address for the target system and may also include part of the redirected datagram.

23.4   RECOMMENDED READING

ICMPv6 is described in RFC 1885. At the time of this writing, the Neighbor Discovery protocols still were in draft form.

IP Security

24.1 INTRODUCTION

The need to develop a new version of IP created additional stimulus to the effort to solve TCP/IP's security problems. The proposed mechanisms introduce security into the IP layer. They are designed to be used with both IP versions 4 and 6. For simplicity, the scenarios in this chapter will refer to version 4 addresses.

Everyone agrees that there is a need for security, but why build it into the IP layer? Why not use the application layer? In fact, it is likely that many applications *will* add their own security mechanisms. But in an environment where it is easy for snoopers to capture traffic, use all or part of it for later replay, and forge their IP addresses when they do it, you cannot be sure of the validity of any datagram.

Why not use the physical layer? All link traffic could be encrypted. This would solve link eavesdropping problems, but the traffic would need to be automatically decrypted at each router. Today, there is no particular reason to trust all routers.

Also this would not solve authentication problems. It also might cause severe bottlenecks for high-speed traffic, even when encryption and decryption are performed in hardware. Furthermore, every LAN interface card would need to be capable of encryption and decryption, and this would be very costly.

24.2 SECURITY ISSUES

Recall that in Chapter 3 we discussed three security attributes:

Authentication	Validating the identity of a user, client process, or server application
Integrity	Assuring that data has not been changed
Confidentiality	Preventing unwanted disclosure of information

In Chapter 3, we presented several mechanisms for implementing these attributes. In the sections that follow, we'll see how these mechanisms have been adopted to provide security at the IP layer.

24.3 SECURITY STRATEGY

The integration of security into IP is one of the thorniest jobs that is being tackled by the Internet Engineering Task Force (IETF). The need for authentication, data integrity, and confidentiality is immediate and widespread. The security strategy is:

- To promote interworking, start out with well-known, already implemented mechanisms for authentication, integrity, and confidentiality
- Design the security framework so that it is possible to switch to other mechanisms

The startup default mechanisms selected are:

- Message Digest 5 (MD5) for authentication and data integrity.[1]
- Symmetric encryption using the Cipher Block Chaining mode of the U.S. Data Encryption Standard (CBC-DES) for confidentiality

Public key encryption may be used for key distribution.

24.4 SCENARIOS FOR SECURITY

There are many different ways to use the security facilities that we will be describing a little later. Let's look at some scenarios in order to understand at least some of the choices.

Scenario 1.   Company XYZ wants to safeguard its internal client/server communications. They want to eliminate the possibility that someone could compromise their data by forging source IP addresses or altering data in transit.

Scenario 2.   There is an administrator at company XYZ who copies highly sensitive files between hosts. Only this administrator is allowed to perform these transfers. It also is important to prevent an eavesdropper from capturing and using these files.

Scenario 3.   Company XYZ connects its manufacturing division to its remote headquarters location via the Internet. The company wishes to make all of its communications opaque to the outside world.

---

[1]Currently, a problem arises when using MD5 with very high-speed communications because of the time required to perform the calculation.

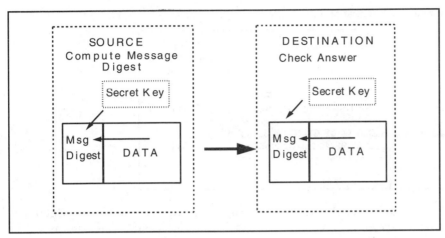

Figure 24.1 Using a message digest.

For simplicity, you can think of each client and server host as having a single interface and single IP address. However, all of the security mechanisms work when a system has multiple interfaces and multiple IP addresses.

24.4.1 Scenario 1

Message Digest technology is used to satisfy the requirements of scenario 1—that is, authenticate senders and detect whether data has been changed. Let's review how message digests work (see Figure 24.1):

- The source and destination know a secret key.
- The source performs a calculation, using the data and the secret key as input.
- The source sends the answer along with the data.
- The destination performs the same calculation and compares the answers.

24.4.2 Configuring Authentication Information for Scenario 1

Suppose company XYZ has an important server at IP address 130.15.20.2. The server's security administrator numbers the client hosts and assigns a secret authentication key to each client IP address.

The server needs to store this security information. A table such as the one shown in Table 24.1 could be used to store security parameters. The table is indexed by a number assigned to each client host—more formally, the number is called the *Security Parameters Index,* or SPI.[2]

---

[2]If the server has multiple IP addresses, the table also is indexed by its destination IP addresses.

TABLE 24.1   Security Information at Destination 130.15.20.2

SPI (for client host)	Source IP address	Client's authentication key	Client's authentication method
301	130.15.24.4	X'2E-41-43-11-5A-5A-74-53-E3-01-88-55-10-15-CD-23	MD5
302	**130.15.60.10**	**X'35-14-4F-21-2B-2C-12-34-82-22-98-44-C0-1C-33-56**	**MD5**
. . .	. . .	. . .	. . .

Of course, each client would need to be configured with the Security Parameters Index and secret key to be used when accessing this server. Table 24.2 shows configuration data at the second client. Note that the client needs separate entries for each destination that it accesses.

What happens when the client host wants to send an authenticated datagram to the server?

- The client looks up the destination IP address in its table.
- The authentication key is used to calculate a message digest for the datagram.
- The SPI number and message digest answer are put into the Authentication header.
- The datagram is sent.

When the server receives the datagram:

- The server uses the SPI in the Authentication header to look up the client entry in the table.
- The source IP address of the message is compared to the source address in the table.
- The message digest is calculated using the authentication key in the table entry.
- The answer is compared with the value in the Authentication header.

TABLE 24.2   Security Information at Source 130.15.60.10

Destination IP address	SPI	Client source IP address	Client's authentication key	Client's authentication method
130.15.20.2	**302**	**130.15.60.10**	**X'35-14-4F-21-2B-2C-12-34-82-22-98-44-C0-1C-33-56**	**MD5**
130.15.65.4	. . .	. . .	. . .	. . .

24.4.3   One-Way Security Association

Note that we really have done only half of the job. We have set up authentication in *one direction only*. Datagrams sent from the client to the server are authenticated.

The information that we have described is said to define a one-way *Security Association*. At both the source and destination, the combination of the *destination* IP address for this association and the SPI is sufficient to identify the entry to be used. Thus, a Security Association corresponds to a destination and an SPI.

In order to authenticate data flowing from the server to the client, *we need a separate set of table entries that define the authentication keys for the Security Association in the reverse direction*. That is, each host needs:

- A security table used when the host is the source of datagrams.
- A security table used when the host is the destination of datagrams.

Figure 24.2 shows a pair of Security Associations.

24.4.4   How Many Authentication Keys?

How many authentication keys need to be used by a server when it is sending datagrams to its clients? Intuitively, it might seem natural to assign a server a single MD5 authentication key, which the server would use to say to all clients "I am server so-and-so."

But then all clients would know that key. A client might use a forged IP address and masquerade as the server. To prevent this from happening, a separate authentication key could be assigned for each client host. The total number of keys could be reduced by using a same key for client-to-server and server-to-client authentication.

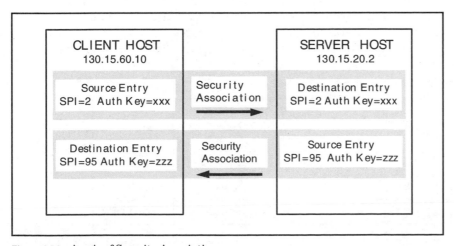

Figure 24.2   A pair of Security Associations.

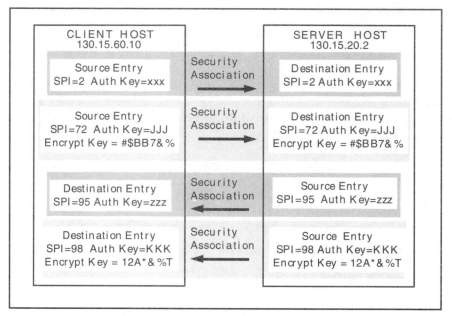

Figure 24.3 Multiple Security Associations for a client and server.

24.4.5 Scenario 2

In scenario 1, security was imposed at the host level. But suppose that there is a user or role that requires a different level of security. The security framework provides for user, role, or information sensitivity-based security.

Suppose that the client host that we discussed in scenario 1 is a multiuser system. For scenario 2, a shared host-based authentication key is sufficient for the ordinary users at the client host 130.15.60.10. However, the system administrator's file transfers to the server will need special authentication and will need to be encrypted. Figure 24.3 illustrates the Security Associations that are created.

Let's look at the Security Association tables when they are augmented with an additional entry for the administrator and with encryption keys. Table 24.3 shows destination information at the server, and Table 24.4 shows source

TABLE 24.3   Destination Security Information at 130.15.20.2

SPI	Source IP address	Client's authentication key	Client's authentication method	Client's encryption key	Client's encryption method
301	130.15.24.4	. . .	MD5	None	None
2	130.15.60.10	..xxx..	MD5	None	None
72	130.15.60.10	..JJJ..	MD5	#$BB7&%	CBC-DES
. . .	. . .	. . .	. . .	. . .	. . .

TABLE 24.4   Source Security Information at 130.15.60.10

Destination IP address	Role or userid	SPI	Source IP address	Client's authentication key	Client's authentication method	Client's encryption key	Client's encryption method
130.15.20.2	Host	2	130.15.60.10	..xxx..	MD5	None	None
130.15.20.2	Admin.	72	130.15.60.10	..JJJ..	MD5	#$BB7&%	CBC-DES
130.15.65.4	Host	...	...	...	MD5		...
...	...	...	...	...	...	...	...

information at the client. There are now separate SPIs for the ordinary users at 130.15.60.10 and for the administrator at that address.

Tables 24.3 and 24.4 include security parameters for the one-way Security Associations with the source at client 130.15.60.10 and the destination at server 130.15.20.2. A separate set of parameters would be defined in the reverse direction, with the server acting as the source and the client acting as the destination. Here again, local planners need to decide whether to use the same keys in both directions, or assign different keys for client-to-server and server-to-client traffic.

24.4.6   Scenario 3

Scenario 3 is illustrated in Figure 24.4. The goal is to make all of the traffic that company XYZ sends across an untrusted network opaque to the outside world. *Tunnel-mode* encapsulation is used. This means that datagrams are encrypted and encapsulated inside other datagrams.

As shown in the figure, when a datagram whose destination is in network 193.40.3 arrives at the boundary router for network 130.15, the router encrypts the entire datagram, including its headers. The router prepends a temporary (cleartext) IP header[3] and forwards the datagram across the Service Provider network to the boundary router for network 193.40.3. There, the temporary header is removed, the datagram is decrypted, and then the datagram is forwarded to its true destination. In this case, Security Associations are defined between the two boundary routers.

24.4.7   Generalizing

We have looked at some specific examples in order to become acquainted with the basic security framework. It is easy to see that in general, a common set of mechanisms can be used to secure traffic when it is transmitted:

- Host to host

---

[3]In addition to a main header, other headers could also be prepended. For example, a separate Authentication header could be used to authenticate the router-to-router transfer.

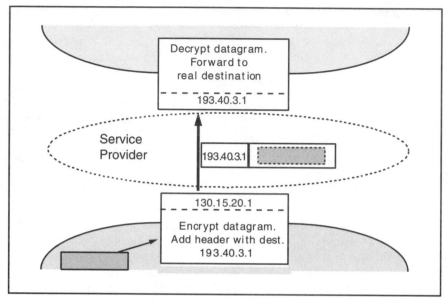

Figure 24.4   Tunneling traffic between networks.

- Router to router
- Host to router
- Router to host

If a destination host has more than one IP address, separate Security Association parameter sets could be defined for each destination address. And, there is no barrier to providing authentication, data integrity, and confidentiality for multicast destination addresses.

In scenario 2, we saw that security can be defined at a user- or role-based level. This can be made as fine grained as may be needed. Furthermore, security parameters could be configured based on the sensitivity of information (e.g., unclassified or top secret). Of course, the maintenance of many different parameter sets will depend on having a very good key distribution application.

24.5   SECURITY PROTOCOL ELEMENTS

Now we are ready to take a more formal look at how security is implemented.

24.5.1   Security Associations

As we have seen, security is handled one direction at a time. To enable a source to communicate securely with a destination, both the source and destination need to store a set of parameters, such as:

- Source address

- The authentication and integrity algorithm(s) to be used
- The confidentiality algorithm to be used
- Secret keys that will be used and any other information needed for the algorithms
- The lifetime limit for the keys
- The lifetime limit for the Security Association
- Sensitivity level (e.g., unclassified or top secret)

A *Security Association* is formally defined as the set of security parameters that supports secure one-way communication between a source and destination. From the scenarios above we can see that:

- A source host might use a single set of parameters when sending data to a destination.
- Alternatively, a host might have several Security Associations that it uses to send data to a given destination host. The association that is selected might be based on source userid, role, or sensitivity.

Recall that a numeric identifier called a Security Parameter Index is assigned to each distinct parameter set *for a given destination.*[4]

The same SPI numbers can be reused for different destinations. The parameter sets for (Destination = A, SPI = 300) and (Destination = B, SPI = 300) are very likely to be different. In other words, sets of parameters are indexed by both destination and SPI.

The IP *Authentication Header* and the IP *Encapsulating Security Payload Header* are used to implement IP version 4 and version 6 security.

24.5.2 Authentication Header

When a message digest is used for authentication, the *Authentication* Header serves a double purpose:

- It validates the sender because the sender knows the secret key used to create the computed message digest result.
- It indicates that data has not been changed in transit.

The Authentication Header has the format shown in Figure 24.5. The receiver uses the Security Parameters Index to look up the authentication protocol and authentication key. The receiver uses the authentication key to perform the MD5 calculation.

The MD5 authentication calculation is performed against all fields in the IP datagram that do not change in transit. (Fields that change, such as the

---

[4]Some standard parameter sets will be assigned Security Parameter Indices by the Internet Assigned Numbers Authority (IANA).

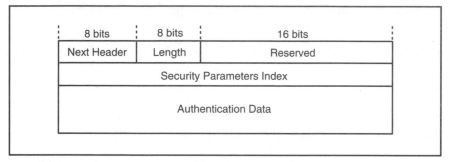

Figure 24.5   Format of the Authentication Header.

hop counter or IP version 6 routing pointer are treated as zero for the calculation.) The receiver's answer is compared to the value in the *Authentication Data* field. If they differ, the datagram is discarded.

24.5.3   Transport-Mode and Tunnel-Mode

Now let's take a look at the way that confidentiality is implemented. The format of an IP version 6 datagram whose upper-layer payload has been encrypted is shown in Figure 24.6. This format is called *Transport-mode*.

Figure 24.7 shows a *Tunnel-mode* format. An entire datagram, including all of its headers, is encrypted. A new header is prepended in order to forward the datagram. Note that tunneling from host to host could run into trouble if the path between them includes a filtering firewall router. The firewall router may wish to check information such as the source and destination IP addresses and ports, which will be hidden inside the encrypted message.

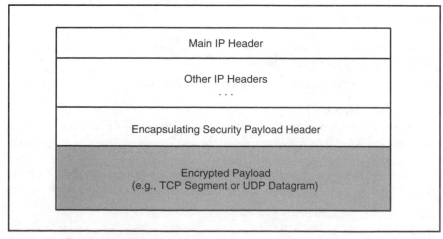

Figure 24.6   Transport-mode encryption.

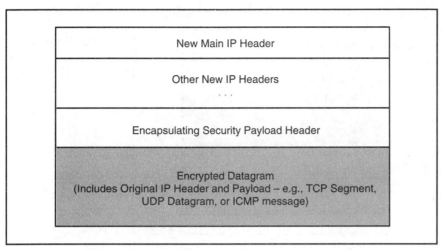

Figure 24.7   Tunnel-mode encryption.

24.5.4 Encapsulating Security Payload

The IP Encapsulating Security Payload header is used to implement both Transport-mode and Tunnel-mode encryption.

The format of the Encapsulating Security Payload header is shown in Figure 24.8. The recipient will use the Security Parameter Index to look up the algorithm and key(s) to be used. The remainder of the data depends on the algorithm choice.

When using CBC-DES, the format of the Encapsulating Security Payload header and the remainder of the message is as shown in Figure 24.9.

The *Initialization Vector* is a block of data needed to start off the CBC-DES algorithm. The shaded area is transmitted in encrypted form. Type = 4 means that the payload encapsulates a complete datagram (Tunnel-mode).

Although systems are expected to use CBC-DES initially, future Encapsulating Security Payload protocols could combine authentication and data integrity with encryption.

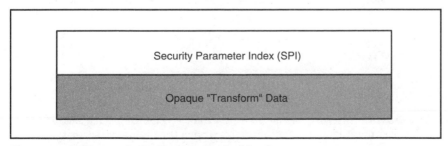

Figure 24.8   An Encapsulating Security Payload header.

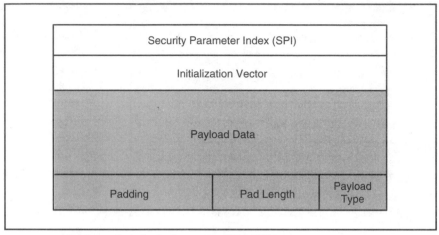

Figure 24.9   Header and payload when using CBC-DES.

24.5.5   Using Authentication with Tunnel-Mode

Two separate Authentication headers might be included when Tunnel-mode encryption is used between boundary routers. One would be within the *original* datagram header, which would be encrypted and hidden for all or part of the journey. This header would provide end-to-end authentication. The other Authentication header would be part of the cleartext IP header that is used between boundary routers. This header would provide boundary-to-boundary authentication.

24.5.6   Key Management

As we have seen, wide use of IP security will require the distribution of many secret keys to a large number of nodes. Keys need to be changed periodically, and the use of matching keys needs to be synchronized.

There is a growing literature on key management. No single standard method of key management is mandated, and much experimentation is likely.

The use of asymmetric public/private key pairs instead of symmetric CBC-DES could significantly reduce the number of keys that need to be administered.

24.6   RECOMMENDED READING

The following list of RFCs was current at the time of this writing. Check the RFC index for updates.

RFC 1825   *Security Architecture for the Internet Protocol.* The Reference section of this document lists many other publications that relate to security.

RFC 1826   *IP Authentication Header*

RFC 1828   *IP Authentication using Keyed MD5*

RFC 1321   *The MD5 Message-Digest Algorithm*

RFC 1827   *IP Encapsulating Security Payload (ESP)*

RFC 1829   *The ESP DES-CBC Transform*

Abbreviations and Acronyms

AAL	ATM Adaptation Layer
ACK	An Acknowledgment
AF	Address Family
AH	Authentication Header
ANSI	American National Standards Institute
API	Application Programming Interface
ARP	Address Resolution Protocol
ARPA	Advanced Research Projects Agency
ARPANET	Advanced Research Projects Agency Network
AS	Autonomous System
ASA	American Standards Association
ASCII	American National Standard Code for Information Interchange
ASN.1	Abstract Syntax Notation 1
ATM	Asynchronous Transfer Mode
BBN	Bolt, Beranek, and Newman, Incorporated
BCP	Best Current Practices
BECN	Backward Explicit Congestion Notification (Frame Relay)
BER	Basic Encoding Rules
BGP	Border Gateway Protocol
BIND	Berkeley Internet Name Domain
BOOTP	Bootstrap Protocol
BPDU	Bridge Protocol Data Unit
BRI	Basic Rate Interface
BSD	Berkeley Software Distribution
CBC	Cipher-Block Chaining

CCITT	International Telegraph and Telephone Consultative Committee, now ITU-T (Comite Consultatif International de Telegraphique et Telephonique)
CERT	Computer Emergency Response Team
CHAP	Challenge Handshake Authentication Protocol
CIDR	Classless Inter-Domain Routing
CLNP	Connectionless Network Protocol
CMIP	Common Management Information Protocol
CMIS	Common Management Information Services
CMOT	Common Management Information Services and Protocol over TCP/IP
CPU	Central Processing Unit
CR	Carriage Return
CRC	Cyclic Redundancy Check
CSLIP	Compressed SLIP
CSMA/CD	Carrier Sense Multiple Access with Collision Detection
CSO	Computer Services Organization
CSU	Channel Service Unit
DAP	Directory Access Protocol
DARPA	Defense Advanced Research Projects Agency
DCA	Defense Communications Agency
DCE	Data Circuit-terminating Equipment
DCE	Distributed Computing Environment
DDN	Defense Data Network
DDN NIC	Defense Data Network Network Information Center
DE	Discard Eligibility (Frame Relay)
DEC	Digital Equipment Corporation
DES	Data Encryption Standard
DEV	Deviation
DFS	Distributed File Service
DHCP	Dynamic Host Configuration Protocol
DISA	Defense Information Systems Agency
DIX	Digital, Intel, and Xerox Ethernet protocol
DLCI	Data Link Connection Identifier
DLL	Dynamic Link Library
DME	Distributed Management Environment
DMI	Desktop Management Interface
DMTF	Desktop Management Task Force
DNS	Domain Name System
DOD	Department of Defense
DOS	Disk Operating System

DSA	Directory System Agent
DSAP	Destination Service Access Point
DSU	Data Service Unit
DTE	Data Terminal Equipment
DUA	Directory User Agent
DUAL	Diffusing Update Algorithm
DXI	Data Exchange Interface
EBCDIC	Extended Binary-Coded Decimal Interchange Code
EGP	Exterior Gateway Protocol
EIGRP	Enhanced Internet Gateway Routing Protocol
EOF	End of File
EOR	End of Record
ESMTP	Extensions to SMTP
ESP	Encapsulating Security Payload
FAQ	Frequently Asked Questions
FCS	Frame Check Sequence
FDDI	Fiber Distributed Data Interface
FECN	Forward Explicit Congestion Notification (Frame Relay)
FIN	Final Segment
FIPS	Federal Information Processing Standard
FTAM	File Transfer, Access, and Management
FTP	File Transfer Protocol
FYI	For Your Information
GGP	Gateway-to-Gateway Protocol
GIF	Graphics Interchange Format, used for graphic files
GMT	Greenwich Mean Time
GOSIP	Government Open Systems Interconnection Profile
GUI	Graphical User Interface
HDLC	High Level Data Link Control Protocol
HIPPI	High Performance Parallel Interface
HTML	Hypertext Markup Language
HTTP	Hypertext Transfer Protocol
IAB	Internet Architecture Board (Internet Activities Board)
IAC	Interpret As Command
IANA	Internet Assigned Numbers Authority
IBM	International Business Machines
ICMP	Internet Control Message Protocol
ID	Identifier
IDRP	OSI Inter-Domain Routing Protocol

IEEE	Institute of Electrical and Electronics Engineers
IEN	Internet Engineering Notes
IESG	Internet Engineering Steering Group
IETF	Internet Engineering Task Force
IGMP	Internet Group Management Protocol
IGP	Interior Gateway Protocol
IGRP	Internet Gateway Routing Protocol (Cisco proprietary)
ILMI	Interim Local Management Interface
IMAP	Internet Mail Access Protocol
I/O	Input/Output
IP	Internet Protocol
IPng	IP next generation (version 6)
IPSO	IP Security Option
IPX	Internetwork Packet eXchange (for NetWare)
IRQ	Interrupt Request
IRTF	Internet Research Task Force
ISDN	Integrated Services Digital Network
IS-IS	Intermediate System to Intermediate System
ISN	Initial Sequence Number
ISO	International Organization for Standardization
ISOC	Internet Society
ISODE	ISO Development Environment
ISP	Internet Service Provider
ITU	International Telecommunications Union
ITU-T	Telecommunication Standardization Sector of the ITU
JPEG	Joint Photographic Experts Group
LAN	Local Area Network
LAPB	Link Access Procedures Balanced
LAPD	Link Access Procedures on the D-channel
LF	Line Feed
LLC	Logical Link Control
MAC	Media Access Control
MAN	Metropolitan Area Network
MD5	Message Digest 5
MIB	Management Information Base
MIME	Multipurpose Internet Mail Extensions
ms	Millisecond
MSS	Maximum Segment Size
MTA	Message Transfer Agent

MTU	Maximum Transmission Unit
MX	Mail Exchanger
NAP	Network Access Point
NCSA	National Center for Supercomputing Applications
NDIS	Network Device Interface Specification
NETBIOS	Network Basic Input Output System
NFS	Network File System
NIC	Network Information Center
NIS	Network Information System
NISI	Network Information Service Infrastructure
NIST	National Institute of Standards and Technology
NLPID	Network Level Protocol ID
NNTP	Network News Transfer Protocol
NOC	Network Operations Center
NREN	National Research and Education Network
NS	Name Server
NSAP	Network Service Access Point
NSF	National Science Foundation
NTP	Network Time Protocol
NVT	Network Virtual Terminal
ODI	Open Device Interface
ONC	Open Network Computing
OSF	Open Software Foundation
OSI	Open Systems Interconnect
OSPF	Open Shortest Path First
OUI	Organizationally Unique Identifier
PAD	Packet Assembler/Disassembler
PAP	Password Authentication Protocol
PC	Personal Computer
PDU	Protocol Data Unit
PGP	Pretty Good Privacy
PI	Protocol Interpreter
PING	Packet Internet Groper
POP	Point Of Presence
POP	Post Office Protocol
PPP	Point-to-Point Protocol
PTT	Postal Telegraph and Telephone
QoS	Quality of Service
RA	Routing Arbiter

RARP	Reverse Address Resolution Protocol
RFC	Request For Comments
RIF	Routing Information Field
RIP	Routing Information Protocol
RIPE	Reseaux IP Europeens
RMON	Remote Network Monitor
ROM	Read Only Memory
RPC	Remote Procedure Call
RR	Resource Record
RR	Routing Registry
RST	Reset
RTO	Retransmission Timeout (for TCP)
RTT	Round-Trip Time
SDEV	Smoothed Deviation
SDLC	Synchronous Data Link Control
SEI	Software Engineering Institute
SGML	Standard Generalized Markup Language
SIP	SMDS Interface Protocol
SLIP	Serial Line Interface Protocol
SMDS	Switched Multimegabit Data Service
SMI	Structure of Management Information
SMTP	Simple Mail Transfer Protocol
SNA	Systems Network Architecture
SNAP	Sub-Network Access Protocol
SNMP	Simple Network Management Protocol
SOA	Start of Authority
SONET	Synchronous Optical Network
SPF	Shortest Path First
SPI	Security Parameters Index
SPX	Sequenced Packet Exchange (for Netware)
SRTT	Smoothed Round-Trip Time
SSAP	Source Service Access Point
SSL	Secure Sockets Layer
SWS	Silly Window Syndrome
SYN	Synchronizing Segment
TCB	Transmission Control Block
TCP	Transmission Control Protocol
TCU	Trunk Coupling Unit
TELNET	Terminal Networking

TFTP	Trivial File Transfer Protocol
TLI	Transport Layer Interface
TOS	Type of Service
TP4	OSI Transport Class 4
TSAP	Transport Service Access Point
TTL	Time-To-Live
UA	User Agent
UDP	User Datagram Protocol
ULP	Upper Layer Protocol
URI	Universal Resource Identifier
URL	Uniform Resource Locator
URN	Uniform Resource Name
UTC	Universal Time Coordinated
VCC	Virtual Channel Connection
VLSM	Variable Length Subnet Masks
VPC	Virtual Path Connection
W3	World Wide Web
WAIS	Wide Area Information Service
WAN	Wide Area Network
WWW	World Wide Web
WYSIWYG	What You See is What You Get
XDR	eXternal Data Representation
XNS	Xerox Network Systems

RFCs and Other
TCP/IP Documents

B.1 LOCATION OF RFC DOCUMENTS

At the time of this writing, Request For Comments (RFC) documents could be obtained from the InterNIC Directory and Database Services (operated by AT&T), which is accessed from:

http://www.internic.net/

by choosing *DIRECTORY AND DATABASE SERVICES* and following the pointers to RFC documents. RFC documents also can be accessed at:

ftp://ftp.internic.net/

in directory */rfc*.

Consult the InterNIC document */rfc/rfc-index.txt* for a complete and up-to-date list of RFC documents.

B.2 ASSIGNED NUMBERS

The Internet Assigned Numbers Authority (IANA) coordinates the assignment of unique parameter values for Internet protocols. The IANA is chartered by the Internet Society (ISOC) and the Federal Network Council.

Periodically, the IANA publishes an RFC titled *Assigned Numbers* which reports the currently assigned parameters and their values. New parameter assignments are made available immediately at the public file transfer archive:

ftp:// ftp.isi.edu/in-notes/iana/assignments

B.3 REGISTRATION FORMS

Internet name and address registration forms can be obtained from the InterNIC Registration Services, which can be accessed from:

http://www.internic.net/

by choosing *REGISTRATION SERVICES* and then selecting *Templates.*

B.4 DOMAIN NAME SYSTEM

The Registration Service provides Domain Name System (DNS) information in their file transfer archive, which can be reached by choosing *FTP Archive* from their home page and then selecting the *domain* directory, or via:

ftp://rs.internic.net/domain/

B.5 RFC STANDARDS

Periodically, the official list of standards and their current status is published in an RFC document. This list is itself a standard (STD 1). The information in Tables B.1 to B.5 was taken from RFC 1920 published in March of 1996, and reflects the status and state of the Internet Standards at the time of this writing. Check the file *rfc-index.txt* for the most current official Standards document.

The protocol *state* indicates its progress through the review process. After

TABLE B.1   Standard Protocols

Protocol	Name	Status*	RFC	STD
	Internet Official Protocol Standards	Req.	1920	1
	Assigned Numbers	Req.	1700	2
	Host Requirements—Communications	Req.	1122	3
	Host Requirements—Applications	Req.	1123	3
IP	Internet Protocol	Req.	791	5
	Amended by: IP Subnet Extension	Req.	950	5
	IP Broadcast Datagrams	Req.	919	5
	IP Broadcast Datagrams with Subnets	Req.	922	5
ICMP	Internet Control Message Protocol	Req.	792	5
IGMP	Internet Group Multicast Protocol	Rec.	1112	5
UDP	User Datagram Protocol	Rec.	768	6
TCP	Transmission Control Protocol	Rec.	793	7
TELNET	Telnet Protocol	Rec.	854 855	8
FTP	File Transfer Protocol	Rec.	959	9

TABLE B.1   Standard Protocols

Protocol	Name	Status*	RFC	STD
SMTP	Simple Mail Transfer Protocol	Rec.	821	10
SMTP-SIZE	SMTP Service Ext for Message Size	Rec.	1870	10
SMTP-EXT	SMTP Service Extensions	Rec.	1869	10
MAIL	Format of Electronic Mail Messages	Rec.	822	11
CONTENT	Content Type Header Field	Rec.	1049	11
NTPv2	Network Time Protocol (Version 2)	Rec.	1119	12
DOMAIN	Domain Name System	Rec.	1034 1035	13
DNS-MX	Mail Routing and the Domain System	Rec.	974	14
SNMP	Simple Network Management Protocol	Rec.	1157	15
SMI	Structure of Management Information	Rec.	1155	16
Concise-MIB	Concise MIB Definitions	Rec.	1212	16
MIB-II	Management Information Base-II	Rec.	1213	17
NETBIOS	NetBIOS Service Protocols	Ele.	1001 1002	19
ECHO	Echo Protocol	Rec.	862	20
DISCARD	Discard Protocol	Ele.	863	21
CHARGEN	Character Generator Protocol	Ele.	864	22
QUOTE	Quote of the Day Protocol	Ele.	865	23
USERS	Active Users Protocol	Ele.	866	24
DAYTIME	Daytime Protocol	Ele.	867	25
TIME	Time Server Protocol	Ele.	868	26
TFTP	Trivial File Transfer Protocol	Ele.	1350	33
TP-TCP	ISO Transport Service on Top of the TCP	Ele.	1006	35
ETHER-MIB	Ethernet MIB	Ele.	1643	50
PPP	Point-to-Point Protocol	Ele.	1661	51
PPP-HDLC	PPP in HDLC Framing	Ele.	1662	51
IP-SMDS	IP Datagrams over the SMDS Service	Ele.	1209	52

*Ele. = elective; Rec. = recommended; Req. = required.

TABLE B.2   Network-Specific Standard Protocols

Protocol	Name	State*	RFC	STD
IP-ATM	Classical IP and ARP over ATM	Prop.	1577	
IP-FR	Multiprotocol over Frame Relay	Draft	1490	
ATM-ENCAP	Multiprotocol Encapsulation over ATM	Prop.	1483	
IP-TR-MC	IP Multicast over Token-Ring LANs	Prop.	1469	
IP-FDDI	Transmission of IP and ARP over FDDI Net	Std.	1390	36
IP-HIPPI	IP and ARP on HIPPI	Prop.	1374	
IP-X.25	X.25 and ISDN in the Packet Mode	Draft	1356	
IP-FDDI	Internet Protocol on FDDI Networks	Draft	1188	
ARP	Address Resolution Protocol	Std.	826	37
RARP	A Reverse Address Resolution Protocol	Std.	903	38
IP-ARPA	Internet Protocol on ARPANET	Std.	BBN 1822	39
IP-WB	Internet Protocol on Wideband Network	Std.	907	40
IP-E	Internet Protocol on Ethernet Networks	Std.	894	41
IP-EE	Internet Protocol on Exp. Ethernet Nets	Std.	895	42
IP-IEEE	Internet Protocol on IEEE 802	Std.	1042	43
IP-DC	Internet Protocol on DC Networks	Std.	891	44
IP-HC	Internet Protocol on Hyperchannel	Std.	1044	45
IP-ARC	Transmitting IP Traffic over ARCNET Nets	Std.	1201	46
IP-SLIP	Transmission of IP over Serial Lines	Std.	1055	47
IP-NETBIOS	Transmission of IP over NETBIOS	Std.	1088	48
IP-IPX	Transmission of 802.2 over IPX Networks	Std.	1132	49

*Prop. = proposed; Std. = Standard.

initial screening, a standard is given *proposed* state. After further examination, improvement, and review, it may be advanced to *draft* state. The protocol becomes a *standard* after test, usage, and final review.

The protocol status indicates whether it is *required, recommended, elective, limited use,* or *not recommended.*

TABLE B.3   Telnet Options*

Protocol	Name	Option	State	Status	RFC	STD
TOPT-BIN	Binary Transmission	0	Std.	Rec.	856	27
TOPT-ECHO	Echo	1	Std.	Rec.	857	28
TOPT-RECN	Reconnection	2	Prop.	Ele.		
TOPT-SUPP	Suppress Go Ahead	3	Std.	Rec.	858	29
TOPT-APRX	Approx Message Size Negotiation	4	Prop.	Ele.		
TOPT-STAT	Status	5	Std.	Rec.	859	30
TOPT-TIM	Timing Mark	6	Std.	Rec.	860	31
TOPT-REM	Remote Controlled Trans and Echo	7	Prop.	Ele.	726	
TOPT-OLW	Output Line Width	8	Prop.	Ele.		
TOPT-OPS	Output Page Size	9	Prop.	Ele.		
TOPT-OCRD	Output Carriage-Return Disposition	10	Prop.	Ele.	652	
TOPT-OHT	Output Horizontal Tabstops	11	Prop.	Ele.	653	
TOPT-OHTD	Output Horizontal Tab Disposition	12	Prop.	Ele.	654	
TOPT-OFD	Output Formfeed Disposition	13	Prop.	Ele.	655	
TOPT-OVT	Output Vertical Tabstops	14	Prop.	Ele.	656	
TOPT-OVTD	Output Vertical Tab Disposition	15	Prop.	Ele.	657	
TOPT-OLD	Output Linefeed Disposition	16	Prop.	Ele.	658	
TOPT-EXT	Extended ASCII	17	Prop.	Ele.	698	
TOPT-LOGO	Logout	18	Prop.	Ele.	727	
TOPT-BYTE	Byte Macro	19	Prop.	Ele.	735	
TOPT-DATA	Data Entry Terminal	20	Prop.	Ele.	1043	
TOPT-SUP	SUPDUP	21	Prop.	Ele.	736	
TOPT-SUPO	SUPDUP Output	22	Prop.	Ele.	749	
TOPT-SNDL	Send Location	23	Prop.	Ele.	779	
TOPT-TERM	Terminal Type	24	Prop.	Ele.	1091	
TOPT-EOR	End of Record	25	Prop.	Ele.	885	

*Ele. = elective; Exp. = experimental; Hist = historic; Prop. = proposed; Rec. = recommended; Std. = standard.

TABLE B.3   Telnet Options*

Protocol	Name	Option	State	Status	RFC	STD
TOPT-TACACS	TACACS User Identification	26	Prop.	Ele.	927	
TOPT-OM	Output Marking	27	Prop.	Ele.	933	
TOPT-TLN	Terminal Location Number	28	Prop.	Ele.	946	
TOPT-3270	Telnet 3270 Regime	29	Prop.	Ele.	1041	
TOPT-X.3	X.3 PAD	30	Prop.	Ele.	1053	
TOPT-NAWS	Negotiate About Window Size	31	Prop.	Ele.	1073	
TOPT-TS	Terminal Speed	32	Prop.	Ele.	1079	
TOPT-RFC	Remote Flow Control	33	Prop.	Ele.	1372	
TOPT-LINE	Linemode	34	Draft	Ele.	1184	
TOPT-XDL	X Display Location	35	Prop.	Ele.	1096	
TOPT-ENVIR	Telnet Environment Option	36	Hist.	None	1408	
TOPT-AUTH	Telnet Authentication Option	37	Exp.	Ele.	1416	
TOPT-ENVIR	Telnet Environment Option	39	Prop.	Ele.	1572	
TOPT-EXTOP	Extended-Options-List	255	Std.	Rec.	861	32

*Ele. = elective; Exp. = experimental; Hist = historic; Prop. = proposed; Rec. = recommended; Std. = standard.

TABLE B.4   Draft Standards

Protocol	Name	Status*	RFC
COEX-MIB	Coexistence between SNMPv1 & SNMPv2	Ele.	1908
SNMPv2-MIB	MIB for SNMPv2	Ele.	1907
TRANS-MIB	Transport Mappings for SNMPv2	Ele.	1906
OPS-MIB	Protocol Operations for SNMPv2	Ele.	1905
CONF-MIB	Conformance Statements for SNMPv2	Ele.	1904
CONV-MIB	Textual Conventions for SNMPv2	Ele.	1903
SMIV2	SMI for SNMPv2	Ele.	1902
CON-MD5	Content-MD5 Header Field	Ele.	1864
OSPF-MIB	OSPF Version 2 MIB	Ele.	1850
STR-REP	String Representation of Distinguished Names	Ele.	1779
X.500syn	X.500 String Representation of Standard Attribute Syntaxes	Ele.	1778
X.500lite	X.500 Lightweight Directory Access Protocol	Ele.	1777
BGP-4-APP	Application of BGP-4	Ele.	1772
BGP-4	Border Gateway Protocol 4	Ele.	1771
PPP-DNCP	PPP DECnet Phase IV Control Protocol	Ele.	1762
RMON-MIB	Remote Network Monitoring MIB	Ele.	1757
802.5-MIB	IEEE 802.5 Token Ring MIB	Ele.	1748
BGP-4-MIB	BGP-4 MIB	Ele.	1657
POP3	Post Office Protocol, Version 3	Ele.	1725
RIP2-MIB	RIP Version 2 MIB Extension	Ele.	1724
RIP2	RIPVersion 2-Carrying Additional Info.	Ele.	1723
RIP2-APP	RIP Version 2 Protocol App. Statement	Ele.	1722
SIP-MIB	SIP Interface Type MIB	Ele.	1694
	Def Man Objs Parallel-printer-like	Ele.	1660
	Def Man Objs RS-232-like	Ele.	1659
	Def Man Objs Character Stream	Ele.	1658

Ele. = elective; Rec. = recommended.

TABLE B.4   Draft Standards (Continued)

Protocol	Name	Status*	RFC
SMTP-8BIT	SMTP Service Ext or 8bit-MIMEtransport	Ele.	1652
OSI-NSAP	Guidelines for OSI NSAP Allocation	Ele.	1629
OSPF2	Open Shortest Path First Routing V2	Ele.	1583
ISO-TS-ECHO	Echo for ISO-8473	Ele.	1575
DECNET-MIB	DECNET MIB	Ele.	1559
	Message Header Ext. of Non-ASCII Text	Ele.	1522
MIME	Multipurpose Internet Mail Extensions	Ele.	1521
802.3-MIB	IEEE 802.3 Repeater MIB	Ele.	1516
BRIDGE-MIB	BRIDGE-MIB	Ele.	1493
NTPv3	Network Time Protocol (Version 3)	Ele.	1305
IP-MTU	Path MTU Discovery	Ele.	1191
FINGER	Finger Protocol	Ele.	1288
BOOTP	Bootstrap Protocol	Rec.	951, 1497
NICNAME	WhoIs Protocol	Ele.	954

Ele. = elective; Rec. = recommended.

TABLE B.5   Proposed Standards

Protocol	Name	Status*	RFC
WHOIS++M	How to Interact with a Whois++ Mesh	Ele.	1914
WHOIS++A	Architecture of Whois++ Index Service	Ele.	1913
DSN	Delivery Status Notifications	Ele.	1894
EMS-CODE	Enhanced Mail System Status Codes	Ele.	1893
MIME-RPT	Multipart/Report	Ele.	1892
SMTP-DSN	SMTP Delivery Status Notifications	Ele.	1891
RTP-AV	RTP Audio/Video Profile	Ele.	1890
RTP	Transport Protocol for Real-Time Apps	Ele.	1889
DNS-IPv6	DNS Extensions to Support IPv6	Ele.	1886

Ele. = elective.

TABLE B.5 Proposed Standards (Continued)

Protocol	Name	Status*	RFC
ICMPv6	ICMPv6 for IPv6	Ele.	1885
IPv6-Addr	IPv6 Addressing Architecture	Ele.	1884
IPv6	IPv6 Specification	Ele.	1883
HTML	Hypertext Markup Language - 2.0	Ele.	1866
SMTP-Pipe	SMTP Serv. Ext. for Command Pipelining	Ele.	1854
MIME-Sec	MIME Object Security Services	Ele.	1848
MIME-Encyp	MIME: Signed and Encrypted	Ele.	1847
WHOIS++	Architecture of the WHOIS++ service	Ele.	1835
	Binding Protocols for ONC RPC Version 2	Ele.	1833
XDR	External Data Representation Standard	Ele.	1832
RPC	Remote Procedure Call Protocol V. 2	Ele.	1831
	ESP DES-CBC Transform	Ele./Req.	1829
	IP Authentication Using Keyed MD5	Ele./Req.	1828
ESP IP	Encapsulating Security Payload	Ele./Req.	1827
IPv6-AH	IP Authentication Header	Ele./Req.	1826
	Security Architecture for IP	Ele./Req.	1825
RREQ	Requirements for IP Version 4 Routers	Ele.	1812
URL	Relative Uniform Resource Locators	Ele.	1808
CLDAP	Connectionless LDAP	Ele.	1798
OSPF-DC	Ext. OSPF to Support Demand Circuits	Ele.	1793
TMUX	Transport Multiplexing Protocol	Ele.	1692
TFTP-Opt	TFTP Options	Ele.	1784
TFTP-Blk	TFTP Blocksize Option	Ele.	1783
TFTP-Ext	TFTP Option Extension	Ele.	1782
OSI-Dir	OSI User Friendly Naming ...	Ele.	1781
MIME-EDI	MIME Encapsulation of EDI Objects	Ele.	1767
Lang-Tag	Tags for Identification of Languages	Ele.	1766

Ele. = elective. Req. = required.

TABLE B.5   Proposed Standards (Continued)

Protocol	Name	Status*	RFC
XNSCP	PPP XNS IDP Control Protocol	Ele.	1764
BVCP	PPP Banyan Vines Control Protocol	Ele.	1763
Print-MIB	Printer MIB	Ele.	1759
ATM-SIG	ATM Signaling Support for IP over ATM	Ele.	1755
IPNG	Recommendation for IP Next Generation	Ele.	1752
802.5-SSR	802.5 SSR MIB using SMIv2	Ele.	1749
SDLCSMIv2	SNADLC SDLC MIB using SMIv2	Ele.	1747
BGP4/IDRP	BGP4/IDRP for IP/OSPF Interaction	Ele.	1745
AT-MIB	Appletalk MIB	Ele.	1742
MacMIME	MIME Encapsulation of Macintosh files	Ele.	1740
URL	Uniform Resource Locators	Ele.	1738
POP3-AUTH	POP3 AUTHentication Command	Ele.	1734
IMAP4-AUTH	IMAP4 Authentication Mechanisms	Ele.	1731
IMAP4	Internet Message Access Protocol V4	Ele.	1730
PPP-MP	PPP Multilink Protocol	Ele.	1717
RDBMS-MIB	RDMS MIB - using SMIv2	Ele.	1697
MODEM-MIB	Modem MIB - using SMIv2	Ele.	1696
ATM-MIB	ATM Management Version 8.0 using SMIv2	Ele.	1695
SNANAU-MIB	SNA NAUs MIB using SMIv2	Ele.	1665
PPP-TRANS	PPP Reliable Transmission	Ele.	1663
BGP-4-IMP	BGP-4 Roadmap and Implementation	Ele.	1656
	Postmaster Convention X.400 Operations	Ele.	1648
TN3270-En	TN3270 Enhancements	Ele.	1647
PPP-BCP	PPP Bridging Control Protocol	Ele.	1638
UPS-MIB	UPS Management Information Base	Ele.	1628
AAL5-MTU	Default IP MTU for use over ATM AAL5	Ele.	1626
PPP-SONET	PPP over SONET/SDH	Ele.	1619

Ele. = elective.

TABLE B.5 Proposed Standards (Continued)

Protocol	Name	Status*	RFC
PPP-ISDN	PPP over ISDN	Ele.	1618
DNS-R-MIB	DNS Resolver MIB Extensions	Ele.	1612
DNS-S-MIB	DNS Server MIB Extensions	Ele.	1611
FR-MIB	Frame Relay Service MIB	Ele.	1604
PPP-X25	PPP in X.25	Ele.	1598
OSPF-NSSA	The OSPF NSSA Option	Ele.	1587
OSPF-Multi	Multicast Extensions to OSPF	Ele.	1584
SONET-MIB	MIB SONET/SDH Interface Type	Ele.	1595
RIP-DC	Extensions to RIP to Support Demand Cir.	Ele.	1582
	Evolution of the Interfaces Group of MIB-II	Ele.	1573
PPP-LCP	PPP LCP Extensions	Ele.	1570
X500-MIB	X.500 Directory Monitoring MIB	Ele.	1567
MAIL-MIB	Mail Monitoring MIB	Ele.	1566
NSM-MIB	Network Services Monitoring MIB	Ele.	1565
CIPX	Compressing IPX Headers Over WAN Media	Ele.	1553
IPXCP	PPP Internetworking Packet Exchange Control	Ele.	1552
DHCP-BOOTP	Interoperation Between DHCP and BOOTP	Ele.	1534
DHCP-BOOTP	DHCP Options and BOOTP Vendor Extensions	Ele.	1533
BOOTP	Clarifications and Extensions BOOTP	Ele.	1532
DHCP	Dynamic Host Configuration Protocol	Ele.	1541
SRB-MIB	Source Routing Bridge MIB	Ele.	1525
CIDR-STRA	CIDR Address Assignment...	Ele.	1519
CIDR-ARCH	CIDR Architecture...	Ele.	1518
CIDR-APP	CIDR Applicability Statement	Ele.	1517
	802.3 MAU MIB	Ele.	1515
HOST-MIB	Host Resources MIB	Ele.	1514
	Token Ring Extensions to RMON MIB	Ele.	1513

Ele. = elective.

TABLE B.5 Proposed Standards (Continued)

Protocol	Name	Status*	RFC
FDDI-MIB	FDDI Management Information Base	Ele.	1512
KERBEROS	Kerberos Network Authentication Ser (V5)	Ele.	1510
GSSAPI	Generic Security Service API: C-bindings	Ele.	1509
GSSAPI	Generic Security Service Application...	Ele.	1508
DASS	Distributed Authentication Security...	Ele.	1507
	X.400 Use of Extended Character Sets	Ele.	1502
HARPOON	Rules for Downgrading Messages...	Ele.	1496
Mapping	MHS/RFC-822 Message Body Mapping	Ele.	1495
Equiv	X.400/MIME Body Equivalences	Ele.	1494
IDPR	Inter-Domain Policy Routing Protocol	Ele.	1479
IDPR-ARCH	Architecture for IDPR	Ele.	1478
PPP/Bridge	MIB Bridge PPP MIB	Ele.	1474
PPP/IP MIB	IP Network Control Protocol of PPP MIB	Ele.	1473
PPP/SEC MIB	Security Protocols of PPP MIB	Ele.	1472
PPP/LCP MIB	Link Control Protocol of PPP MIB	Ele.	1471
X25-MIB	Multiprotocol Interconnect on X.25 MIB	Ele.	1461
PEM-KEY	PEM - Key Certification	Ele.	1424
PEM-ALG	PEM - Algorithms, Modes, and Identifiers	Ele.	1423
PEM-CKM	PEM - Certificate-Based Key Management	Ele.	1422
PEM-ENC	PEM - Message Encryption and Auth	Ele.	1421
SNMP-IPX	SNMP over IPX	Ele.	1420
SNMP-AT	SNMP over AppleTalk	Ele.	1419
SNMP-OSI	SNMP over OSI	Ele.	1418
FTP-FTAM	FTP-FTAM Gateway Specification	Ele.	1415
IDENT-MIB	Identification MIB	Ele.	1414
IDENT	Identification Protocol	Ele.	1413
DS3/E3-MIB	DS3/E3 Interface Type	Ele.	1407

Ele. = elective.

TABLE B.5   Proposed Standards (Continued)

Protocol	Name	Status*	RFC
DS1/E1-MIB	DS1/E1 Interface Type	Ele.	1406
BGP-OSPF	BGP OSPF Interaction	Ele.	1403
	Route Advertisement In BGP2 And BGP3	Ele.	1397
SNMP-X.25	SNMP MIB Extension for X.25 Packet Layer	Ele.	1382
SNMP-LAPB	SNMP MIB Extension for X.25 LAPB	Ele.	1381
PPP-ATCP	PPP AppleTalk Control Protocol	Ele.	1378
PPP-OSINLCP	PPP OSI Network Layer Control Protocol	Ele.	1377
TABLE-MIB	IP Forwarding Table MIB	Ele.	1354
TOS	Type of Service in the Internet	Ele.	1349
PPP-AUTH	PPP Authentication	Ele.	1334
PPP-LINK	PPP Link Quality Monitoring	Ele.	1333
PPP-IPCP	PPP Control Protocol	Ele.	1332
	X.400 1988 to 1984 downgrading	Ele.	1328
	Mapping between X.400(1988)	Ele.	1327
TCP-EXT	TCP Extensions for High Performance	Ele.	1323
FRAME-MIB	Management Information Base for Frame	Ele.	1315
NETFAX	File Format for the Exchange of Images	Ele.	1314
IARP	Inverse Address Resolution Protocol	Ele.	1293
FDDI-MIB	FDDI-MIB	Ele.	1285
	Encoding Network Addresses	Ele.	1277
	Replication and Distributed Operations	Ele.	1276
	COSINE and Internet X.500 Schema	Ele.	1274
BGP-MIB	Border Gateway Protocol MIB (Version 3)	Ele.	1269
ICMP-ROUT	ICMP Router Discovery Messages	Ele.	1256
IPSO	DoD Security Options for IP	Ele.	1108
OSI-UDP	OSI TS on UDP	Ele.	1240
STD-MIBs	Reassignment of Exp MIBs to Std MIBs	Ele.	1239

Ele. = elective.

TABLE B.5   Proposed Standards (Continued)

Protocol	Name	Status•	RFC
IPX-IP	Tunneling IPX Traffic through IP Nets	Ele.	1234
GINT-MIB	Extensions to the Generic-Interface MIB	Ele.	1229
IS-IS	OSI IS-IS for TCP/IP Dual Environments	Ele.	1195
IP-CMPRS	Compressing TCP/IP Headers	Ele.	1144
NNTP	Network News Transfer Protocol	Ele.	977

Ele. = elective.

Network Information Centers and Other Services

C.1 REGISTRATION

Before an organization can connect its network to the Internet, it needs to obtain one or more blocks of IP addresses—either from its Service Provider or directly from an Internet registration service. The organization must register its Domain Name and identify its Domain Name Servers (DNSs). The organization might need an *Autonomous System number,* which is a unique integer assigned to the network.

C.1.1 Primary Registration NIC

The primary Internet Registration Service currently is funded by the National Science Foundation (NSF). It provides a worldwide coordination function and delegates registration services for North and South America. The service is furnished by:

Network Solutions
Attn: InterNIC Registration Services
505 Huntmar Park Drive
Herndon, Virginia 22070

Via electronic mail: hostmaster@internic.net

Registration of hosts and domains, and updates of registration information, can be performed via electronic mail. As noted in Appendix B, registration forms are available via World Wide Web (WWW) by following selections from:

http://www.internic.net/

or via file transfer from:

ftp://ftp.internic.net/templates

C.1.2   European NIC

The primary European NIC is the:

RIPE Network Coordination Centre (RIPE NCC)
(Registry for the European Region)

Via electronic mail: hostmaster@ripe.net, ncc@ripe.net

Via telephone: + 31 20 592 5065

Via fax: + 31 20 592 5090

Via postal mail: RIPE NCC
Kruislaan 409
1098 SJ Amsterdam
The Netherlands

The RIPE Network Coordination Centre can be accessed at:

http://www.ripe.net/

C.1.3   Asia-Pacific NIC

The NIC for the Asia-Pacific region is:

Asia Pacific Network Information Center
c/o Internet Initiative Japan, Inc.
Sanbancho Annex Bldg.
1-4 Sanbancho, Chiyoda-ku
Tokyo 102, Japan
Via electronic mail: ip-request@rs.apnic.net
Via telephone: + 81-3-5276-3973
Via fax: + 81-3-5276-6239

The Asia Pacific Network Information Center can be accessed at:

http://www.apnic.net/

ftp://archive.apnic.net/apnic/docs/

These three major Network Information Centers (NICs) delegate address registration to national and Service Provider NICs within their regions of the world.

C.2   FINDING OTHER NICS

The AT&T Data services NIC publishes a list of NICs at:

http://ds.internic.net/pub/niclocator/

The list is maintained by the Network Information Services Infrastructure (NISI) working group of the Internet Engineering Task Force (IETF).

C.3 FINDING ADMINISTRATORS VIA WHOIS

An organization's registration information includes the names of its administrative and technical points of contact, and information on how to reach them.

This information is made available on-line in a database that can be accessed via the *whois* application. Below, we ask about the domain *yale.edu*. The first response gives us Yale's "handle," YALE-DOM, which is used to get more information about the domain.

```
> whois -h rs.internic.net yale.edu
Yale University (YALE-DOM)                                         YALE.EDU
Yale University (YALE)         YALE.EDU         128.36.0.1, 130.132.1.1

The InterNIC Registration Services Host contains ONLY Internet Information
(Networks, ASN's, Domains, and POC's).
Please use the whois server at nic.ddn.mil for MILNET Information.

> whois -h rs.internic.net yale-dom
Yale University (YALE-DOM)
 Yale University Computing & Information Systems
 Mail Stop 2112
 New Haven, CT 06520

 Domain Name: YALE.EDU

 Administrative Contact:
   Paolillo, Joseph (JP218) joseph_paolillo@yale.edu
   (203) 432 6673
 Technical Contact, Zone Contact:
   Long, Morrow H. (HML1) LONG-MORROW@CS.YALE.EDU
   (203) 432-1254

 Record last updated on 15-Dec-93.
 Record created on 17-Mar-87.

 Domain servers in listed order:

 SERV1.NET.YALE.EDU             130.132.1.9
 SERV2.NET.YALE.EDU             130.132.1.10
 SERV3.NET.YALE.EDU             130.132.1.11
 YALE.EDU                       128.36.0.1, 130.132.1.1
 NIC.NEAR.NET                   192.52.71.4

The InterNIC Registration Services Host contains ONLY Internet Information
(Networks, ASN's, Domains, and POC's).
Please use the whois server at nic.ddn.mil for MILNET Information.
```

C.4 IPv6 REGISTRY IDENTIFIERS

The Internet Assigned Numbers Authority (IANA) coordinates the use of IPv6 addresses. Current registry identifiers for Provider-Based IPv6 addresses are:

Regional registry	Registry ID
Multiregional (IANA)	10000
RIPE NCC	01000
INTERNIC	11000
APNIC	10100

C.6   CERT SECURITY FUNCTIONS

The CERT Coordination Center was established in 1988 and is located at the Software Engineering Institute (SEI), Carnegie Mellon University, Pittsburgh, PA. CERT is an acronym for Computer Emergency Response Team.

CERT publishes notices of security problems found in operating systems or software packages and provides pointers to the solutions. CERT coordinates responses to security attacks upon the Internet. CERT information is available at:

http://www.sei.cmu.edu/technology/cert.cc.html

ftp://cert.org/

CERT can be reached at:

CERT Coordination Center
Software Engineering Institute
Carnegie Mellon University
Pittsburgh, Pennsylvania 15213-3890

Via electronic mail: cert@cert.org

Via telephone: + 1-412-268-7090 (24-hour hotline)

Via fax: + 1-412-268-6989

CERT advisories are published in the newsgroup:

comp.security.announce

Advisories are sent to a mailing list that can be joined by sending mail to:

cert-advisory-request@cert.org

Variable-Length Subnet Masks

D.1 INTRODUCTION

The Internet address format has caused many problems for network administrators. The 32-bit address space is just too small and cramped.

Computers work with these addresses using bit boundaries. A computer is happy to accept configurations such as a 16-bit network number, 7-bit subnet number, and 9-bit host number. Humans are not very comfortable with batches of bits.

To add to the confusion, we *write* addresses by translating *bytes* to decimal numbers—for example, 130.15.1.2. When our subnet boundaries do not fall on byte boundaries, we need to engage in some mental arithmetic to extract the subnet and host parts of an address.

It is the purpose of this appendix to make it easier to work with subnet masks that are not aligned on a byte boundary. We will look at several examples in which we define subnets of the Class B network 130.15. See Table 5.2 in Chapter 5 for a complete list of Class B subnet masks. *Although we have not included all-zero subnets in the examples, keep in mind that many sites use them successfully.*

D.1.1 Seven-Bit Subnets

When the subnet part of your address has fewer than 8 bits, you are opting for fewer subnets with more hosts on them. For example:

Subnet bits	Subnets	Host bits	Hosts
7	128	9	510
6	64	10	1022

With a 7-bit subnet, the first host on your first subnet has binary and dot addresses:

```
10000010 00001111 00000010 0000001
 130  .  15  .  2  .  1
```

Note that the host part of the address is in boldface type. The last host on your first subnet has binary and dot addresses:

```
10000010 00001111 00000011 11111110
 130  .  15  .  3  .  254
```

That is, when translated to dot notation, your first subnet includes addresses:

```
130.15.2.1 to 130.15.2.255
130.15.3.0 to 130.15.3.254
```

All addresses starting with 130.15.2 and 130.15.3 are on the same subnet. Note that host address 130.15.2.255 is legal. This address ends with an all-1s *byte* but not with an all-1s *host field* because of the 0 in the previous byte. Similarly, 130.15.3.0 is legal because we end with an all-0s byte but not with an all-0s host field.

The second subnet will include addresses:

```
130.15.4.1 to 130.15.4.255
130.15.5.0 to 130.15.5.254
```

Now we can see the pattern. Each subnet will be introduced by an adjacent pair of even and odd numbers. New subnets start at each even number.

D.1.2 Six-Bit Subnets

Next, let's look at 6-bit subnets. With a 6-bit subnet, the first host on your first subnet has binary and dot addresses:

```
10000010 00001111 00000100 00000001
 130  .  15  .  4  .  1
```

The last host on your first subnet has binary and dot addresses:

```
10000010 00001111 00000111 11111110
 130  .  15  .  7  .  254
```

That is, when translated to dot notation, your first subnet includes addresses:

```
130.15.4.1 to 130.15.4.255
130.15.5.0 to 130.15.5.255
130.15.6.0 to 130.15.6.255
130.15.7.0 to 130.15.7.254
```

All addresses starting with 130.15.4, 130.15.5, 130.15.6, and 130.15.7 are on the same subnet. As before, host addresses 130.15.4.255, 130.15.5.255, and 130.15.6.255 are legal. These addresses end with an all-1s *byte* but not an all-1s *host field* because of zeros in the previous byte. Similarly, 130.15.5.0, 130.15.6.0, and 130.15.7.0 are legal because we end with an all-0s byte but not an all-0s host field.

The second subnet will include addresses:

```
130.15.8.1   to 130.15.8.255
130.15.9.0   to 130.15.9.255
130.15.10.0 to 130.15.10.255
130.15.11.0 to 130.15.11.254
```

Now we can see the pattern. Each subnet will be introduced by four adjacent numbers. New subnets start at each multiple of 4.

If we looked at 5-bit subnets, the first subnet would include the addresses 130.15.8.1 to 130.15.15.254. New subnets would start at each multiple of 8. Now that we have mastered small subnet fields, let's look at large subnet fields.

D.1.3 Nine-Bit Subnets

It is easier to understand large subnets if we start from 130.15.1 instead of starting from the very first subnet. With a 9-bit subnet, the first host on the subnet has binary and dot addresses:

```
10000010 00001111 00000001 00000001
  130  .   15   .   1   .   1
```

The last host on your subnet has binary and dot addresses:

```
10000010 00001111 00000001 01111110
  130  .   15   .   1   .   126
```

That is, when translated to dot notation, your subnet includes addresses:

```
130.15.1.1 to 130.15.1.126
```

The first host on your next subnet has address:

```
10000010 00001111 00000001 10000001
  130  .   15   .   1   .   129
```

The last host on this subnet has binary and dot addresses:

```
10000010 00001111 00000001 11111110
  130  .   15   .   1   .   254
```

The subnet will include addresses:

130.15.1.129 to 130.15.1.254

The first host on your next subnet has address:

10000010 00001111 00000010 00**000001**
130 .   15 .   2 .   1

The last host on your next subnet has binary and dot addresses:

10000010 00001111 00000010 01**111110**
130 .   15 .   2 .   126

Thus, the next subnet will include addresses:

130.15.2.1 to 130.15.2.126

Now we can see the pattern. The last byte is used to construct two subnets, each with 126 addresses. The host numbers for one of them range from 1 to 126. The host numbers for the other range from 129 to 254.

D.1.4   Ten-Bit Subnets

As before, it is easier to understand 10-bit subnets if we start from 130.15.1 instead of starting from the very first subnet. The first host has binary and dot addresses:

10000010 00001111 00000001 00**000001**
130 .   15 .   1 .   1

The last host on this subnet has binary and dot addresses:

10000010 00001111 00000001 00**111110**
130 .   15 .   1 .   62

That is, when translated to dot notation, your subnet includes the 62 addresses:

130.15.1.1 to 130.15.1.62

The first host on your next subnet has address:

10000010 00001111 00000001 01**000001**
130 .   15 .   1 .   65

The last host on your second subnet has binary and dot addresses:

10000010 00001111 00000001 01**111110**
  130 . 15 . 1 . 126

That is, when translated to dot notation, your subnet includes the 62 addresses:

130.15.1.65 to 130.15.1.126

The last byte is used to construct four subnets, each with 62 addresses. The
last byte will hold the ranges:

    1 to 62
   65 to 126
  129 to 190
  193 to 254

D.2 VARIABLE-LENGTH SUBNET MASKS

It can be difficult to choose a single subnet mask for an organization. Many
enterprise networks are made up of a mixture of communications facilities—
long distance lines or frame relay circuits, large headquarters LANs, and
small branch office LANs. Fortunately, today you can assign addresses efficiently by using variable-length subnet masks. In other words, use different
mask sizes that fit each of your subnetworks.

The only reason that this was not done in the past was that subnet mask
information was not passed between routers by the old routing protocols. For
example, classic Routing Information Protocol (RIP) routers exchanged messages that identified:

- A destination network, subnet, or host
- The hop count metric to the destination

Entries did not include any mask information. Implementers dealt with this
limitation by insisting that a single mask must be used throughout a network. Organizations with a Class B network address usually made the conventional choice and used 8 bits to number subnets and 8 bits to number
hosts. This limited them forever to at most 254 subnets, each with at most
254 hosts.

RIP version 2, Open Shortest Path First (OSPF), and Cisco Enhanced
Internet Gateway Routing Protocol (EIGRP) support variable-length subnet
masks. This means that the routers include a mask with each destination
that they describe.

We'll continue to work with the sample Class B network, 130.15.0.0. The
easiest way to work with variable-length masks is to set aside ranges of numbers for each size.

D.2.1 Assigning Masks for Point-to-Point Links

Let's start with Point-to-Point links.[1] Only two addresses are needed for any Point-to-Point circuit. A 14-bit mask produces subnets that support two systems. If we use 14-bit masks for addresses starting with 130.15.251, we get the following 64 subnets:

130.15.251.1 to 130.15.251.2
130.15.251.5 to 130.15.251.6
130.15.251.9 to 130.15.251.10
...
130.15.250.253 to 130.15.250.254

If we use 14-bit masks for all addresses in the range:

130.15.251.0 to 130.15.255.255

we get 5 times as many, that is, 320 subnets.

D.2.2 Small Branch Office LANs

Suppose that an organization has 100 branch offices and each needs 30 or 40 addresses. To be safe, let's use a 10-bit subnet mask, which will support 62 hosts at each site. If we use 10-bit masks for addresses starting with 130.15.101, we get the following four subnets:

130.15.101.1   to 130.15.101.62
130.15.101.65   to 130.15.101.126
130.15.101.129 to 130.15.101.190
130.15.101.193 to 130.15.101.254

If we need 100 of these subnets, we could apply 10-bit masks to the range:

130.15.101.0 to 130.15.125.255

We could reserve a somewhat larger range to allow more sites to be added in the future.

D.2.3 Large LANs

Finally, suppose that we have six large LANs. We want to allow up to 500 hosts to be connected to each. A 7-bit subnet mask will do the job. We looked

---

[1]Some sites prefer another approach. They do not assign IP addresses to interfaces that connect to Point-to-Point lines. Many routers support this option.

at 7-bit subnets in Section D.1.1. Recall that a typical 7-bit subnet includes an address range such as:

130.15.2.1 to 130.15.2.255
130.15.3.0 to 130.15.3.254

Since we need six of these LANs, we could apply 7-bit masks to the range:

130.15.2.0 to 130.15.13.255

We could reserve a somewhat larger range to allow for future needs.

D.2.4   Summary

Variable-length masks support the efficient assignment of IP addresses. The first step in using them is to examine a network and identify the subnet sizes that are needed. Next, range of numbers can be set aside for use with each mask size. It is a good idea to leave gaps between these ranges, to allow for future expansion.

Bibliography

Albitz, Paul, and Cricket Liu, *DNS and BIND*, O'Reilly & Associates, Sebastopol, Calif., 1993.

American National Standards Institute, *Fiber Distributed Data Interface (FDDI)—Token-Ring Physical Layer Protocol (PHY)*, ANS X3.148-1988, (also ISO 9314-1, 1989).

———, *Fiber Distributed Data Interface (FDDI)—Token-Ring Media Access Control(MAC)*, ANS X3.139-1987, (also ISO 9314-2, 1989).

———, *T1.602—Telecommunications—ISDN—Data Link Layer Signalling Specification for Application at the Network Interface*, 1990.

———, *T1.606—Frame Relaying Bearer Service—Architectural Framework and Service Description*, 1990.

———, *T1S1/90-175—Addendum to T1.696—Frame Relaying Bearer Service—Architectural Framework and Service Description*, 1990.

———, *T1S1/90-214—DSS1—Core Aspects of Frame Protocol for Use with Frame Relay Bearer Service—Architectural Framework and Service Description*, 1990.

Bellcore TA TSV 00160, *Exchange Access SMDS Service Generic Requirements*, December 1990

Bellovin, S., and M. Merritt, "Limitations of the Kerberos Authentication System," *Computer Communications Review*, October 1990.

Black, Uyless D., *"Data Communications," Networks, and Distributed Processing*, Reston, 1983.

Bolt, Beranek, and Newman, *A History of the ARPANET: The First Decade*, Technical Report, 1981.

Borman, D., "Implementing TCP/IP on a Cray Computer," *Computer Communication Review*, April 1989.

Brand, R., *Coping with the Threat of Computer Security Incidents: A Primer from Prevention through Recovery*, at cert.sei.cmu.edu in /pub/info/primer, June 1990.

Callon, Ross, "An Overview of OSI NSAP Addressing in the Internet," *ConneXions, The Interoperability Report*, December 1991.

CCITT Recommendation I.22, *Framework for providing additional packet mode bearer services*, Blue Book, ITU, Geneva, 1988

CCITT Recommendation X.25, *Interface between data terminal equipment (DTE) and data-circuit-terminating equipment (DCE) for terminals operating in the packet mode on public data networks*, 1980 and 1984.

CCITT Recommendation X.400, *Message Handling System*, 1984 and 1988.

CCITT Recommendation X.500, *The Directory*, 1988.

Cerf, V., "A History of the ARPANET," *ConneXions, The Interoperability Report*, October 1989.

——— and R. Kahn, "A Protocol for Packet Network Intercommunication," *IEEE Transactions on Communication*, May 1974.

Cheswick, B., "The Design of a Secure Internet Gateway," *Proc. of the Summer Usenix Conference*, Anaheim, Calif., June 1990.

Cheswick, William R., and Steven M. Bellovin, *Firewalls and Internet Security*, Addison-Wesley, Reading, Mass., 1994.

Cisco Systems, StrataCom, Digital Equipment Corporation, *Frame Relay Specification with Extensions*, Draft, 1990.

Cisco Systems, *Gateway System Manual*, 1991.

Coltun, Rob, "OSPF: An Internet Routing Protocol," *ConneXions*, August 1989.

Comer, Douglas E., *Internetworking with TCP/IP, Volume I, Principles, Protocols, and Architecture*, 2d ed., Prentice-Hall, Englewood Cliffs, N.J., 1991.

——— and David L. Stevens, *Internetworking with TCP/IP, Volume II, Design, Implementation, and Internals*, Prentice-Hall, Englewood Cliffs, N.J., 1991.

Cooper, J., *Computer and Communications Security: Strategies for the 1990s*, McGraw-Hill, New York, 1989.

Deering, S., "IP Multicasting," *ConneXions*, February 1991.

Dern, Daniel P., "Standards for Interior Gateway Routing Protocols," *ConneXions*, July 1990.

Digital Equipment Corporation, Intel Corporation, and XEROX Corporation, *The Ethernet: A Local Area Network Data Link Layer and Physical Layer Specification,* September 1980.

Frey, Donnalyn, and Rick Adams, *!%@:: A Directory of Electronic Mail Addressing and Networks,* 2d ed., O'Reilly & Associates, Sebastopol, Calif., 1989.

FRICC, *Program Plan for the National Research and Education Network,* Federal Research Internet Coordinating Committee, U.S. Department of Energy, Office of Scientific Computing Report ER-7, May 1989.

FTP Software, *PC/TCP Kernel Installation and Reference Guide,* Version 2.05 for DOS, 1990.

———, *PC/TCP User's Guide,* Version 2.05 for DOS, 1990.

Garcia-Luna-Aceves, J. J., *A Unified Approach to Loop-Free Routing using Distance Vectors or Link States,* ACM 089791-332-9/89/0009/0212, pp. 212–223, 1989.

———, "Loop-Free Routing using Diffusing Computations," *IEEE/ACM Transactions on Networking,* vol. 1, no. 1, 1993.

GOSIP, *U.S. Government Open Systems Interconnection Profile Version 2.0,* Advanced Requirements Group, National Institute of Standards and Technology (NIST), April 1989.

Green, James Harry, *The Dow Jones-Irwin Handbook of Telecommunications,* Dow Jones-Irwin, Homewood, Ill., 1986.

Hedrick, Charles L., *Introduction to Administration of an Internet-based Local Network,* Rutgers, The State University of New Jersey, 1988, at cs.rutgers.edu, in /runet/tcp-ip-admin.doc.

———, *Introduction to the Internet Protocols,* Rutgers, The State University of New Jersey, 1987, host cs.rutgers.edu, /runet/tcp-ip- intro.doc.

Hoffman, L., *Rogue Programs: Viruses, Worms, and Trojan Horses,* Van Nostrand Reinhold, New York, 1990.

Huitema, Christian, "Routing in the Internet," Prentice-Hall PTR, Englewood Cliffs, N.J., 1995.

IBM GG24-3442, *IBM AS/400 TCP/IP Configuration and Operation,* 1991.

IBM GG24-3696, *Managing TCP/IP Networks Using NetView and the SNMP Interface,* 1991.

IBM GG24-3816, *High-Speed Networking Technology, An Introductory Survey,* 1992.

IBM SC31-6081, *TCP/IP Version 2 Release 2 for VM: User's Guide,* 1991.

IBM SC31-6084, *TCP/IP Version 2 Release 2 for VM: Programmer's Reference,* 1991.

IBM, *Vocabulary for Data Processing, Telecommunications, and Office Systems,* 1981.

Institute of Electrical and Electronics Engineers, *Draft Standard P802.1A—Overview and Architecture,* 1989.

———, *Local Area Networks—CSMA/CD Access Method,* ANSI/IEEE 802.3, (ISO 8802-3).

———, *Local Area Networks—Distributed Queue Dual Bus (DQDB) Subnetwork of a Metropolitan Area Network (MAN),* ANSI/IEEE 802.6 (ISO DIS 8802-6, 1991).

———, *Local Area Networks—Higher Layers and Interworking,* ANSI/IEEE 802.1, 1990 (ISO DIS 8802-1D, 1990).

———, *Local Area Networks—Logical Link Control,* ANSI/IEEE 802.2, 1989 (ISO 8802-2, 1989).

———, *Local Area Networks—Network Management. Draft IEEE 802.1B,* 1990.

———, *Local Area Networks—Token-Bus Access Method,* ANSI/IEEE 802.4, (ISO 8802-3).

———, *Local Area Networks—Token Ring Access Method,* ANSI/IEEE 802.5, 1989 (ISO 8802-5,1989).

International Organization for Standardization, *Information Processing Systems—Common Management Information Protocol (CMIP),* ISO 9596, 1990.

———, *Information Processing Systems—Common Management Information Service (CMIS),* ISO 9595, 1990.

———, *Information Processing Systems—Data Communications—Addendum to the Network Service Definition,* ISO 8348 AD1.

———, *Information Processing Systems—Data Communications—High-Level Data Link Control Procedures—Consolidation of Classes of Procedures,* ISO 7809.

———, *Information Processing Systems—Data Communications—High-Level Data Link Control Procedures—Consolidation of Elements of Procedures,* ISO 4335.

———, *Information Processing Systems—Data Communications—High-Level Data Link Control Procedures—Frame Structure,* ISO 3309.

———, *Information Processing Systems—Data Communications—Network Service Definition,* ISO 8348.

———, *Information Processing Systems—Data Communications—Protocol for Providing the Connectionless-Mode Network Service,* ISO 8473.

———, *Information Processing Systems—Open Systems Interconnection—Basic Connection Oriented Session Protocol Specification,* ISO 8327.

————, *Information Processing Systems—Open Systems Interconnection—Basic Connection Oriented Session Service Definition,* ISO 8326.

————, *Information Processing Systems—Open Systems Interconnection—Connection Oriented Presentation Protocol Specification,* ISO 8823.

————, *Information Processing Systems—Open Systems Interconnection—Connection Oriented Presentation Service Definition,* ISO 8822.

————, *Information Processing Systems—Open Systems Interconnection—Connection Oriented Transport Protocol,* ISO 8073.

————, *Information Processing Systems—Open Systems Interconnection—Intermediate System to Intermediate System Intra-Domain Routing Exchange Protocol for use in Conjunction with the Protocol for Providing the Connectionless-Mode Network Service,* ISO DIS 10589.

————, *Information Processing Systems—Open Systems Interconnection—Message Handling System,* ISO 10021/CCITT X.400.

————, *Information Processing Systems—Open Systems Interconnection—Protocol Specification for the Association Control Service Element,* ISO 8650.

————, *Information Processing Systems—Open Systems Interconnection—Remote Operations: Model, Notation, and Service Definition,* ISO 9072-1.

————, *Information Processing Systems—Open Systems Interconnection—Remote Operations: Protocol Specification,* ISO 9066-2.

————, *Information Processing Systems—Open Systems Interconnection—Service Definition for the Association Control Service Element,* ISO 8649.

————, *Information Processing Systems—Open Systems Interconnection—Specification of Abstract Syntax Notation One (ASN.1),* ISO 8824.

————, *Information Processing Systems—Open Systems Interconnection—Specification of Basic Encoding Rules for Abstract Syntax Notation One (ASN.1),* ISO 8825.

————, *Information Processing Systems—Open Systems Interconnection—Transport Service Definition,* ISO 8072.

————, *OSI Routing Framework,* ISO TC97/SC6/N4616, June 1987.

Jacobson, V., "Berkeley TCP Evolution from 4.3-Tahoe to 4.3-Reno," *Proceedings of the Eighteenth Internet Engineering Task Force.*

————, "Congestion Avoidance and Control," *ACM SIGCOMM-88,* August 1988.

Jain, R., K. Ramakrishnan, and D-M Chiu, *Congestion Avoidance in Computer Networks With a Connectionless Network Layer,* Technical Report, DEC-TR-506, Digital Equipment Corporation, 1987.

Kapoor, Atul, *SNA, Architecture, Protocols, and Implementation,* McGraw-Hill, New York, 1992.

Karn, P., and C. Partridge, "Improving Round-Trip Time Estimates in Reliable Transport Protocols," *Proceedings of the ACM SIGCOMM,* 1987.

Kernighan, Brian W., and Dennis M. Ritchie, *The C Programming Language,* 2d ed., Prentice-Hall, Englewood Cliffs, N.J., 1988.

Kessler, Gary C., and Train, David A., *Metropolitan Area Networks,* McGraw-Hill, New York, 1992.

————, *ISDN,* McGraw-Hill, New York, 1990.

Kochan, Stephen G., and Patrick H. Wood (consulting eds.), *UNIX Networking,* 1989.

Laquey, T. L., *User's Directory of Computer Networks,* Digital Press, Bedford, Mass., 1989.

Lippis, Nick, and James Herman, "Widening Your Internet Horizons," *ConneXions,* October 1991.

Liu, Cricket, Jerry Peek, Russ Jones, Bryan Buus, and Adrian Nye, *Managing Internet Information Services,* O'Reilly & Associates, Sebastopol, Calif., 1995.

Malamud, Carl, *DEC Networks and Architectures,* McGraw-Hill, New York, 1989.

————, *STACKS—The INTEROP Book,* Prentice-Hall, Englewood Cliffs, N.J., 1991.

McKenney, P., "Congestion Avoidance," *ConneXions,* February 1991.

Medin, Milo, "The Great IGP Debate—Part Two: The Open Shortest Path First (OSPF) Routing Protocol," *ConneXions,* October 1991.

Mills, D., and H-W. Braun, "The NSFNET Backbone Network," *Proceedings of the ACM SIG-COMM,* 1987.

Mogul, Jeffrey C., "Efficient Use Of Workstations for Passive Monitoring of Local Area Networks," *Proceedings of SIGCOMM '90 Symposium on Communications Architectures and Protocols,* September 1990.

Narten, T., "Internet Routing," *Proceedings of the ACM SIGCOMM,* 1989.

Nemeth, Evi, Garth Snyder, and Scott Seebass, *UNIX System Administration Handbook,* Prentice-Hall, Englewood Cliffs, N.J., 1989.

Odlyzko, A. M., "The future of integer factorization," *CryptoBytes* (The technical newsletter of RSA Laboratories), 1994.

Perlman, Radia, and Ross Callon, "The Great IGP Debate—Part One: IS-IS and Integrated Routing," *ConneXions,* October 1991.

Pfleeger, C., *Security in Computing,* Prentice-Hall, Englewood Cliffs, N.J., 1989.

Postel, J. B., "Internetwork Protocol Approaches," *IEEE Transactions on Communications,* 1980.

——, C. A. Sunshine, and D. Chen, "The ARPA Internet Protocol," *Computer Networks,* vol. 5, no. 4, July 1981.

Quarterman, John S., "The Matrix," *Computer Networks and Conferencing Systems Worldwide,* Digital Press, Bedford, Mass., 1990.

—— and Hoskins, J. C., "Notable Computer Networks," *Communications of the ACM,* October, 1986.

Romkey, John, "The Packet Driver," *ConneXions,* July 1990.

Rose, Marshall T., *The Little Black Book: Mail Bonding with OSI Directory Services,* Prentice-Hall, Englewood Cliffs, N. J., 1990.

——, *The Open Book: A Practical Perspective on OSI,* Prentice-Hall, Englewood Cliffs, N.J., 1990.

——, *The Simple Book: An Introduction to Management of TCP/IP-based Internets,* Prentice-Hall, N.J., 1990.

Sackett, George C., IBM's Token-Ring Networking Handbook, McGraw-Hill, New York, 1993.

St. Amand, Joseph V., *A Guide to Packet-Switched, Value-Added Networks,* Macmillan, New York, 1986.

Schwartz, Michael F., "Resource Discovery and Related Research at the University of Colorado," *ConneXions,* May 1991.

Seeley, D., "A Tour of the Worm," *Proceedings of 1989 Winter USENIX Conference,* Usenix Association, San Diego, Calif., February 1989.

Sijan, Karanjit, and Hare, Chris, *Internet Firewalls and Network Security,* New Riders Publishing.

Simmons, G. J., ed., *Contemporary Cryptology,* IEEE, 1991.

Spafford, E., "The Internet Worm Program: An Analysis," *Computer Communication Review,* vol. 19, no. 1, ACM SIGCOM, January 1989.

Stallings, William, *Data and Computer Communications,* Macmillan, New York, 1984.

——, *Handbook of Computer Communications Standards,* Department of Defense Protocol Standards, 1988.

Stern, Hal, *Managing NFS and NIS,* O'Reilly and Associates, Sebastopol, Calif., 1991.

Stevens, W. Richard, *TCP/IP Illustrated,* vol. 1, Addison Wesley, Reading, Mass., 1994.

——, *UNIX Network Programming,* Prentice-Hall, Englewood Cliffs, N.J., 1990.

Stoll, C., *The Cuckoo's Egg,* Doubleday, New York, 1989.

Tannenbaum, Andrew S., *Computer Networks,* Prentice-Hall, Englewood Cliffs, N.J., 1981.

Vitalink, *Building and Managing Multivendor Networks using Bridge and Router Technologies,* 1990.

Tsuchiya, Paul F., "Inter-domain Routing in the Internet," *ConneXions,* January 1991.

XEROX, *Internet Transport Protocols, Report XSIS 028112,* Xerox Corporation, 1981.

X/Open specification, *X/Open CAE Specification: Protocols for X/Open Internetworking: XNFS,* X/Open Company, Ltd., 1991.

Glossary

Abstract Syntax Notation One (ASN.1) A language used for defining datatypes. ASN.1 is used in OSI standards and also is used in TCP/IP network management specifications.

Access Control A facility that defines each user's privileges to access computer data.

Acknowledgment TCP requires that data be acknowledged before it can be considered to have been transmitted safely.

Active Open Action taken by an application to initiate a TCP connection.

Address Mask A 32-bit binary number used to identify the parts of an IP address that are used for network and subnet numbers. Every bit in the network and subnet fields is set to 1.

Address Resolution Protocol (ARP) A protocol that dynamically discovers the physical address of a system, given its IP address.

Agent In the Simple Network Management Protocol, the process within a device that responds to get and set requests and sends trap messages.

American National Standards Institute (ANSI) Organization responsible for coordinating United States standardization activities. ANSI is a member of ISO.

AppleTalk A networking protocol developed by Apple Computer for use with its products.

Application Programming Interface (API) A set of routines that enable a programmer to use computer facilities. The socket programming interface and the Transport Layer Interface are APIs used for TCP/IP programming.

Archie A server that gathers and indexes the locations of files at public file transfer archives and supports user searches.

ARPANET The world's first packet-switching network, which for many years functioned as an Internet backbone.

ASCII American National Standard Code for Information Interchange. Seven of the eight bits in an octet are required to define an ASCII character.

Asynchronous Transfer Mode A switch-based technology that transports information in 53-octet cells. ATM may be used for data, voice, and video.

Authentication Verification of the identity of a communications partner.

Authentication Header (AH) An IP-layer header which authenticates a source and protects the integrity of the data. An Authentication Header normally is inserted after the main IP header and before the other information being authenticated.

Autonomous System (AS) A collection of routers under the control of a single administrative authority, and using a common Interior Gateway Protocol. More recently, one or more networks which have a single exterior routing policy.

Bandwidth   The quantity of data that can be sent across a link, typically measured in bits per second.

Basic Encoding Rules (BER)   The rules for encoding datatypes specified using ASN.1 into their transmission format.

Baud   A unit of signaling speed equal to the number of times per second that a signal changes state. If there are exactly two states, the baud rate equals the bit rate.

Berkeley Software Distribution (BSD)   Unix software from the University of California at Berkeley that included TCP/IP support.

Best Current Practices (BCP)   A classification applied to a useful RFC that does not define a protocol standard.

Big Endian   A format for the storage or transmission of data that places the most significant byte (or bit) first.

BIND Software   Domain Name server software from Berkeley university.

Bootstrap Protocol (BOOTP)   Protocol that can be used by booting systems to obtain network configuration information.

Border Gateway protocol (BGP)   A protocol used to advertise the set of networks that can be reached within an Autonomous System. BGP enables this information to be shared with other Autonomous Systems. BGP is newer than EGP, and offers a number of improvements.

Bounce   The return of a piece of mail that cannot be delivered.

Bridge   A device that connects two or more physical segments of a LAN and forwards frames which have source and destination addresses on different segments.

Broadcast   A link frame addressed to all systems on the link.

Brouter   A device that performs both bridging and routing functions. Some traffic is selected for routing, while the rest is bridged.

Buffer   An area of storage used to hold input or output data.

Canonical Name   A host's unique true name.

Carrier Sense Multiple Access with Collision Detection (CSMA/CD)   A simple Media Access Control protocol. All stations listen to the medium. A station wanting to send may do so if there is no signal on the medium. When two stations transmit simultaneously, both back off and retry after a random time period.

Cipher-Block Chaining   A popular option for DES encryption. A block of already encrypted data is fed into the algorithm as it encrypts the next block.

Classless Inter-Domain Routing (CIDR).   A method of routing used to enable the network part of IP addresses to consist of a specified number of bits.

Common Management Information Protocol (CMIP)   A central OSI network management protocol.

Common Management Information Services and Protocol over TCP/IP (CMOT)   A historic (nonrecommended) specification for using OSI management protocols on a TCP/IP network.

Confidentiality   Protection of information from disclosure to unintended parties.

Connection   A logical communication path between TCP users.

Core Gateway Historically, a router on the Internet backbone. Core gateways distributed reachability information among the Autonomous Systems attaching to the Internet backbone.

Cracker Someone who attempts to break into computer systems, often with malicious intent.

Cyclic Redundancy Check (CRC) The value obtained by applying a mathematical function to the bits in a frame and appended to the frame. The CRC is recalculated when the frame is received. If the result differs from the appended value, the frame is discarded.

Data Circuit-terminating Equipment (DCE) Equipment required to connect a DTE to a line or to a network.

Data Encryption Standard (DES) A symmetric encryption protocol officially sanctioned by the United States government. There are several options for the manner in which DES is applied. (See Cipher Block Chaining.)

Data Terminal Equipment (DTE) A source or destination for data. Often used to denote terminals or computers attached to a wide area network.

DECnet Digital Equipment Corporation's proprietary network protocol. Versions are identified by their Phase number—such as Phase IV and Phase V.

Directory Access Protocol (DAP) Client-to-server protocol used to access an X.500 directory service.

Directory System Agent (DSA) A server that accepts queries from Directory User Agents and extracts information from a database. A DSA interacts with a Directory User Agent by means of the X.500 Directory Access Protocol.

Directory User Agent (DUA) A client enabling a user to send queries to an X.500 directory server. A DUA interacts with a Directory Service Agent (DSA) via the X.500 Directory Access Protocol.

Distributed Computing Environment (DCE) A set of technologies selected by the Open Software Foundation to support distributed computing.

Distributed File Service (DFS) A file server technology adopted by the Open Software Foundation.

Distributed Management Environment (DME) A set of technologies selected for network and system management by the Open Software Foundation.

DIX Ethernet Version of Ethernet developed by Digital, Intel, and Xerox.

Domain Name System (DNS) A set of distributed databases providing information such as translation between system names and their IP addresses and the location of mail exchangers.

DS1 A frame and interface specification for synchronous T1 lines.

DS3 A frame and interface specification for synchronous T3 lines.

Encryption Transformation of information into a form that cannot be understood without possession of a secret ("decryption") key.

Encapsulating Security Payload (ESP) A protocol designed to provide confidentiality (and optionally, authentication and integrity) to IP datagrams. ESP can be used between a pair of hosts, between a pair of routers, or between a host or router and multiple other hosts and routers.

Exterior Gateway Protocol (EGP)   Routers in neighboring Autonomous Systems use this protocol to identify the set of networks that can be reached within or via each Autonomous System. EGP is being supplanted by BGP.

eXternal Data Representation (XDR)   A standard developed by Sun Microsystems to define datatypes used as parameters and to encode these parameters for transmission.

Fiber Distributed Data Interface (FDDI)   A standard for high-speed data transfer across a dual fiber-optic ring.

File Transfer, Access, and Management (FTAM)   The OSI file transfer and management protocol. FTAM allows users to copy whole files or part of a file, such as an individual record.

File Transfer Protocol (FTP)   The TCP/IP protocol that enables users to copy files between systems and perform file management functions, such as renaming or deleting files.

Finger   A program that displays information about one or more remote users.

Flow Control   A mechanism that allows a receiver to limit the amount of data that a sender may transmit at any time. Flow control prevents a sender from exhausting the receiver's memory buffers.

For Your Information (FYI)   A set of documents including useful information, such as answers to frequently asked questions about TCP/IP. FYI documents also are published as RFCs.

Fragmentation   Partitioning of a datagram into pieces. This is done when a datagram is too large for a network technology that must be traversed to reach the destination.

Frame   A link layer Protocol Data Unit.

Frame Check Sequence (FCS)   A mathematical function applied to the bits in a frame and appended to the frame. The FCS is recalculated when the frame is received. If the result differs from the appended value, the frame is discarded.

Frequently Asked Questions (FAQ)   A document in the form of questions and answers that summarizes information for a newsgroup or mailing list.

Gateway   An IP router. Many RFC documents use the term *gateway* rather than *router*.

Gateway-to-Gateway Protocol (GGP)   A protocol formerly used to exchange routing information between Internet core routers.

Gopher   A protocol that enables clients to access data at a server by means of a series of menus.

Government Open Systems Interconnection Profile (GOSIP)   Specification of a set of OSI protocols to be preferred in government procurements of computer equipment.

Graphics Interchange Format (GIF)   A popular format for graphical image files.

High Level Data Link Control Protocol (HDLC)   A standard that is the basis for several link layer protocols.

High Performance Parallel Interface (HIPPI)   A high-speed communications technology defined by an ANSI standard. Devices communicate via HIPPI across short distances at speeds of 800 and 1600 megabits per second.

Hypertext Markup Language (HTML)   A markup language used to write hypertext documents. Tags in the document identify elements such as headers, paragraphs, and lists.

Initial Sequence Number (ISN) A sequence number defined during TCP connection setup. Data octets sent over the connection will be numbered starting from this point.

Integrated Services Digital Network (ISDN) A telephony technology that provides digital voice and data services.

Interior Gateway Protocol (IGP) Any routing protocol used within an Autonomous System.

Intermediate System to Intermediate System Protocol (IS-IS) A protocol that can be used to route both OSI and IP traffic.

International Organization for Standardization (ISO) An international body founded to promote international trade and cooperative progress in science and technology.

International Telecommunications Union (ITU) A body that oversees several international organizations devoted to communications standards and cooperation.

International Telecommunications Union Telecommunication Standardization Sector (ITU-T) Presides over study groups and writes "Recommendations" for international communications standards. Formerly CCITT.

International Telegraph and Telephone Consultative Committee (CCITT) Former name of an organization formed to facilitate connecting communications facilities into international networks.

Internet A set of networks connected by IP routers and appearing to its users as a single network.

Internet The world's largest network, the Internet is based on the TCP/IP protocol suite.

Internet Architecture Board (IAB) Formerly the Internet Activities Board. An Internet Society group responsible for promoting protocol development, selecting protocols for Internet use, and assigning state and status to protocols.

Internet Assigned Numbers Authority (IANA) The authority responsible for controlling the assignment of a variety of parameters, such as well-known ports, multicast addresses, terminal identifiers, and system identifiers.

Internet Control Message Protocol (ICMP) A protocol that is required for implementation with IP. ICMP specifies error messages to be sent when datagrams are discarded or systems experience congestion. ICMP also provides several useful query services.

Internet Engineering Notes (IEN) An early set of documents discussing features of the TCP/IP suite. These documents are available on-line at the Network Information Center.

Internet Engineering Steering Group (IESG) A group that coordinates the activities of the IETF working groups and performs technical reviews of standards.

Internet Engineering Task Force (IETF) A set of working groups made up of volunteers who develop and implement Internet protocols.

Internet Group Management Protocol (IGMP) A protocol that is part of the multicast specification. IGMP is used to carry group membership information.

Internet Protocol (IP)   The TCP/IP layer 3 protocol responsible for transporting datagrams across an internet.

Internet Research Task Force (IRTF)   A group directed by the IAB, charged with long-term research on Internet protocols.

Internet Service Provider (ISP)   An organization that sells Internet connectivity services.

Internet Society (ISOC)   An international organization formed to promote the growth and continued technical enhancement of the Internet.

IP Address   A 32-bit quantity that identifies a network interface.

IP Datagram   The unit of data routed by IP.

IP Security Option   In version 4, an optional field in the IP header that contains a security label. The option was designed for use by military and government agencies.

ISO Development Environment (ISODE)   A research effort that has produced software enabling OSI protocols to run on top of TCP/IP.

Joint Photographic Experts Group (JPEG)   A specification for an image compression scheme.

Kerberos   An authentication service developed at the Massachusetts Institute of Technology. Kerberos uses encryption to prevent intruders from discovering passwords and gaining unauthorized access to files or services.

Link   A medium over which nodes can communicate using a link layer protocol.

Little Endian   A format for the storage or transmission of data that places the least significant byte (or bit) first.

Local Area Network (LAN)   A data network intended to serve an area of only a few square kilometers or less and consisting of a single subnetwork.

Logical Byte   A logical byte is a specified number of bits in length. In a file transfer, it is sometimes necessary to specify a logical byte size in order to preserve the integrity of data that is transferred.

Logical Link Control (LLC)   A layer 2 (data link layer) protocol that governs the exchange of data between two systems connected to the same physical segment or residing on segments that are connected via one or more bridges.

MAC Address   A physical address assigned to a LAN interface.

MAC Protocol   A Media Access Control protocol defines the rules that govern a system's ability to transmit and receive data on a medium.

Mail Exchanger   A system used to relay mail into an organization's network.

Mail Gateway   A system that performs a protocol translation between different electronic mail delivery protocols.

Management Information Base (MIB)   The set of all definitions of network-manageable objects. Also, the configuration, status, and performance information that can be retrieved from a network device.

Maximum Segment Size   The maximum permissible size for the data part of any segment sent on a particular connection.

Maximum Transmission Unit (MTU)   The largest datagram that can be sent across a particular network medium, such as an Ethernet or Token-Ring.

Media Access Control (MAC) A protocol governing a station's access to a network. For example, CSMA/CD provides a set of MAC rules for sending and receiving data across a local area network.

Message Digest 5 (MD5) An algorithm that combines a secret key with message or file information and produces a 16-octet answer. Its purpose is to detect at a later time whether the information has been altered.

Message Transfer Agent (MTA) An entity that moves messages (such as electronic mail) between computers.

Metropolitan Area Network (MAN) A technology supporting high-speed networking across a metropolitan area. IEEE 802.6 defines a MAN protocol.

Multicast IP Address An IP address that can be adopted by multiple hosts. Datagrams sent to a multicast IP address will be delivered to all hosts in the group.

Multihomed Host A host that has multiple network interfaces and therefore requires multiple IP addresses.

Multipurpose Internet Mail Extensions (MIME) Extensions to Internet mail that enable messages to be made of one or more parts, each of which can contain a variety of content types, such as text, image, sound, or application data.

National Education and Research Network (NREN) A planned high-capacity network to be used as part of a future backbone for the Internet.

National Institute of Standards and Technology (NIST) A United States standards organization that has promoted communications standards. NIST formerly was the National Bureau of Standards.

National Science Foundation Network (NSFnet) A network used as part of the current Internet backbone.

Neighbors Nodes attached to the same link.

NETBIOS A network programming interface and protocol developed for IBM-compatible personal computers.

Network Address The 32-bit IP address of a system.

Network File System (NFS) A set of protocols introduced by Sun Microsystems; it enables clients to mount remote directories onto their local file systems and use remote files as if they were local.

Network Information Center (NIC) An Internet administration facility that supervises network names and network addresses and can provide other information services.

Network Information Service (NIS) A set of protocols introduced by Sun Microsystems, used to provide a directory service for network information.

Network Service Access Point (NSAP) An identifier used to distinguish the identity of an OSI host and to point to the transport layer entity at that host to which traffic is directed.

Network Virtual Terminal (NVT) A set of rules defining a very simple virtual terminal interaction. The NVT is used at the start of a *telnet* session, but a more complex type of terminal interaction can be negotiated.

Nonrepudiation The ability to prove that a source sent specific data, even if the source later tries to deny that fact.

Open Shortest Path First (OSPF)   An internet routing protocol that scales well, can route traffic along multiple paths, and uses knowledge of an internet's topology to make accurate routing decisions.

Open Software Foundation (OSF)   A consortium of computer vendors cooperating to produce standard technologies for open systems. The MOTIF user interface and Distributed Computing Environment (DCE) are OSF technologies.

Open Systems Interconnection (OSI)   A set of ISO standards relating to data communications.

Packet   Originally, a unit of data sent across a packet-switching network. Currently, the term may refer to a communications Protocol Data Unit at any layer.

Packet Assembler/Disassembler (PAD)   Software that converts between a terminal's stream of traffic and X.25 packet format.

Page Structure   A file organization supported in FTP for use with older Digital Equipment Corporation computers.

Passive Open   Action taken by a TCP/IP server to prepare to receive requests from clients.

Pathname   The character string which must be input to a file system by a user in order to identify a file.

Payload   The information carried in a Protocol Data Unit.

Physical Address   An address assigned to a network interface.

Point-to-Point Protocol (PPP)   A protocol for data transfer across serial links. PPP supports authentication, link configuration, and link monitoring capabilities and allows traffic for several protocols to be multiplexed across the link.

Port Number   A 2-octet binary number identifying an upper-level user of TCP or UDP.

Post Office Protocol (POP)   A protocol used to download electronic mail from a server to a client (usually at a desktop system).

Protocol Data Unit (PDU)   A generic term for the protocol unit (e.g., a header and data) used at any layer.

Protocol Interpreter (PI)   An entity that carries out FTP functions. FTP defines two PI roles: user and server.

Protocol State   Position on the standards track or classification as informational, experimental, or historic.

Protocol Status   Requirement level.

Proxy ARP   Use of a router to answer ARP requests. This will be done when the originating host believes that a destination is local, when in fact it lies beyond a router.

Push Service   A service provided by TCP that lets an application specify that some data should be transmitted and delivered as soon as possible.

Receive Window   The valid range of sequence numbers that a sender may transmit at a given time during the connection.

Record Structures   Common structure for data files. During a transfer of a file that is organized as a sequence of records, records can be delimited by End-of-Record markers.

Remote Network Monitor (RMON) A device that collects information about network traffic.

Remote Procedure Call (RPC) A protocol that enables an application to call a routine that executes at a server. The server returns output variables and a return code to the caller.

Request For Comments (RFC) A document describing an Internet protocol or related topics. RFC documents are available on-line at various Network Information Centers.

Reseaux IP Europeens (RIPE) Coordination center for network registration for Europe.

Resolver Software that enables a client to access the Domain Name System databases.

Retransmission Timeout If a segment is not ACKed within the period defined by the retransmission timeout, TCP will retransmit the segment.

Reverse Address Resolution Protocol (RARP) A protocol that enables a computer to discover its IP address by broadcasting a request on a network.

Round-Trip Time (RTT) The time elapsed between sending a TCP segment and receiving its ACK.

Router A system that forwards layer 3 traffic not explicitly addressed to itself. A router is used to connect separate LANs and WANs into an internet and to forward traffic between the constituent networks.

Routing Information Field (RIF) A field in a Token-Ring frame used to identify the path to a destination that is reached via one or more bridges.

Routing Information Protocol (RIP) A simple protocol used to exchange information between routers. The original version was part of the XNS protocol suite.

Routing Policy The sets of sources and destinations for which an Autonomous System is willing to route traffic.

Routing Registry A database containing route information, used to forward data along a path that traverses two or more Autonomous Systems.

Security Association A communication protected by a specific selection of security parameters.

Security Gateway A system that provides security to datagrams sent between internal systems and untrusted external systems.

Segment A Protocol Data Unit consisting of a TCP header and optionally, some data.

Send Window The range of sequence numbers between the last octet of data that already has been sent and the right edge of the receive window.

Sequence Number A 32-bit field of a TCP header. If the segment contains data, the sequence number is associated with the first octet of the data.

Serial Line Interface Protocol (SLIP) A very simple protocol used for transmission of IP datagrams across a serial line.

Service Provider An organization that provides TCP/IP connectivity services to a set of customers. Some Service Providers support customers in a small, local area, while others have national or international scope.

Shortest Path First   A routing algorithm that uses knowledge of a network's topology in making routing decisions.

Silly Window Syndrome   Inefficient data transfer that results when a receiver reports small window credits and a sender transmits correspondingly small segments. This problem is easily solved using algorithms cited in RFC 1122.

Simple Mail Transfer Protocol (SMTP)   A TCP/IP protocol used to transfer mail between systems.

Simple Network Management Protocol   A protocol that enables a management station to monitor network systems and receive trap (alarm) messages from network systems.

Smoothed Deviation   A quantity that measures deviations from the smoothed round-trip time and is used to calculate the TCP retransmission timeout.

Smoothed Round-Trip Time (SRTT)   An estimate of the current round-trip time for a segment and its ACK, used in calculating the value of the TCP retransmission timeout.

Socket Address   The full address of a communicating TCP/IP entity, made up of a 32-bit network address and a 16-bit port number.

Socket Descriptor   An integer that an application uses to identify a connection. Socket descriptors are used in the Berkeley socket programming interface.

Source Quench   An ICMP message sent by a congested system to the sources of its traffic.

Source Route   A sequence of IP addresses identifying the route a datagram must follow. A source route may optionally be included in an IP datagram header.

Standard Generalized Markup Language (SGML)   A powerful markup language used to describe elements in portable documents.

Stub Network   A network that does not carry transit traffic between other networks.

Subnet Address   A selected number of bits from the local part of an IP address, used to identify a set of systems connected to a common link.

Subnet Mask   A 32-bit quantity, with 1s placed in positions covering the network and subnet part of an IP address.

Switched Multimegabit Data Service (SMDS)   A data transfer service developed by Bellcore, whose access protocol is based on the IEEE 802.6 Metropolitan Area Network protocol.

SYN   A segment used at the start of a TCP connection. Each partner sends a SYN containing the starting point for its sequence numbering and, optionally, the size of the largest segment that it is willing to accept.

Synchronous Data Link Protocol (SDLC)   A protocol similar to HDLC that is part of IBM's SNA communications protocol suite. SDLC is used for point-to-point and multipoint communications.

Synchronous Optical Network (SONET)   A telephony standard for the transmission of information over fiber-optic channels.

Systems Network Architecture (SNA)   The data communications protocol suite developed and used by IBM.

T1 A digital telephony service that operates at 1.544 megabits per second. DS1 framing is used.

T3 A digital telephony service that operates at 44.746 megabits per second. DS3 framing is used.

Telnet The TCP/IP application protocol that enables a terminal attached to one host to login to other hosts and interact with their applications.

Time-To-Live (TTL) A limit on the length of time that a datagram can remain within an internet. The TTL usually is specified as the maximum number of hops that a datagram can traverse before it must be discarded.

Tn3270 Telnet, used with options that support IBM 3270 terminal emulation.

Token-Ring A local area network technology based on a ring topology. Stations on the ring pass a special message, called a token, around the ring. The current token holder has the right to transmit data for a limited period of time.

Transmission Control Block (TCB) A data structure used to hold information about a current TCP or UDP communication.

Transmission Control Protocol (TCP) TCP provides reliable, connection-oriented data transmission between a pair of applications.

Transport Class 4 (OSI TP4) An OSI transport layer protocol that is functionally similar to TCP.

Transport Layer Interface (TLI) An application programming interface introduced by AT&T that interfaces to both TCP/IP and OSI protocols.

Transport Service Access Point (TSAP) An identifier that indicates the upper-layer protocol entity to whom a Protocol Data Unit should be delivered.

Trivial File Transfer Protocol (TFTP) A very basic TCP/IP protocol used to upload or download files. Typical uses include initializing diskless workstations or downloading software from a controller to a robot.

Trojan Horse A program that appears to do useful work but also includes secret routines that the perpetrator can use to access the victim's data or to open up access to the victim's computer.

Trunk Coupling Unit (TCU) A hardware element connecting a Token-Ring station to the backbone of a ring.

Unicast Address An address assigned uniquely to a single interface.

Uniform Resource Locator (URL) An identifier for an item that can be retrieved by a World Wide Web browser, which provides a specific location.

Uniform Resource Name (URN) An identifier for an item that can be retrieved by a World Wide Web browser, which provides a generic name. This may map to several locations from which the item may be retrieved.

Universal Resource Identifier (URI) An identifier for an item that can be retrieved by a World Wide Web browser. The identifier may be a Uniform Resource Locator or a Uniform Resource Name.

Universal Time Coordinated (UTC) Formerly known as Greenwich Mean Time.

Urgent Service A service provided by TCP that lets an application indicate that specified data is urgent and should be processed by the receiving application as soon as possible.

Usenet   Thousands of bulletin-board-like newsgroups whose information is available on the Internet.

User Agent (UA)   An electronic mail application that helps an end user to prepare, save, and send outgoing messages and view, store, and reply to incoming messages.

User Datagram Protocol   A simple protocol enabling an application to send individual messages to other applications. Delivery is not guaranteed, and messages need not be delivered in the same order as they were sent.

Virtual Circuit   A term derived from packet-switching networks. A virtual circuit is supported by facilities which are shared between many users, although each circuit appears to its users as a dedicated end-to-end connection.

Virus   A routine that attaches to other, legitimate programs and usually harms local data or program execution.

Wide Area Network (WAN)   A network that covers a large geographical area. Typical WAN technologies include point-to-point, X.25, and frame relay.

Well-Known Port   A TCP or UDP port whose use is published by the Internet Assigned Numbers Authority.

World Wide Web   A set of Internet servers that enable clients to access many types of information, including documents that include images, sounds, and links to other documents.

Worm   A program that replicates itself at other networked sites.

X11   A windowing system invented at MIT.

X.121   A CCITT standard describing the assignment of numbers to systems attached to an X.25 network. These numbers are used to identify a remote system so that a data call can be set up over a virtual circuit.

X.25   A CCITT standard for connecting computers to a network that provides reliable, virtual circuit-based data transmission.

X.400   A series of protocols defined by the CCITT for message transfer and interpersonal messaging. These protocols were later adopted by ISO.

Xerox Network System (XNS)   A suite of networking protocols developed at Xerox Corporation.

X/Open   A consortium of computer vendors that cooperate to provide a common application environment.

X-Window System   A set of protocols developed at MIT that enable a user to interact with applications which may be located a several different computers. The input and output for each application occurs in a window at the user's display. Window placement and size are controlled by the user.

Index